MW00911378

The Warrior's Handbook

A Volume Containing:

Warrior's Heart Revealed
The Art of War
The Sayings of Wutzu
Tao Te Ching
The Book of Five Rings
Behold, The Second Horseman: Quotes on War

Compiled by Joseph Lumpkin

Dedicated to Melanie Chesnut, United States Marine

Joseph Lumpkin

The Warrior's Handbook

Copyright © 2010 by Joseph Lumpkin

All rights reserved.

Printed in the United States of America. No part of this book may be used or reproduced in any manner whatsoever without written permission except in the case of brief quotations embodied in critical articles and reviews.

Fifth Estate, Post Office Box 116,

Blountsville, AL 35031

First Edition

Cover Designed by An Quigley

Printed on acid-free paper

Library of Congress Control No: 2010937471

ISBN: 9781933580999

Fifth Estate, 2010

Joseph Lumpkin

Table of Content

Editor's Note:
"The Tao Te Ching," by Lao Tsu, has been translated by Joseph Lumpkin. "Behold the Second Horseman" is a book of quotes and essays compiled by Joseph Lumpkin. "Warrior's Heart Revealed" is the textbook of Shinsei Hapkido (martial art) written and compiled Joseph Lumpkin and Daryl Covington with contributions by many masters and grandmasters. Edition of "The Art of War", "The Sayings of Wutzu", and "The Book of Five Rings" are in Public Domain.

Dedications

"War is an ugly thing, but not the ugliest of things. The decayed and degraded state of moral and patriotic feeling which thinks that nothing is worth war is much worse." - John Stuart Mill

---

This book is dedicated to Melanie Chesnut and to all those who would test themselves in the fires of war. There is war in our streets, at our borders, and across our lands. There is war across the seas and in the deserts of the world. We salute you who would dare to make a difference by risking your lives to protect us – the ones you love.

Let us not forget the greatest war of all – that which rages within the human soul. To conquer there is to conquer the world.

On August 1st, 2010, Melanie Chesnut was sworn in as a Marine. Her love of her family, her country, her martial art, her fellow students, and her instructor enriched our lives. We are proud of her lion's heart, but saddened by her absence.
We await her return to our family.

This book was conceived as a gift to her, in the hope that the centuries of wisdom found within these pages would be her shield and strength, no less effective than our arms around her and our prayer for her safety.

"Choosing a path, others lost
Regret for what will be missed
Joy in the journey."

Joseph Lumpkin
Shinsei Hapkido

Dear Melanie,

I've had the pleasure of watching you grow from a girl, to a young lady and into a warrior; but your journey is not over; and as we said when you first got your black belt, "Welcome to the Starting Line". Your now back at the "Starting Line", at a commencement within your life, the end of one era and the beginning of another. When starting any new leg of life, it is important to clearly define your goals (as I am sure you have); but also, continue to evaluate them over time to ensure life's path has not changed without you.

I struggle wanting to share everything I have come to learn within my Journey; however, wisdom is not knowledge that I can pass along, but the action for which you live, e.g. I have the knowledge to loose weight; but struggle with the wisdom to bring this knowledge into action. With this understanding, keep in mind sometimes the action you must take is patience and waiting or the action of inaction.

Many individuals at your age seek to live the "American Dream", but I caution against this. Its a never ending path of chasing material possession in an attempt to find happiness. Instead of the "American Dream", seek to live "Your Dreams".

Happiness is only found in the here and the now. Keep in mind life is a mystery that does not require to be "figured out". Just live, stay light-hearted and always keep a sense of humor about yourself. You can find strength through humor. As you learn, live and love remember change is inevitable; and paradoxically it is the only constant. This is best stated in the phrase, "And this too shall pass". While this phrase will bring you comfort during times of need; it also serves as a reminder to live in the here and the now, to savor every moment.

Life is composed of a series of moments; past and future does not exist. They are only abstract concepts of the present. Therefore do not seek to live to "Carpe Diem", "Seize the Day"; but to "Carpe Temporis Punctum", which is to "Seize the Moment".

May God Bless You.
Your Friend in Life and the Arts,
Rick

Go with God speed, serve with pride, and come home safely.
-        Daryl Ray Covington

Melanie,
 It is my honor to know you, to train and spend time with you. I pray
that all of your life you will be blessed as you have blessed me. I
cannot accurately express just how much you mean to me and how
you've changed my life but I want you to know that you are a light in
the darkness wherever you go. You are my inspiration, my best
friend, and more than that...my sister. I love you girl.
        Love,
        Elisabethe Dorning

Few things in life truly matter and even fewer people, not because
people don't matter but because they choose not to. People and
events fly in and out of our lives so fast it's often like trying to catch
the wind. But for those people who do matter, those who do make a
difference in our lives, whether in be in word, in deed, in song, or in
friendship, those are the people we should always take time to thank
God for, because he's allowed them as a blessing in our lives.
Melanie, never slow down or let the world around you hold you
back, but always make time to be still. The time I've known you I've
been blessed to see glimpses into your life and its been those small
joys we often don't take time to experience. Keep the faith, Run the
good race and know your Shensei Family will never be too far away
to find you. There is no place you will go that God will not be there.
Keep God first, stay constant in prayer and meditation, and never
forget who you are because that's the Melanie that has impacted my
life and been a blessing in it.
Josh Buckelew

The Seasons of the Warrior
By Ed Byers
For Melanie Chestnut

In Spring, the Warrior trains. It is time to learn new techniques that encourage additional personal growth. The Warrior's physical movement is made to fit the technique. The Warrior approaches the new concept with an open mind. Training becomes an exploration of new ideas. And Spring becomes Summer.

In Summer, the Warrior trains. It is time to fit the new technique into the Warrior's mindset and physical skill level. The Warrior adapts the new technique by changing it to become comfortable physical movement. The Warrior's existing skills are enhanced. Training becomes the blending of new concepts and old movements into a new reality. And Summer becomes Fall.

In Fall, the Warrior trains. It is time to experiment with the new technique. The Warrior teaches students and peers the new technique. The Warrior observes how others adapt the movement to fit themselves. Training becomes a refinement through exposure to others. Concepts are reborn and techniques age properly. And Fall becomes Winter.

In Winter, the Warrior trains. It is time to master the technique by removing unnecessary movement and ideas. Homage is paid to ancient Warrior Priests by tempering the technique with justice, honor, and respect. Wisdom is gained through patience and labor. The Warrior begins to see the smallest hint of the technique's potential. And Winter becomes Spring.
And the Warrior trains......

"For evil to flourish, all that is needed is for good men to do nothing." Edmund Burke

Martial artists typically fall into two categories, those who view their training as a hobby or exercise program and the much smaller group who consider it an integral aspect of a way of life. This small percentage has chosen to train their mind and body from the Warrior's perspective. Trophies and rank are of little concern to them while enhanced awareness, improved mental and physical fitness and the ability to protect themselves and their loved ones is of paramount importance.

As a martial artist who has decided to follow the Warrior's path, we come to view the world differently from those around us. Situational awareness is heightened as is our acceptance of personal responsibility. We come to realize that our desire and willingness to better ourselves and our environment is our greatest resource. Regardless of opportunities or obstacles, a Warrior knows that success or failure hinges on our own merit and we take responsibility for failures and learn from them for future conflict.

Realistic training entails pain and hardship. A Warrior accepts this and understands that nothing of value is gained without sincere effort. Whatever rank a Warrior carries has been earned, never given.

It can be a lonely journey. Few will understand our viewpoint and many will openly ridicule it. The Warrior is sometimes ostracized and even feared for their willingness to stand apart and alone. Understand that the Warrior is only truly appreciated when the timid cry out for protection. In a world largely populated by sheep, the sheep dog are all too often only welcome when the wolves attack.

While not all Warriors chose to serve in the military or as Law Enforcement Officers, those that do deserve a special appreciation. By placing themselves in harm's way each day, they are the most obvious and praiseworthy of Warriors. As an instructor, I often find myself pushing them especially hard because I know that what they take from the Dojo could one day mean the difference between life and death for them, their comrades or the populace they protect. It is a duty I do not take lightly and I believe that I honor their commitment by expecting more from them.

Originally, the word samurai meant servant or one who serves. I like to think of the modern Warrior as serving country, community, clan and self. Whether in business, social interaction or charging into danger, we see and appreciate the Big Picture and thereby consider more than just personal gain or safety.

Along with tempering the body and spirit, expand your view and knowledge of the world, both around you and beyond your borders. Commit to study history, anatomy, psychology and the strategy and tactics of the Warriors who proceeded you. Learn from both great victories and tragic defeats.

Finally, know your strengths and weaknesses. The oak and willow both have their place in the world. Rather than bemoan a lack of physical strength, make use of flexibility or stamina. Take an honest assessment of what you bring to the battle and choose your weapons wisely.

Mark Barlow
Akayama Ryu Jujutsu
Bon Secour, AL
September 2010

Melanie,
I had the pleasure of meeting you at the Akayama-Ryu winter camp. At that time I shared with you how your teachers were so proud of you when you made black belt. I just heard you are joining the Marines and want to wish you the best of luck. All of us patriots are proud of you. Thank you for serving our country.
Semper fi
David Dunn
Springfield Missouri Shinsei Hapkido

Melanie,

Your time, like all others, is not guaranteed on this earth, and it is limited, so don't waist it living someone else's life or dreams. Don't be converted or ensnared into living the results of someone else's viewpoint, or for that matter the world's viewpoint. Don't be drowned by the clatter of someone else's opinion or the world's opinion. Doing so will extinguish that inner witnessing voice. Be courageous enough to follow your heart and dreams wherever they may take you. Remember, courage often doesn't roar. it's the quite inner witnessing voice at the end of the day saying "I will try again tomorrow." So, live life each and every day as though it the last one because your time isn't guaranteed on this earth and it is limited!!
Blessings,
Eddie Graydon

The Great Answers

There are questions that we ask,
As we learn and grow;
But they are small and weak,
And do not shake the ground,
Upon which, sure-footed
We erect our banners of truth.

There are other questions,
That give no safety in the asking.

Their answers lie,
As always…
At the frontiers of our lives;
Be they in personal endurance or meditation
Or in science, or in faith
Or yet again in art
Or in the horrors of our wars.

The answers carry all before them;
And we either shrink away,

To some dim-lit corner,
And pretend we never saw;

Or we too are swept away,
To new realities;
And more noble truths.

Where few have been,
And fewer still would choose to go;
This is where the Great Answers lie.

And if you look,
They lie there,
Waiting at your feet.

Chris Dewey 08/09

Wishing you all the best on you journey, I have been traveling 33 years and in my brief time on earth I can truly say that I have never seen the righteous forsaken or his children begging for bread. I aspire to be more inspirational but I'll leave you with a Psalm.

Yours in the Way,

David Lyons

Psalm 33: 13-22 (KJV)

13 The LORD looketh from heaven; he beholdeth all the sons of men.

14 From the place of his habitation he looketh upon all the inhabitants of the earth.

15 He fashioneth their hearts alike; he considereth all their works.

16 There is no king saved by the multitude of an host: a mighty man is not delivered by much strength.

17 An horse is a vain thing for safety: neither shall he deliver any by his great strength.

18 Behold, the eye of the LORD is upon them that fear him, upon them that hope in his mercy;

19 To deliver their soul from death, and to keep them alive in famine.

20 Our soul waiteth for the LORD: he is our help and our shield.

21 For our heart shall rejoice in him, because we have trusted in his holy name.

22 Let thy mercy, O LORD, be upon us, according as we hope in thee.

All the best - David Lyons

To Every Thing There is a Season
Ecclesiastes 3

To everything there is a season,
and a time to every purpose under the heaven:

A time to be born, and a time to die;
a time to plant, and a time to pluck up that which is planted;

A time to kill, and a time to heal;
a time to break down, and a time to build up;

A time to weep, and a time to laugh;
a time to mourn, and a time to dance;

A time to cast away stones,

and a time to gather stones together;
a time to embrace, and a time to refrain from embracing;

A time to get, and a time to lose;
a time to keep, and a time to cast away;

A time to rend, and a time to sew;
a time to keep silence, and a time to speak;

A time to love, and a time to hate;
a time of war, and a time of peace.

Ecclesiastes 3:1-8

I have watched you grow from a cute young lady to a beautiful woman and
a new season is upon you. Remember the Lord thy God first and foremost
and return to us safely. With all my hope and love,
David Williams.

Melanie, We are so proud and stand behind you in your support of our country by joining our military service and the U.S. Marine Corps at that -- **"Oorah"!!!!** Here's something to keep in mind, SEMPER-FI = "Semper Fidelis" which means, "ALWAYS FAITHFUL." Thus, always remain faithful to God, family, friends, and country. May God bless and be with you on this phase of your life's journey.

-Mike Lewis-sensei,
Seishin-Do Karate Seishoku-Kai. U.S. HQ. Karate for Christ Intl.

Quotes:

"Creed of the Christian warrior, 'For we wrestle not against flesh and blood, but against principalities, against powers, against the rulers of the darkness of this world, against spiritual wickedness in high places.'" [Ephesians 6:12 (KJV).] -Mike Lewis.

"In being a Christian warrior, you must insure that your spirit is never broken while in the service of Christ, and for this country." -Mike Lewis

"The warrior, who is of the world, is aware of everything only when he thinks he should be -- But the mind of the Christian warrior, however, is to be aware of everything at all times." -Mike Lewis.

"Marine Corps integrity is doing that thing which is right, when no one is looking." -Col. Colin Lampard, USMC.

"I can never again see a UNITED STATES MARINE without experiencing a feeling of reverence." -GEN. JOHNSON, U.S. ARMY.

"There are only two kinds of people that understand Marines: Marines and those who have met them in battle. Everyone else has a second-hand opinion." -Unknown.

"Panic sweeps my men when they are facing the AMERICAN MARINES." -CAPTURED NORTH KOREAN MAJOR.

"The man who will go where his colors go without asking, who will fight a phantom foe in a jungle or a mountain range, and who will suffer and die; in the midst of incredible hardship, without complaint, is still what he has always been, from Imperial Rome to sceptered Britain to democratic America. He is the stuff of which legends are made. His pride is his colors and his regiment, his training hard and thorough and coldly realistic, to fit him for what he must face, and his obedience is to his orders. As a legionnaire, he held the gates of civilization for the classical world...today he is called United States Marine." -LTCOL FEHRENBACH, USA, in "This Kind of War."

"You cannot exaggerate about the Marines. They are convinced to the point of arrogance, that they are the most ferocious fighters on earth - and the amusing thing about it is that they are."- Father Kevin Keaney, 1st MarDiv Chaplain, Korean War.

Everyone wants to be strong like the mighty oak, but when the great winds come the oak falls over. Everyone should strive to be like bamboo, because bamboo is strong like the oak, but when the great winds come the bamboo bends, adjusts, and lets it pass. You have made a big difference in my life, Melanie, thank you for that.

We love you.

Galen Dorning

Melanie Chesnut-
The principles you have learned in Shinsei Hapkido about hard work, balance, strength, and others will serve you well, don't forget them. Love the Lord with all your heart and He will be there for you... even in the times of darkness.

Jim Lindsey, Sensei
Paradise Hill Karate for Christ Certified School.

THE WARRIOR'S HEART REVEALED

The book, "Warrior's Heart Revealed," is the textbook of a martial art called Shinsei Hapkido. It is presented here, not because it is the greatest art, but because the art draws on a deep well of wisdom from many modern day masters and thus reflects some of the best martial art philosophy of today. The system stresses a balance within the student and their life. Not only must the body be trained, but also the emotional, mental, and spiritual aspects of a student must be addressed in order to produce a fully capable person. Melanie Chesnut is a student of Shinsei Hapkido.

FOREWORD, BY ERLE MONTAIGUE

Far too many books written about Karate and the martial arts place emphasis only upon the killing aspects and worse still upon the 'seeing who is better' aspects of the art. This book does not. If fact, it reflects upon the martial arts from a different time and place, as it was hundreds of years ago, before it became popular in the West and people were awarded ranks in astonishingly short periods of time.

This book speaks about the true spirit of the martial arts ...when it should be used and when it shouldn't; it accurately recounts the history of a number of mainstream martial arts and then goes on to put forth the philosophy of 'Christians In The Martial Arts'.

For those who wish to read something very unusual about the martial arts from a totally different point of view, I highly recommend this book. It will make a valuable addition to your library.

Erle Montaigue

BEGINNINGS

The Beginning for Joseph Lumpkin - Winter of 1972

We were led out of a small, secluded house that sat miles off any
main road. There we gathered and dressed for war...with ourselves.
We were quiet and reverent at heart, awaiting the beginning of class.
Our minds became as cathedrals whose doors were open wide,
beckoning, empty, eager. He entered the room and left the house,
silent as a breeze and we followed him into the virgin snow.

The cold air came as a bracing rush into our lungs as we breathed in
the ozone rich twenty-degree tonic. The snow bit at our feet as we
walked into the woods along a narrow path, twisting between large
trees and clumped brush. The forest closed in behind us and the dark
canopy held our presence secret until we came to the clearing. The
sun cast a spotlight marking the place of enlightenment. The old barn
had stood for a hundred and sixty years. Its beams had been cut by
hand and were a foot and a half square and twenty feet long, each
spanning the length of the place. We walked in and bowed to the
raised mat made of straw and quilts. It was time for the warrior
within to awaken. We felt the groan and stir of anticipation deep
within. Our souls were agape. It was time for church. It was time for
class. It was time to fly.

He said nothing but instead settled on his haunches facing us. Class
would be conducted in silence, him leading, we following his every
movement. He sat transfixed and prayed silently. He slowly arose
and climbed to the top of a homemade ladder. Stepping onto the
center beam he strode like a cat, stopping at a point that stretched
above the mat. He leaped, tucked, and for a moment seemed to
pause in mid air, hanging there, waiting for the command of gravity
to overtake him, waiting to ride it down like a horse under him. He
hung, looking up at the roof, his back flat as if he were laying on
some invisible surface.

For a time it seemed he had surprised nature and caught the force off
guard, then it snatched him out of the air and down to the mat as if it
was angry at his taunting. A ribbon of light separated him from the
mat when he slapped the surface with his hands as if striking for his

life. He landed on his back like a pillow dropped from the bed. As he got up, we were climbing the ladder behind him. For an atom of time, for a twinkling of an eye, he had shown us we could fly and we believed. In believing and in trusting that he had prepared us, we also took to the air and our souls laughed at the tickle of gravity's greedy hands around us.

The only words I remember being said for those two hours were his closing prayer: "Father, as they have faith in me, and I have faith in you, so let my faith guide these students to you."

His name was James Hiner, student of Byung Kyu Park, student of Choi. Although he taught for 15 years, I am one of only three people Mr. Hiner ever promoted to black belt. My name is Joseph Lumpkin. I am a life-long student. I am a teacher.

You may view faith as trusting in the outcome even though you cannot see the path. In practicing breakfalls we jump over a rope suspended in mid-air at the height of our shoulders. If you attempt to watch where you are going you will spear the mat and break your neck. You must tuck and flip. This makes it impossible to see where you are going or where you are in space. You must trust your technique in the blindness. This is faith. We build up our level of faith by seeing the techniques work over and over again. I have lost count of how many times God has protected and led me in the darkness, in my blindness. I have faith in Him although I do not see Him. Each time I feel His hand on my life my faith increases.

As teachers, we must live the life and communicate the faith.

The Beginning for Daryl Covington

There was nothing fancy. A barren area in the grass near the side of a cedar-sided home served as the dojang. Suspended from a limb of an old oak tree by a worn rope was an Army issue duffle bag full of rags, wrapped in duct tape. It was while practicing on this piece of homemade equipment that I first fell in love with the martial arts.

Years later, after earning my *4th Dan* in Hapkido and a *2nd Dan* in TaeKwonDo under Master Knight and Dyung Vu, I was introduced to wrestling, kickboxing, and finally under the direction of a true living legend, I obtained my *3rd Dan* in Kodokan Judo from Phil Porter.

Unlike Do Ju Nim (the founder and head of a system) Joseph Lumpkin, there was no Christian emphasis in the arts I practiced. I grew up in church, and accepted the Lord's saving grace into my life at a young age, but never saw how the two could fit together until the early 1990's.

I spent my martial arts career competing and fighting. Although I had trained a few students, and taught self-defense to many groups, it was not until Do Ju Nim encouraged and persuaded me that I began to teach the arts as part of my Christian outreach ministry.

After years of training, I realized there are many approaches to the practice and teaching of martial arts. Some teach "self-defense", while others claim to teach for other reasons, including health and fitness, well-being, competitive sport, and performance. The way the student thinks about the art, his way of learning precisely what the instructor is communicating, as well as the student's ability to implement that knowledge is directly proportional to the relationship the student develops with his art, his mind, and his body. What a student gains from the teacher will often depend upon the quality of effort and instruction put forth by the teacher.

While modern America has made martial arts into a form of fitness conditioning, recreation, and self-defense, the pure study of martial arts is not by nature a casual activity, nor should it be taught as such.

Instructors must bring a serious attitude and respect for the art they are teaching to the dojang; else the students will treat "martial arts night", like "soccer practice" or "T-Ball night". To slow the demise of martial arts in the United States, we must teach the arts as something more than some new work out program. We are not aerobics instructors. We are artists.

How can you be more than simply a good martial arts instructor? You must teach your students how to learn. Teaching "how to learn" is a common practice in traditional martial arts. Teaching the student how to learn while you are teaching them the arts has been overlooked in this country. Unfortunately, some so-called martial arts schools have discarded "learning to learn" and put their emphasis on "how to defend yourself if you ever find yourself being attacked by point sparring bandits!"

In the Shinsei Hapkido system, our students are not simply taught the "basics". They learn to think, analyze, and apply. The problem with most teachers in the United States is the fact they have focused on "tournament" success to the point of disassociating the mind from the body in their training. When the student learns how to "score", they are not being taught martial arts, they are being taught how to compete. By teaching in this way, the instructors have lost focus on teaching the students how to learn the arts. Instead, the students are learning how to collect useless trophies and ribbons.

Many students drop out of classes because they are looking for the quick fix. They are seeking the "ten months to black belt plan", and trust me, they are out there. In those classes the instructors are putting the emphasis on achieving rather than learning. The rewards of teaching the students to learn are immense. Students who are taught to learn will have a lifelong loyalty to the instructor, and to the arts. Those not taught to achieve will be of the many who say, "Oh yeah, I USED to take martial arts".

Many times we must put the student in a place to learn. Other times, we must allow them to fail. To be a good martial artist, one must be able to admit mistakes, be smart enough to profit from them, and strong enough to correct them. Students will make mistakes…sometimes grave mistakes. We, as instructors, must be there for them. Not to insulate them from failures, but to help them learn from them. Tom Watson of IBM was once faced with a dilemma. An employee had made a mistake that cost IBM over $600,000.00. When asked if he was going to fire the employee, he responded, "No way. I just spent $600,000.00 training him. Why would I want to fire him and let another company get his experience"?

As a final thought on being a good instructor, I would like to remind the teacher he has to spend time with his students in order to be able to know their problems, but he must spend time with God in order to solve them.

Beginnings Collide – Winter of 2003

In January 2003, an unlikely band of Christian warriors swept the World Cup Martial Arts Competitions in Cancun, Mexico. Most were trained in the southern delta of Mississippi, in the poorest county in the United States. Most had trained for only three years. All were students of a little known style simply referred to as "Shinsei". Nineteen Shinsei students took back to Mississippi and Tennessee 46 gold medals, 11 silver medals, and 16 bronze medals. The feat caught the martial arts community by surprise and was reported in the May 2003 issues of Black Belt Magazine and TaeKwonDo Times. Such an accomplishment reflects an astounding shift in internal and physical training and philosophy from that of martial arts seen in the West today.

Plaudits for the style and philosophy were echoed in the acclaim issued by Master Kang Rhee when he awarded recognition to Shinsei Hapkido as a separate and unique Kwan. Shinsei Hapkido is also recognized by the World Taiji Boxing Association, and is a member and recognized Hapkido Kwan of the World Kido Federation guided by In Sun Seo.

The roots of Shinsei reach far back into history - over 1,000 years, to the beginnings of codified martial combat, taking its linage from Hapkido, Jujitsu, Judo, Shotokan Karate, wrestling, Muay Thai, and Pa Sa Ryu TaeKwonDo. Shinsei is a distilled and purified form of effective and applicable martial combat.

Here before you, is the history and philosophy of this remarkable form of martial art written by its founder and head instructor. The insight of these two men is presented in conjunction with essays on teaching, application, and personal insights from some of the greatest men in today's international martial arts community. Essays were solicited from a number of highly-ranked teachers in various styles that are compatible with the application and philosophy of Shinsei Hapkido. In these pages, we will examine the various components of form and philosophy that compose a martial art. We will then discuss how to teach these components. Major contributors are masters highly ranked in various arts. The insights they provide will inspire all teachers and students alike. Anyone interested in the philosophy, history, techniques, and teaching tools of martial arts will be delighted.

Every Culture Has Its Own Art

If one were to travel to any continent, or examine any culture, there would be an obvious difference in the ways the people cook, create, dress, and fight. These differences are formed from the physical attributes of the people, their environment, terrain, and the various ways a distinct culture views its world. Some traits of societies are obvious, such as food choices. People living in a cold climate will not have many fruits from which to choose. Some characteristics of a culture are not so obvious, such as how a form of combat evolves. For tall and thin or short and stocky people living in pastoral settings as opposed to those in the woods and underbrush the strengths of the people and the restrictions of the environment are different. Yet, if we examine each culture, there will be punches, kicks, or throws, all with their own unique flavor of execution.

The arts can be compared according to their function. Arts of self-defense will differ in function and form from styles used for sport and contests. The sport of Judo in Japan is not dissimilar to the Greco-Roman wrestling of the West. The arts of self-defense in Japan are very similar to those in China, Korea, and even Israel. The martial art of Jujitsu used in Japan looks and functions like the art of Krav Maga used in Israel.

There is boxing in the West and Karate in the East, wrestling in the West and Judo in the East. The French have Savat and the Philippines have Kali. Each culture has its own martial system and in all of them, a punch is just a punch and a kick is just a kick.

There may be no direct connections between a martial art and religion but it should be noted the culture and society spawning the art will also have a central belief system, which would have colored everything in it. From the time any art reached the masses what seems to influence it is the physical body type, way of life, environment, and the philosophy of the people practicing the art.

According to Sifus Derek Prout and David Kash of the Cloud-Forrest Chin Woo Association, "Terrain was not so much an issue as the body type of people from these regions. The northern Chinese were taller and therefore developed a system of martial arts that incorporated the strengths and weaknesses of their body types. It may or may not be coincidence that the terrain in that area is more flat and open. It is debated that they may have developed that way from an influence and/or migration of Mongolian Nordic races.

The southern Chinese were shorter and more compact and therefore developed a system of martial arts, which incorporated their strengths and/or weaknesses. It is also true that the terrain in that area has less open space. I will note that when this topic is brought up in the Chinese martial arts circles that these styles are usually attributed to the body types and not necessarily the terrain of the areas. In other words, a person with longer legs and longer arms is going to capitalize on these issues when in personal combat with someone who is shorter. The reach is longer so they use it to keep the opponent at bay while striking him as often as possible. Likewise, a shorter, stockier person is going to develop a close-in style of martial arts in order to close the distance as quickly as possible and stay there while wrapping up and nailing the opponent."

One of the major concerns during the writing of this book was to maintain a distinct separation between the Christian religion and the various philosophies of the cultures contributing to the development of martial philosophy. The relevance and genius of Sun Tzu, Miyamoto Musashi, Alexander the Great, or Genghis Kahn must not be expunged from the annals of martial arts simply because of a difference in religions. There is the wisdom of man and then there is the wisdom of God.

As long as these two points are not in conflict and one does not worship the wisdom of man, there will be balance and all will be well.

A History of Shinsei Hapkido

A Dragon emerges from a ditch (Kae-ch'eon-e-seo young nan-da) – A Korean Proverb

The combined backgrounds of Shinsei Hapkido Do Ju Nim Joseph Lumpkin, and Kwan Jang Nim Daryl Covington were synergistic in the development of Shinsei.

Joseph Lumpkin is the C.E.O. of Karate for Christ International. He has training in Hapkido (6th Dan), Shotokan Karate (2nd Dan), as well as Judo (2nd Dan). His love for the arts and his desire to train in what works motivated him to continue training and earn a 3rd Dan in Akayama-Ryu Jujitsu under Nidai Soke Mark Barlow. Lumpkin's Hapkido lineage passes from its founder, Choi, to Park, to Lumpkin. His Akayama-Ryu Jujitsu linage comes from its founder, Alex Marshal to Mark Barlow, who is now the head of the system, to Lumpkin. His Shotokan Karate as well as his Judo lineage comes from Clifton Green to James Hiner to Lumpkin. The contributions of Lumpkin include Hapkido, Shotokan Karate, Kodokan Judo, Akayama-Ryu Jujitsu, and practical applications, as well as law enforcement tactics.

Joseph Lumpkin began his training in 1971. He trained in Hapkido under James Hiner, who was trained by Park, who was trained by Choi, the founder of Hapkido. Lumpkin's relationship with Hiner and Park provided insight into the roots of Hapkido. Byung Kyu Park began teaching Hapkido in the United States in the early 1970's, but only to those already holding Dan ranking in other arts. Although Mr. Park kept secret the real reason he had come to the United States, evidence leans to the fact that he fled Korea or was exiled due to his alleged affiliation with those involved in an assassination plot against the President of Korea. It should be stated emphatically there was never any evidence that would implicate Park in the assassination. Upon arriving in the United States, Park moved to Indianapolis, Indiana, possibly becoming one of the first teachers of Hapkido in this country. It was here that James Hiner met Master Park. Hiner had already earned his black belts in both Shotokan Karate and Kodokan Judo. The Korean government responded to Park's absence by placing restrictions on his wife and children. Being in the United State alone, unable to bring his family to the United States, and unable to go home, Mr. Park's emotional and physical health suffered. In the mid to late 70's Mr. Park closed his studio and moved to California to teach acupuncture in a small university. He never published any material, claiming his art was too dangerous and not meant for tournaments or demonstrations. When asked about those who trained him, he would smile and say, "the mountain peaks." Because of the flowing motion and similarity of his techniques we assume we was also taught by or trained with Myung Nam.

He is remembered as a calm and quiet man, quick to smile, and warm of heart. After the death of Park, Lumpkin continued to train under Hiner until he retired in 1982. In the early 1980's, Joseph Lumpkin took the first steps in the creation of a more effective and balanced martial art. He pursued an educated and painstaking study to determine and remove any techniques considered ineffective or dangerous to execute. By removing slow and ineffective techniques and by eliminating or simplifying techniques that were overly complicated or risky, a system was created containing only the best and most effective techniques from the wide variety of arts. The creation of Shinsei owes its origin to a synthesis of Hapkido, Akayama-Ryu Jujitsu, Kodokan Judo, Shotokan Karate, Wrestling, and Shoot Fighting. Lumpkin blended the systems, discarding what did not work, and integrating the rest into a tightly meshed system with a well-articulated martial philosophy.

As a result, many kicks and several other techniques were eliminated, having been proven to be risky in combat situations or difficult to perform under combat conditions. The outcome of this objective re-creation was an art suitable and usable for a wide range of ages, sizes, and body types. In 2006 the style was recognized by the World Kido Federation and Lumpkin was awarded a 7th dan under In Sun Seo. As a side note, Seo commented that the Shinsei techniques looked like those of a dear desceded friend of his; Nam.

The Shinsei system has more kicks and strikes than most modern Jujitsu styles but fewer high and aerial kicks than modern Hapkido. Shinsei's kicking philosophy was adapted from the temples of Korea and from Shotokan Karate, which emphasizes powerful, simple, and direct kicks. Although the aerial kicks of TaeKwonDo were never included, five supplemental kicks remain but are executed in a very straightforward approach.

Shotokan's philosophy may be stated as, "Enter at one stroke", indicating a more direct and no-nonsense approach to combat. This philosophy and application fit perfectly with the Shinsei philosophy which is stated as "Evade, Invade, Control". The invasion principal must be a strong and direct movement. If punches or kicks are used for this purpose, the linear and forceful aspect of Shotokan or Pa Sa Ryu, its Korean counterpart, is helpful.

Owing to the fact that the majority of all fights end up on the ground, Shinsei added many ground techniques and a solid arsenal of grappling movements from wrestling and shoot fighting, as well as Judo and Jujitsu. Many techniques were taken from the direct approach of Akayama-Ryu Jujitsu, Brazilian Jujitsu, and Wha Moo Hapkido, as techniques were stripped down to the minimum number of essential movements. Shinsei was developed to supply an effective attack and defense strategy over the three ranges of combat: close range, mid-range, and long-range hand-to-hand. These three distances are clearly visible in the effective extents between the long range of the kicks, the medium striking range of the punches and backfist, and the close range of the elbow strike, kneeing, throws, and grappling. Shinsei also has special techniques to defend against weapons such as the knife and gun. Shinsei has been as tried and tested as effective in street combat and to date has been taught to police officers, sheriff's deputies, and even fugitive recovery agents and has proven itself in a wide range of situations.

Life and war are played out in the same watchful and strategic ways. The philosophy of life and combat developed by Shinsei can and should be used in everyday life. It is applicable to life situations and to life and death matters. When training, in research, and in life's experiences, always take the useful, reject the useless, and realize your own potential.

Daryl Covington began his training in TaeKwonDo. He currently holds a 4th Dan in Pa Sa Ryu TaeKwonDo from the system's founder, Master Kang Rhee, and is a certified branch school instructor of the style. At a young age, Covington desired to learn more self-defense oriented arts as well, and began his quest in Hapkido, simultaneously training in Kodokan Judo, earning his

Kodokan Judo 3rd Dan directly under 9th Dan Phillip Porter. Covington is also a Level 2 Certified Coach with the United States Judo Association. Covington's Hapkido training was from Dyung Vu, at the time, a 5th Dan. Dyung Vu was a Vietnamese gentleman that was trained by a student of Ji, Han Jae, in the country of Vietnam, thus linking Covington's roots indirectly to Do Ju Nim, Ji, Han Jae. Master Knight, a student of Vu, was also involved in the process. Vu passed away after Covington had received his 4th Dan, the highest rank to which Master Vu would promote students. Covington was later awarded the 5th Dan in Hapkido from the Grandmasters Council, and aligned himself under Joseph Lumpkin as his instructor, receiving his 6th and 7th Dan promotions in Shinsei Hapkido. This is the lineage of Hapkido Covington Brings to the Shinsei Hapkido System.

Covington's Kodokan Judo lineage is from Jigoro Kano to Kotani Sensei to Phil Porter. His TaeKwonDo lineage is directly from Pa Sa Ryu system founder, Kang Rhee. These elements, along with Covington's extensive knowledge of grappling, all flowed into the Shinsei system.

Covington is also a student of Dim Mak, and is one of very few American's hold actual instructor status in the art. Covington trained and holds his Instructor Certification (Level 1) directly under Erle Montaigue, bringing knowledge of meridians, knockouts, antidotes, etc., into the upper levels of the Shinsei system.

By the mid 1980's, Joseph Lumpkin established the art of Shinsei and was teaching the art to a select group of students in the Alabama area. At that time the art was not named and no written curriculum was established except a manual, which Mr. Lumpkin began some ten years earlier. When Joseph Lumpkin took over the presidency of Karate for Christ in the mid 1990's, it was decided the ministry should have an in-house system in which members could be trained in and promoted on a worldwide basis. In Korean "Shinse" means a debt of gratitude owed. In Japanese the name, "Shinsei" means holy or new. Both meanings represent our Christian belief.

Soon after the official naming and establishment of Shinsei the style began to take on national dimensions. Seminars were held, classes were taught, and the style began to grow. It was at this time that Daryl Covington contacted Mr. Lumpkin. Dr. Covington was the founder of Wha Moo Hapkido, a fully conceived and robust art. Covington had established and nurtured several schools, which taught Wha Moo Hapkido. Dr. Covington asked to join Mr. Lumpkin as an active participant of the Karate for Christ ministry. Dr. Covington resigned his position as head of Wha Moo Hapkido and came under Mr. Lumpkin in the rank and organizational structure of Shinsei, bringing all students and teachers under him into the Shinsei family.

The two men began in earnest the monumental task of obtaining recognition from the highest levels for Shinsei Hapkido. Much time, testing, and labor went into this approval. Lumpkin and Covington diligently pursued their quest for the certification of Shinsei Hapkido to the highest levels of the Korean Government. Master Kang Rhee, the National Hapkido Association, Erle Montaigue, the World Taiji Boxing Association, and many others have recognized Joseph Lumpkin as Do Ju Nim and Shinsei became the only Hapkido Kwan to our knowledge that is a true Christian art with such backing.

By establishing a true martial philosophy, then designing and tailoring Shinsei Hapkido accordingly, the system took on a unique quality and became recognized as a Hapkido Kwan. Every technique fits into the philosophy of "evade, invade, and control". In his work with law enforcement agencies, and being a bounty hunter himself, Lumpkin knows what works and what doesn't. Daryl Covington, a U.S. Army war veteran, knows as well. It is the combined 58 years of experience that makes Shinsei Hapkido practical, effective, and one of the best self-defense systems available today.

A Ministry of Discipline

Think of what we could be if we could give it all away.
The Philosophy of Shinsei

Shinsei means – New, sacred, genius, truth, purity, a new face, nova. Shin, in the Korean Language, means "God, or "Deity". Shinsei means to owe a debt of gratitude. Thus the name Shinsei, in Japanese and Korean, represent the Christian nature of the art, and our indebtedness to God for all. Within its teachings are contained not only the traditional martial arts techniques, but also a philosophy by which to live, grow, and succeed in life. It also includes a philosophy that calls for a balanced and educated individual. It demands self-knowledge, self-sufficiency, and self-improvement. It also demands loyalty, strength and humility.

As more and more martial artists seek to escape the death grips of various political martial arts organizations, Shinsei has opened up an oasis of monetary, political, and religious relief. Teachers and students of Shinsei are called upon to teach the art for the love of it and not for the money it can raise. Teachers are expected to be honest and as non-political as possible in the running of their classes.

The wise adapt themselves to circumstances, as water molds itself to the pitcher.-(Chinese proverb).

The philosophy of Shinsei is to **evade, invade, and control**. Evade the attack by never meeting force with force. Invade by penetrating the opponent's defenses. Control the opponent by seizing his balance or foundation. Once contact is made, control is acquired. Once the opponent is seized he should not be released until he is overcome. The philosophy of Shinsei is a well-considered and deeply conceived philosophy that is to be lived out in the lives of its practitioners. The evolution of the art was based on having a robust and effective arsenal of techniques and a philosophy that would allow those techniques, and the traits they developed, to apply to life as well as martial arts.

Thus, to live within the philosophy of Shinsei one must strive be a multi-dimensional person. The philosophy applies to adults and children, to business and social events, and in war and peace. Combat or the ability to fight is but one manifestation of the application of this philosophy. We do not train to fight. We train to live and to do so in a full and productive manner. We train to know ourselves.

The height of cultivation is really nothing special. It is simplicity, the ability to express the utmost with the minimum. It is the halfway cultivation that leads to ornamentation. –Bruce Lee

Shinsei is one of the few martial arts based on Christian ideals. Yet, within the philosophy of Shinsei we do not overlook the knowledge of combat brought to us by those great warriors of the past, who may have been non-Christian, as there is a great difference between a philosophy and religion. Keeping this in mind, we can discern and separate the knowledge of warfare and martial arts from the Buddhist or Shinto beliefs of the generals and teachers of the past.

Let us consider the meaning of "a philosophy" as it differs from "a religion." Simply stated, a philosophy is how one views life. A religion is whom or what one views as God. Thus, in this context, it can be said a philosophy is how to view, and a religion is whom or what to view. This book does not cover the religious approach of Shinsei. The sole purpose of this work is to explain the basic martial and life philosophy held and taught within the Shinsei system. Just as to glean the martial training of the old masters, one must taken into account their background, to fully understand the applications of any tenets expounded upon here, one must lay them over a framework and foundation of the love, mercy, grace, and salvation through Jesus Christ.

The art of Shinsei is based on the water theory. This statement is not meant to be in the least bit enigmatic but is totally and obviously applicable to life in all situations. Water, when meeting an obstacle, tends to roll off and flow around it, never meeting force with force. Water gives way. Water is very difficult to stop. It tends to continue to flow. Water finds the smallest cracks in an object, and when it does, it invades and fills any gaps and vacancies it encounters. Water wears away defenses, goes into places unseen, invades the smallest opening, and eventually overcomes the strongest obstacle. Water can support the largest ship. Water can also overturn it.

Water shapes its course according to the nature of the ground over which it flows; the soldier works out his victory in relation to the foe who he is facing. Therefore, just as water retains no constant shape, so in warfare there are no constant conditions. He who can modify his tactics in relation to his opponent and thereby succeed in winning, may be called a heaven-born captain. Sun Tzu

Higher worth is like water. Water is good at benefiting ten thousand beings without vying for position... In dwelling, be close to the land. In heart and mind value depth. In interacting with others, value kindness. In words, value reliability. In rectifying, value order. In social affairs, value ability. In action, value timing. In general, simply don't fight and hence have no blame. Lao Tzu

As applied to combat, this theory is made up of three important parts: Passively defend based on evasion whenever possible. Advance toward the enemy so as to invade him at his weaker point. Control the enemy in order to prevent further attacks.

Master Knight, instructor of Covington, on the difference in Hapkido and Aikido: "We Hapkidoists tend to run right of center, while the Aikidoist says they are the center of the universe. The Aikidoist says, "Blend with your enemy until he is no longer a threat". Hapkido practitioners say "Killing an enemy today keeps you from having to deal with him tomorrow!"

As applied to life, we should try to maintain peace and cooperation when ever possible. We should strive to continue advancing in our endeavors and in life. We should never stop pursuing resolutions to the problems put before us. Once we conquer a problem, we should master it so as to not repeat the same issues or mistakes. If life is truly threatened, the threat should not be allowed to continue. Above all, we should live as warriors... learned, studied, dedicated, and devoted.

A Code of Honor

MAT 22:37 Jesus said unto him, Thou shalt love the Lord thy God with all thy heart, and with all thy soul, and with all thy mind. 38 This is the first and greatest commandment. 39 And the second is like unto it, Thou shalt love thy neighbour as thyself. 40 On these two commandments hang all the law and the prophets.

On this simple code rests all truths and all paths to honor. Actions based on these two commandments will bring esteem and peace of mind. One may speak of honor, loyalty, courage, respect, and truth, but these things will be lost and hollow without love and mercy. One may call one and one's students to a higher purpose, but what is purpose without people? All endeavors begin and end with people. Our lives are eventually judged on the basis of how people were treated along the way, as a result of our endeavors.

This is the Warrior Honor Code of Shinsei Hapkido:

- *Love the Lord your God with all your heart and honor Him above all things*
- *Believe in Jesus as Son of God, Savior, and resurrected Lord*
- *Be loyal to God, yourself, and those who have benefited you*
- *Love others and treat them as you wish to be treated*
- *Respect others. The greatest folly is not to have respect for your enemy*
- *Whenever possible, show love and mercy*
- *Be courageous in all endeavors*
- *Be ready to defend the weak and oppressed*
- *Speak the truth clearly. Let your yes be yes and your no be no*
- *Train hard and seek the truth in yourself*
- *Never stop learning*
- *Focus your mind. Most problems come from lack of concentration*
- *Be tenacious. Never, ever, ever give up. If you give up you have lost*
- *Expect none of these attributes from others but do them anyway, because it is right*

One could list the desired attributes of students and teachers of Shinsei as:
Faith; Honor; Respect; Mercy; Courage; and Determination.

If one were to look at the codes insisted upon by the great Samurai of the past, we would see great similarities in social and ethical applications of the Shinsei and Christian codes. All they lack is the Christian viewpoint of eternity.

The Code of the Samurai as stated by Musashi:

- *Do not think dishonestly*
- *The Way is in training*
- *Become acquainted with every art*
- *Distinguish between gain and loss in worldly matters*
- *Develop intuitive judgment and understanding for everything*
- *Perceive those things, which cannot be seen*
- *Pay attention even to trifles*
- *Do nothing, which is of no use*

The same attitude is reflected in a general creed stated below.

- *You must know your weakness. Unless you know your weakness, you cannot move yourself forward. This is the way to true strength.*
- *You must know others' pain. Before you get angry, always think of the reason for one's behavior.*
- *A threat to ego is not a threat to life.*
- *You must heighten your sense of values. Consider deeply what is worthwhile in life.*

Train. An unpolished crystal does not shine; an undisciplined Samurai does not have brilliance. A Samurai therefore should cultivate his mind. Anonymous

Teaching Martial Arts Philosophies and Traditions

The martial arts are not only art forms, they are specific fields of scientific study as well. In order to truly instruct, the Sensei must be scientist, psychologist, philosopher and artist. Many students are taught forms and techniques, but if you are not teaching the traditions, theories, principles, and laws upon which these techniques and forms are based, you are robbing the student of the artistry within the martial arts. It is after one is taught all these concepts he is able to practice them with passion and discipline, because he has caught a glimpse of understanding as to what he is doing, why he is doing it, and how it works. It is only then the student may begin to master the arts, and he himself becomes the artist.

A Meeting of Old and New

True wisdom is found in the fear of the Lord, but there is knowledge from the experiences of the world that the Christian warrior can apply. We must keep in mind the distinction between religious or higher wisdom from God and his perfect Word, and the wisdom of men. In these pages the wisdom of the warriors from the past has been brought to bear on the problems of today. Warfare, fighting the good fight, as well as strategies in life, business, and issues of personal combat have not changed. It still takes determination, courage, and intelligence to overcome obstacles, odds, enemies and most importantly, oneself. In the grand scheme of things, it is much more important to know and control oneself than others. To do this it is important to have insight, courage, and discipline. The ancient texts and modern essays we draw on in these pages address these issues.

The words from the warriors of today will resonate with us, but to put the words of leaders and warriors of the distant past into our own modern context, a brief explanation of the major texts cited throughout this book needs to be addressed.

There are many reasons why conflict among people exists. First, disharmony may originate from the ignorance of the theory of extremes and opposites. This theory states as things evolve they will do so in opposite or extreme directions. It is true with most aspects of human nature that we have trouble maintaining the center or balanced line. Most failings of human nature can be summed up as imbalances or extremes. We tend to be passive or aggressive, happy or sad, angry or loving... life is seen in extremes. Lao Tzu says, "those who do not know when to stop or do not know what is enough will encounter troubles". Error occurs in extremes when one strays from the center path. This is either the opposite of a state or an extreme of a state. For example, one could be stoic to the point of brutal, or completely undisciplined. One can be passive to the point of inaction, or aggressive to the point of destructiveness.

This leads to the second theory. People who believe they can manipulate others by a display of emotions are not real, thus their attempts will backfire. Being belligerent, violent, angry, or supercilious may temporarily repress conflicts by emotionally intimidating others but it destroys interpersonal harmonies. Violence, intimidation and any lack of "realness" are contradictory to the Way (Tao).

Thirdly, individuals who are hard, stiff, unbending, and intractable in their attitudes put themselves in predicaments of conflict by interposing themselves on the rights and freedoms of others; just as the egotist does not think of others and offends them by his actions.

To place these "Eastern" concepts within a Western and Christian concept, we will take the approach that the problems is one of opposites and balance. The view of this is extended both microcosmically and macrocosmically. It affects the view of the individual and the universe.

"Te" (as in the Tao Te Ching) can be viewed as virtue or integrity. It is the purity of intent and action used to carry out what is right. To gain integrity one must be exposed to and taught what is right. To walk the path, one must have integrity. To step out on a limb, I might point out that there are non-Christians who have high integrity and live with kindness toward others, and there are Christians whose integrity I have questioned. I continually question my own, for that matter.

As warriors we are called upon to live the Way (Tao) always with integrity (Te).

Over the long and extensive process of research and investigation to complete this work, innumerable volumes of wisdom were studied – both Eastern and Western. A Chinese military philosopher, Sun Tzu, wrote "The Art of War" about 2,400 years ago. It has been applied to life and business, as well as warfare. It is a practical guidebook dealing with human interaction (Li, 1985). "The Art of War" gives insight into interpersonal behavior. According to Sun Tzu, people or groups who want to solve their conflicts with others should make a move that invalidates the aggressors' expectations. This can be done in one of three ways. We acquiesce to the demands. We give them what they want but not how they want it. We produce something that looks like what they want but is not it at all. Obviously, the second and third options are suitable for compromise or war.

To invalidate the opponent's expectation, a person must know others and know the self (Sun Tzu, Chap. 3).

The victory of a military force is determined by the opponent and his reactions (Sun Tzu, Chap. 6).

Miyamoto Musashi wrote "A Book of Five Rings". Musashi was born in 1584, in Japan, at a time Japan was recovering from four centuries of civil unrest. The rule of the emperors had been overthrown in the twelfth century, and although the station of emperor remained, his powers were reduced. Japan had seen almost continuous civil war between the lords and warriors, who were fighting each other for land and power. In the fifteenth and sixteenth centuries the lords built huge stone castles to protect themselves and their troops. Walled, fortress towns began to grow up. Trade and growth were restricted and the economy flagged. The feuding lords had no more resources to keep their troops and the Samurai were released from service. When the armies were gradually disbanded, many out-of-work Samurai roamed the countryside. These out of work Samurai were called Ronin. They had no leader, no lord, and were sometimes viewed as "swords for hire".

Musashi became a Ronin at the time when the Samurai were formally considered to be the elite, but actually had no means of livelihood unless they owned lands and castles. Many Ronin put up their swords and became artisans, but others, like Musashi, pursued the ideal of the warrior searching for enlightenment through the perilous paths of Kendo (The Way of the Sword).

We find the origins of the Samurai system established around 792 AD, when the Japanese were training the ranks of foot soldiers using training officers recruited from among the young sons of the high families. These officers were mounted, wore armor, and used the bow and sword. In 782 the emperor Kammu started building Kyoto, and in Kyoto he built a training hall, which exists to this day called the Butokuden, meaning "Hall of the Virtues of War". The Samurai were considered an upper class due to their training and knowledge and the fact that they were envoys of the lord of the area. In the time of Musashi, the out of work Samurai found themselves living in a society which was completely based on the old chivalry, but there was no place for men at arms. They became an "inverted" class, keeping the old chivalry alive by their devotion to military arts.

In this period of Japanese history, the art of Kendo came into its own. Since the founding of the Samurai class in the eighth century, the military arts had become the highest form of study, inspired by the teachings of the philosophy of Zen. The martial training expanded to include all areas of life, including poetry and writing. The Japanese had a saying, "Pen and sword in accord". Today, prominent businessmen and political figures in Japan still practice the old traditions of the Kendo schools, preserving the forms, strategies, and tactics of several hundred years ago and applying them to business and daily life. There are other resources quoted in this work, however, however these three books have heavily influenced the authors.

A Philosophy of Life

Life is a gestalt of experiences. We put the pieces together by trial and error in an attempt to assemble an understandable image. When the pieces start to fit, it is called wisdom. From this life wisdom we develop a personal philosophy based on how we perceive life to work and how we may manipulate it. The rules of manipulation are based on the mechanics of relationships, natural laws, and what behavior we believe to be socially, morally, and religiously acceptable. When all of these things are brought together into a single visage it is called a personal philosophy.

Concepts and philosophies taught by the martial arts seep into a person's spirit because the philosophy is contained within the movements and techniques. The teacher and senior students pass along attitudes and intangibles such as honor and decorum through constant exposure and guidance. Movements, techniques, and their applications combine with the lessons and attitudes taught within the school to form a force of influence. In this way the art becomes part of the person.

For those who study martial arts, their personal philosophy is made up of both physical and mental lessons. As a person who studies Judo or Jujitsu will learn to "yield and overcome – bend and be straight", or a person who studies Shotokan Karate will learn to "enter at one stroke", attacking problems head on, unrelentingly, so the philosophy of the art permeates the soul and attaches to the heart in ways we cannot suspect. The philosophy of how we fight will become the philosophy of how we live as we grow to understand life is war waged on the self to make it stronger and nobler.

The arts of peace and war are like two wheels of a cart, which, lacking one, will have difficulty in standing. Kuroda Nagamasa 1568-1623 A.D.

For a philosophy to be viable it must be well articulated and applicable to life. This is especially true of a martial philosophy since on it hangs the life and the welfare of others. Life is war. It is a survival game. It is a game of dominance, possession, position, and tactics. This is certainly true in business, but is true to some degree in all aspects of life. We wish to "win" the job. We "beat" others to the draw. We talk of "knocking them dead" at the presentation. We speak of "winning" the heart of our beloved. We "kill" the competition. Even in our language we acknowledge life is war.

In light of this, it is such a sad thing that most people will go through life without a personal philosophy for living, competing, or succeeding. Even sadder are martial artists, those who practice and even teach combat, who have little or no personal or martial philosophy beyond that of striking someone and attempting to avoid being hit. Trying to win and not lose is not a successful philosophy. To live and fight with honor is admirable and necessary but does not articulate a way in which to accomplish the goal. Without a philosophy for living there is no plan by which to win and no rules by which to guide a plan.

Most people realize life is about competition but few take time to formulate a strategy and philosophy that works for them. In martial arts there are complete systems of combat made up of hundreds of individual moves and techniques with no discernable or coherent philosophy of combat. Even fewer have articulated philosophies applicable to life outside the ring. One would think after several thousand years we would come to expect more from the warrior.

There are divisions within the ranks or classes of warriors, just as there are with those who work and live around us. There are those with varying degrees of understanding and vision. There are those who wish to excel, and those whose goals are no more complex than to survive by putting forth the minimal amount of effort to do so. In the higher classes of the warrior one sees vision, strategy, and determination. For those warriors, such as Miyamoto Musashi or Sun Tzu, life and war were seen from a wider perspective, having well thought out and codified strategies, and the spirit to carry them forth. These attributes are as important for our physical and social survival today as they were in the days of the old masters of combat and warfare.

Our society has evolved into a pseudo-warrior society with the C.E.O. as the general and those below him the Samurai class in his fiefdom. War is now about conquering the market. It is a highly complicated, mental game. Some have adopted the ways and philosophies of the Samurai to achieve success. We will explore how the ancient rules apply to modern life, competition, and warfare.

We will lay a solid foundation of martial philosophy and tactics and to draw parallels between combat tactics and life situations so that the warrior may be a better leader and the leader can become the warrior.

In all aspects of life, relationships form the basis of everything. In all things, think with one's starting point in man. Nabeshima Naoshige 1538-1618 A.D.

Unhappy is the fate of one who tries to win his battles and succeed in his attacks without cultivating the spirit of enterprise; for the result is a waste of time and general stagnation. Hence the saying: The enlightened ruler lays his plans well ahead; the good general cultivates his resources. Sun Tzu

A goal is not always meant to be reached, it often serves simply as something to aim at. Bruce Lee

Achieve results, but never glory in them. Achieve results, but do not boast. Achieve results, but do not be proud. Achieve results, because it is the natural way. Achieve results, but not through violence. Lao Tzu

What others teach, I also teach; that is, "A violent man will die a violent death." This is the essence of my teaching. Lao Tsu

Beyond the basic philosophy of Shinsei of "evade, invade, control", there is a higher philosophy of perseverance. As we study the Old Testament, we find it teaches of our need for a savior. The Old Testament promises His coming. It teaches us in symbols and types about His purpose and His nature. The New Testament proclaims

His life, death, resurrection, and imminent return. We see and understand our need for a savior.

Yet, in our study we often skim over those few scriptures containing His ministry and words. Out a volume containing 66 books, there are only four short books that contain His life. Most of those small works repeat one another. So often we overlook the very seed and core of our faith; those books of His life, His ministry, and His sayings. What was Jesus trying to tell us? Certainly He was telling us we need Him and should accept Him as Savior and Lord. We know He came and died, as payment for our sins, but there is more. Jesus also told us how to treat each other and how to live our lives. This message we must shout from the rooftops. With this message we cannot be silent.

To the woman caught in adultery, He did not condemn her but gave her another chance. He said, "GET UP". To the daughter who had died, leaving her father no hope, he gave hope. He said, "GET UP". To Lazarus, his best friend, who lay in the grave three days and was completely without life, he gave him life. He said, "GET UP!"

To Paul, who was knocked down, beaten, thrown in jail, and held in prison, he gave him courage to go on. He said, "GET UP". On the third day the Father told Jesus "GET UP". And when He comes for me, I may be dead, working in the field, laboring on the rooftop, sweating in the mill, or asleep in my bed, but the words will be the same. With the sound of a trumpet, with the voice of an archangel, we who believe will all hear Him say, "GET UP, GET UP, GET UP! You cannot win unless you keep getting up. You cannot be victorious in the battle unless you keep getting up. We will all fall, but he said "GET UP". We will all fail but we must GET UP. The world is not our friend. It would see faith in God fail, but we must keep getting up!

We fight an enemy who wants to knock us down, keep us down, and see us ruined. But Paul said, I may be knocked down but I am not out! Paul proclaimed we were to finish the race. He got back up! Solomon said, Here is a righteous man, a man who although he falls seven times, he keeps getting up! Jesus said to those caught in sin, "GET UP"! To those who have sorrow, "GET UP"! To those who have lost all hope, "GET UP"! The only way to fail is not to get up. YOU MUST GET UP!

He will be there when the race is over. He will meet us when our battle is through. GET UP! It is time for Christians to unite. Tell those you love to GET UP; help those in need to GET UP; Strengthen those who have fallen and HELP LIFT THEM. Tell them to GET UP. Do not punish, do not judge, do not condemn. Our fight is with Satan and those who would keep us down. Our fight is with the world, not with each other. Though the world hates us, politicians loathe us, friends forsake us, judges won't let us pray, laws are made against our will, government steals our land, our money and our rights; through we are tested with sickness, debt, and despair, GET UP! It is not over GET UP! He is coming soon. GET UP! GET UP! GET UP! And teach your children to keep GETTING UP!

If you want to know why we train in martial arts, if you wonder why we worship God, if you are puzzled as to why we fight against our limitations; it is to have the strength in our spirits and the will of mind to keep getting up until Jesus comes again. Shinsei as an art is beyond a physical training system or self-defense. It focuses on "teaching the whole".

Teaching the Whole

There are three aspects of being which Shinsei Hapkido is designed to develop: Body, Soul, and Spirit. These three aspects must be developed in balance for a person to become properly balanced as a martial artist and therefore as a person. This balance is represented by the belt being tied with the knot in the center, and either side of the ends being equal in length.

The first part of this tripartite division of the human is the Body. Most instructors practice this portion of the training. The body is developed through the physical exercises involved in martial arts training. They may include forms, Kata, sparring, grapping, conditioning exercises, stretching, and other drills. Repetition of the basic and advanced techniques develops the proficiency of the practitioner at this level.

The second aspect of this tripartite division is the Soul (including the mental training here as well). The soul is the emotional center of our being. Training must include studies that will help balance the student such as some form of artistic expression such as writing, painting, poetry, or music. It is very important that students have a healthy and adequate emotional expression. Lack of healthy emotional connection produces power without compassion.

Shinsei also emphasizes teaching the student to focus his mind and to coordinate his thinking with his movements. Students are encouraged to listen, read, think, and write. Instructors should never restrict themselves to simply teaching the martial arts, but should study all aspects of history, philosophy, science, medicine and other subjects, as well as spiritual concepts that might have a bearing on the Shinsei system.

The third part, the Spirit, is often overlooked as well. Students need a firm foundation of beliefs to guide them. The true martial artist is in pursuit of personal improvement. It is not enough to have a strong mind and body, the true martial artist should also strive to be strong in spirit. He should have a goal in life and a firm foundation of beliefs to guide him. Shinsei offers a vehicle to reach others with the saving knowledge of Jesus Christ. It is when Jesus becomes one's personal Savior that the Spirit of God takes up residency in the individual's life. When we ask Jesus to save us, the Holy Spirit comes in and "Christ liveth in me". (Galatians 2:20).

To be a good instructor, we must look beyond the obvious, the physical. We must address the reality of the martial arts trinity. We can feed the body, and starve the spirit.

PART THREE – WHAT IS A WARRIOR?

What is a Warrior?

The path of the Warrior is lifelong, and mastery is often simply staying on the path. - Richard Strozzi Heckler, In Search of the Warrior Spirit

The Way of a Warrior is based on humanity, love, and sincerity; the heart of martial valor is true bravery, wisdom, love, and friendship. Emphasis on the physical aspects of warriorship is futile, for the power of the body is always limited. - Ueshiba Morihei, The Art Of Peace

The warrior's intention should be simply to grasp his sword and to die.- Kiyomasa Kato (1562-1611)

A good soldier is not violent. A good fighter does not get angry. A good winner is never vengeful. A good employer is humble. This is the virtue of not striving. It is known as the ability to deal with people. Lao Tzu

There are many ways to describe a warrior. The descriptions may vary widely, so it may be easier to start with what a warrior is not. The warrior is never a brute. He is never reactionary. He is not arrogant or pompous. The warrior should be a balanced and educated person. Most of all, the warrior should have the strength of a strong man, the heart of a compassionate woman, and the open mind of a little child.

The sage takes care of all men, and abandons no one. He takes care of all things and abandons nothing. This is called following the light. - Lao Tzu

I have heard that when a man has literary business, he will always take military preparations; and when he has military business, he will always take literary preparations. - Confucius

Without knowledge of Learning, one will have no military victories. - Imagawa Sadayo 1325-1420 A.D.

Knowing your ignorance is strength. Ignoring knowledge is sickness. When one becomes sick of sickness he is no longer sick. Lao Tzu

It is hardly necessary to record that both learning and the military arts are the Way of the Warrior, for it is an ancient law that one should have learning on the left and martial arts on the right. Hojo Nagauji 1432 – 1519 A.D.

The martial arts considers intelligence most important because intelligence involves the ability to plan and to know when to change effectively. Sun Tzu

Warfare is not about joining the battle, but about timing, reason, and skill. Life and war are about viewing the bigger picture and having a willingness to sacrifice for the goal. The warrior is patient and calm in the midst of chaos. The center of his spirit is quiet amidst the storm. He has devoted his life to understanding himself and others. He is a student of human nature.

The man whose profession is arms (fighting) should calm his own spirit and look into the depths of others. Doing so is likely the best of the martial arts. Shiba Yoshimasa 1350 – 1410 A.D

He is intelligent, caring, and dedicated. He is capable of giving all. However, unless it is a last resort, the sacrifice should never be about dying, it should always be about living fully. The warrior seeks first to save life and do no harm. He lives the way of a minimalist by conserving movement and action. He knows it is better to walk away than to fight. It is better to fight one than to fight many. It is better to hurt than to maim. It is better to maim than to kill. It is better to kill one than to kill many. He seeks to fight intelligently knowing speed and strength will one day fail. The warrior seeks peace above all things, but when the battle must be joined in earnest, he will seek to triumph. He will seek solution and honorable compromise even in the midst of battle.

Yield and overcome. Bend and be straight. Empty and be filled. Wear out and become new. Have little and gain. Have much and be confused. Lao Tsu

If a man becomes alienated from his friends, he should make endeavors in the way of humanity…. One should not turn his back on reproof. Takeda Nobushige 1525 -1561 A.D.

Not being tense, but ready. Not thinking yet not dreaming, not being set, but flexible - it is being wholly and quietly alive, aware and alert, ready for whatever may come. Bruce Lee

A person's character and depth of mind is seen by his behavior. Thus, one should understand that even the walls and fences have eyes… one should not take a single step in vain, or speak a word in a way that others may speak of him as shallow. Shiba Yoshimasa 1350 – 1410 A.D

He does not want war but if war is to be fought it is to be won. True victory is in finding a path that does not lead to conflict. Conflict leads to a lesser victory, which is based on survival and not resolution. Ultimate victory is the saving of all life and honor. This is winning without fighting.

Out of a martial art, out of combat I would feel something peaceful. Something without hostility. Bruce Lee

A brave and passionate man will kill or be killed. A brave and calm man will preserve life. Of these two types, which is good and which does harm? Lao Tzu

Knowing is not enough; we must apply. Willing is not enough; we must do. Bruce Lee

When things dissolve into conflict, harmony and life are lost. The most direct way of viewing this is in the statement made by one of our generals which I will paraphrase: no one ever won a war by dying. The war is won by making the other guy die.

However, if this sacrifice is to be made, let it be made to count.

The defining trait of a warrior is an indomitable spirit. There is a saying, "If you cut off his head, he will not stop until the last move is complete."

Like everyone else, you want to learn the way to win, but never to accept the way to lose. To accept defeat – to learn to die – is to be liberated from it. Once you accept, you are free to flow and to harmonize. Bruce Lee

I will stand off the forces of the entire county here, and die a glorious death. Torii Mototada 1539-1600 A.D.

DEU 20:3 - 4 And shall say unto them, Hear, O Israel, ye approach this day unto battle against your enemies: let not your hearts faint, fear not, and do not tremble, neither be ye terrified because of them; For the LORD your God is he that goeth with you, to fight for you against your enemies, to save you.

This is not to say the warrior blindly follows orders. He has considered the consequences of his decision and is committed fully to each action he undertakes.

Success in warfare is gained by carefully accommodating ourselves to the enemy's purpose. Sun Tzu

I feel I have this great creativity and spiritual force within me that is greater than faith, greater than ambition, greater than confidence, greater then determination, greater than vision. It is all of these combined. ... It is like a strong emotion mixed with faith... Bruce Lee

There is no such thing as an effective segment of totality. Bruce Lee

Teaching the Warrior

Martial arts students love to view themselves as "warriors". In fact, most "wars" today do not take place with hand-to-hand combat. The officers give orders to press buttons, launch missiles, and discuss ceasefires. Yet, within the modern context, there is still a need to teach our students to be warriors. Perhaps the best way to teach one to be a warrior is to define what a warrior is.

The following essay was written by Master Erle Montaigue and is titled, *"The Way of the Warrior"*.

A warrior is not merely a person who has learned some moves, is able to kick at ninety miles an hour or who has won the world championship in kickboxing. A warrior must earn his title. The martial artist is a person who knows things which go far deeper than just self-defense; he is someone who walks into a room full of people and an immediate calm falls upon that room; he is a person who can touch a person's head, arm, or hand and cause an inner stillness and peace to fall upon that person. You know a warrior not from the way he looks, his big biceps, or his rolled-up sleeves revealing a row of tattoos, his shaven head or the fact that he wears his full gi (karate uniform) to parties! We know the warrior by his presence.

The warrior looks upon the earth in a different way than those who are not warriors. Everything, from the smallest insect to the largest mammal, and the most insignificant rock or tree is important and has life; the grass he walks upon he thanks God for softening the rough path he trods; the trees, he thanks God for giving him shade and oxygen. Everything has importance because it was put there by God for some vital reason.

Certainly, he lives in modern times, he drives a motorcar and goes to the supermarket and mows his lawn, but he never loses sight of what he is, and more importantly, where he is. He knows he does not simply emerge from what he has made of himself, but also what has been handed down to him and is accumulated inside the very cells he is made of. Everything his ancestors were is now in him, every bit of information his fathers and mothers gathered is now inside of him. We are God's crowning creation on this earth.

We pass on to our children our own knowledge, together with eons of knowledge accumulated since Genesis 1:2. Everything we are at the conception of our children is passed on to them.

The warrior knows he is an element of change. However, this comes not without payment; for he also knows we cannot receive without first having paid. The whole of the universe is based upon this giving and taking, it is called yin and yang. For every up, there must be a down; for every happiness, there must be a sadness, for every full tummy, there must be an empty one. The warrior knows he must lose in order to gain, and so he sacrifices. He sacrifices his food, he sacrifices his sexual longings, his every day comforts, in order that he may have the power to change and to help others to change specifically. In doing so, he assists them to see where they are, and who they are.

We are not only someone's son or daughter, we are the sons and daughters of God.

Being a martial artist is only one-hundredth of what makes up a warrior; it is only a part of the whole; it is what gives us the confidence to take on life. A warrior knows we do not go through life without teachers, or guides; the people we meet who are able to give us something internal, that something extra to cause us to become our own greatest teachers. By simply being, a guide helps us to realize it is we, ourselves, who teach us, because the warrior also knows locked away inside of everything, is the fact there is a creator.

The warrior communicates without speaking; he knows his physical needs are being looked after, and he need not worry from where the next mortgage payment will come.

The warrior finds his place on the earth and stays there, where the power is. It is not a physical searching, but rather the warrior is 'taken' to where he must be. There he stays, and the whole world will pass by, he need not. Those who will in turn need to seek him out, will do so when their time is right, in just the same manner he did when he found his own guides. They will learn to teach themselves from within, to find their own place...he may never see them again, but it does not worry the warrior.

The warrior is not the master, he is neither the sifu nor the sensei. These are just physical words we place upon ourselves to make us seem important, or better than those we guide. The warrior is a friend to his students, and so cannot be the master. There is only ONE master. The Warrior does not wish to gather students to him, as they will search him out. Those who need to have a master or a sensei will not stay, they will keep searching until they realize what they search is within themselves.

Whenever you meet difficult situations dash forward bravely and joyfully. - Tsunetomo Yamamoto, Hagakure

Take the arrow in your forehead, but never in your back. Hwa Rang maxim

The Warrior's Heart – A Search for Peace

I count him braver who overcomes his desires than he who conquers his enemies: for the hardest victory is the victory over self. - Aristotle

Once I trained because it made me strong. Once I trained because I thought I needed to protect myself. Once I trained because it was my habit to do so. Now I train because it has become a part of me.

The warrior has a diminishing place in our society. His mentality is feared and is misunderstood at times. Society tends to use him for ends and needs of its own. He is "allowed" to protect and serve, but even while he is saving their lives, some look down on him. It is possible this is because of what he is; direct and determined in approach, not easily swayed or distracted. People dislike this, especially if they are trying to hide something. He intimidates some who are unsure of their own abilities, although he does not mean to. Some see the warrior as a threat. Society has trouble separating the warrior from the brute. In times of war, the actions may appear the same, but actions arise from a very different intent. The warrior will only fight when it is necessary. The intent is to serve the common good and re-establish peace. The heart of the brute is self-serving. The brute fights for gain, whether it is for pride or fortune. What is in the warrior's heart? It is called, "the Way".

The Way, or "Do", (pronounced doe), is the path the warrior walks. It is the reason we chose to live the self-disciplined life. Without aggression and discipline nothing would be accomplished. Discipline gives proper direction to aggression. Aggression is masculine energy (yang) applied toward a goal.

Where does this take us? When aggression is applied toward making our actions or ourselves as perfect as possible, it will bring together all parts of us in harmony. Mind, body, and spirit unite in a common goal. At the moment all parts are in fullness and balance, heaven and earth stand still, and for a moment, a brief moment, the angels in heaven stand and applaud and you know you are one.

ISA 40:31 But they that wait upon the LORD shall renew their strength; they shall mount up with wings as eagles; they shall run, and not be weary; and they shall walk, and not faint.

The mind of the warrior remains focused on his own mortality. The sacredness and brevity of life is always in his thoughts. Life is lived to the fullness, moment by moment, when the possibility of death is realized. The warrior's heart is not reserved for those who do battle with others, but is kept secret for those who battle themselves and their own limitations.

JOB 36:10 He openeth also their ear to discipline, and commandeth that they return from iniquity.11 If they obey and serve him, they shall spend their days in prosperity, and their years in pleasures.12 But if they obey not, they shall perish by the sword, and they shall die without knowledge.

The Art calls for complete mastery of techniques and developed reflection within the soul. Bruce Lee

I went to the woods because I wished to live deliberately, to front only the essential facts of life, and see if I could not learn what it had to teach, and not, when I came to die, discover that I had not lived. (Walden, p. 90)

I wanted to live deep and suck out all the marrow of life. (Walden, p. 91)

I learned this, at least, by my experiment; that if one advances confidently in the direction of his dreams, and endeavors to live the life which he has imagined, he will meet with a success unexpected in common hours. He will put some things behind, will pass an invisible boundary; new, universal, and more liberal laws will begin to establish themselves around and within him; or the old laws be expanded, and interpreted in his favor in a more liberal sense, and he will live with the license of a higher order of beings. In proportion as he simplifies his life, the laws of the universe will appear less complex, and solitude will not be solitude, nor poverty poverty, nor weakness weakness. Henry David Thoreau – Walden Pond

What is a warrior but he who, through enormous self-discipline, has trained himself to be concerned with accuracy, strength, and skill, and not to be concerned with self. He is, at times, unaware of self, concerned only with the moment in time and action therein, for death may come in the next instant. But to be unaware of self does not make us free from responsibility for our actions. Indeed, a lifetime of training demands an answer for every action, especially for those, which are wasted or useless. To be free of self makes the warrior loyal and consistent in devotion. In quiet times he plumbs the depths of his soul, finding life precious and finite. In those poignant times, he becomes poet and lover. And so, his heart is broken and complete in a single atom of time. The warrior is not the brute. War makes them look alike. Life separates them fully.

PSA 103:15 - 18 As for man, his days are as grass: as a flower of the field, so he flourisheth.16 For the wind passeth over it, and it is gone; and the place thereof shall know it no more. But the mercy of the LORD is from everlasting to everlasting upon them that fear him, and his righteousness unto children's children; To such as keep his covenant, and to those that remember his commandments to do them.

What is life? It is the flash of a firefly in the night. It is the breath of the buffalo in the wintertime. It is the little shadow, which runs across the grass and loses itself in the sunset. - Crowfoot, a Blackfoot warrior, 1890

The tumult and shouting dies; the captains and kings depart; still stand the sacrifice, a humble and contrite heart. Lord God of Hosts, be with us yet, lest we forget – lest we forget. Rudyard Kipling, 1865 – 1936

The end of our Way of the sword is to be fearless when confronting our inner enemies and our outer enemies. - Tesshu Yamaoka, 19th century Kendo master

It's a difficult thing to truly know your own limits and points of weakness - Hagakure

Licking the Outside of the Watermelon

"Su-bak keot hal-kki" A Korean Proverb

The United States and the world are now inundated with martial arts studios where teachers take money and push students through a routine for one or two years. When the students emerge on the other side of this money-grabbing ride they are awarded a black belt for their troubles. This is called. "licking the outside of the watermelon". This fast food, "McDojo" concept is absolutely contrary to the "Way". It has diluted any concept of what a student of the art should be. This is my thirty-second year as a student of the martial arts, and still when I practice my eyes tear and heart sinks because my art is not quite right. The heart of this quest is one of discovery. How much of oneself and a 2,500 year old art can one explore and discover in such a short time? It is like being given a watermelon. If you lick the outside and do not take time to eat the fruit within, what have you tasted? Three decades and I am into the rind…maybe.

PSA 92:5 - 6 O LORD, how great are thy works! And thy thoughts are very deep. A brutish man knoweth not; neither doth a fool understand this.

Starting is Half

"Shi Jak i pan i da" A Korean Proverb

Out of all the obvious lessons in life, the lesson of inertia is the most obvious but least applied. Remembering our high school physics, the law of inertia states that a body at rest tends to remain at rest and a body in motion tends to remain in motion unless an outside force acts upon it. This is the law of inertia as it applies to the outside or physical world. But, if this law is applied emotionally and spiritually, it is just as true. Our habits, interior journey, and motivation tend to remain in the same state unless acted on by an inside force. The force that changes them is our force of will. As with physical inertia, the greatest amount of energy is spent in getting the object at rest to start moving. This can be seen as we try pushing our car. We strain to begin the first few feet of motion then, life becomes easier as we realize how much less energy it requires to keep the heavy hulk moving. In a lifetime of training, self-discipline, and self-discovery, starting is half the effort.

Remember, things in motion tend to stay in motion. Arise and move!

Know Yourself and Know Others

Tell me – Who do you think I think I am? Don Henley

2CH 16:9 For the eyes of the LORD run to and fro throughout the whole earth, to shew himself strong in the behalf of them whose heart is perfect toward him. Herein thou hast done foolishly: therefore from henceforth thou shalt have wars.

Be honest with yourself about your intentions. It is not enough to know what we ought to be, unless we know what we are. T.S. Elliot

Knowing others is wisdom. Knowing yourself is enlightenment. Mastering others demands force. Mastering self demands inner strength. Lao Tzu

To take responsibility of one's actions, good and bad, is something else. After all, knowledge simply means self-knowledge. Bruce Lee

Fame or self: Which matters more? Self or wealth: Which is more precious? Loss or gain: which is more painful? He who is attached to things will suffer much. Lao Tzu

There is no greater sin than desire, no greater curse than discontent, no greater misfortune than wanting something for oneself. He who knows that enough is enough will always have enough. Lao Tzu

We may think we know ourselves. We seldom do. It is said that between the lightning and the thunder the true face of the man is known. That is to say in times of stress, when one has to put all he is on the line, his true face shows. Those with mild and humble faces can become angry, wild men. It is their true nature they were hiding. If you wish to know what you truly are, look at yourself and your reactions when in unfamiliar circumstances, when angry, when hurt, when troubled. This is the true test of a man. We never know who or what we are until we are living with the unfamiliar; life can become so choreographed that the real and natural mind is hidden.

Know honor yet keep humility. Be the valley of the universe. Being the valley, become true and resourceful. Become as an uncarved block. Lao Tzu

In training, the student first focuses on his own control, actions, and reactions. In the second phase the student begins to learn about others. He begins to learn what makes a formidable enemy or a weak ally. In the third phase the student learns about the interactions between himself and others. It is this final phase we begin to gain insight about ourselves. How we behave and perform with others teaches us about our own weaknesses and strengths. These are not lessons in fighting, but lessons in humanity we learn.

A truly good man is not aware of his goodness and is therefore good. A foolish man tries to do good and is therefore not good. A truly good man does nothing yet nothing is left undone. A fool is always doing and always there is much left undone. Lao Tzu

There is a difference between being a good person and doing a good thing. A good person responds and is kind by nature. He does not notice he is doing good. Good deeds are a burden to those who are not good. Therefore, it is more important to develop a kind and good nature than to be a person who does kind and good acts.

To know oneself is to study oneself in action with another person. Bruce Lee

This is the noble truth of the way, which leads to the cessation of pain. This is the eight-fold path, namely, right views, right intention, right speech, right action, right vocation, right effort, right mindfulness, and right concentration. Gautama Buddha

PSA 112:7 He shall not be afraid of evil tidings: his heart is fixed, trusting in the LORD.

JER 17:9 - 10 The heart is deceitful above all things, and desperately wicked: who can know it? I the LORD search the heart, I try the reins, even to give every man according to his ways, and according to the fruit of his doings.

The art of karate is a never ending quest for perfection....of developing spirit and body to defeat your opponent...one's self.
Tak Kubota, The Art of Karate

Self-Evaluation

The other person's rice cake looks bigger. (Nam-eui tteok-I teo k'eu-ge po-in-da). – A Korean Proverb

MAT 7:3 – 5 And why beholdest thou the mote that is in thy brother's eye, but considerest not the beam that is in thine own eye? Or how wilt thou say to thy brother, Let me pull the mote out of thine eye; and, behold, a beam is in thine own eye? Thou hypocrite, first cast the beam out of thine own eye; and then shalt thou see clearly to cast the mote out of thy brother's eye.

It is said the more a mirror is polished and cleaned the more obvious its imperfections. In this aspect, the martial arts have within them the tools for both self-improvement and self-effacement. Both are contingent upon self-analysis. It is not only the critiquing of technique that is needed, but also the critiquing of self. As technique is polished and polished it begins to look powerful and crisp, but inside, there is the realization of inadequacy and imperfection. This is the time of decision. Here, we can choose to ignore the internal signals of weakness and believe what others may say about our skills, or we can heed the inner voice of our own spirit and continue our search.

We stand on a razor's edge; on the point of a pin; suspended between ego and reality. Tempted as we may be to take the easy way of ego and outward show, it is not the spiritual path. It is not until our physical limits are reached that we begin to develop spiritual strength. This explains why there is such a misunderstanding regarding the meaning of the black belt rank. A first-degree black belt should be considered merely an advanced student. At this level only the physical limits are being pressed and self-discipline has been learned. It is only after this stage the spiritual growth can begin.

He who is aware of his own weakness will remain master of himself in any situation.   Gichin Funakoshi

It is at the point of reaching our physical limits that frustration begins to mount. As we push ourselves to the extremes of our physical limits we attempt to reach beyond our outward strength. We turn inward, becoming frustrated at our limits. Frustration is good. It drives us to find that part of us lying untapped and unused, until now.

Is your technique imperfect because of lack of inner strength? Are physical limitations holding you back? When you are told how strong or fast you are, will you believe what others may say of your technique, or will you dig deeper and look into your heart and find what is lacking?

Knowledge comes from your instructor; wisdom comes from within Dan Inosanto

He Who Rides the Tiger Dares Not Dismount

Whether in everyday life or in battle, a path is chosen. A strategy is sought. The path must contain multiple outlets. The strategy must have contingency plans. More importantly, we must be willing to take the divergent paths. Otherwise, the limitations of choices will entrap us. If your mind is not opened to other possibilities, or your ego will not permit you to let go of one way and embrace another, the limitation may kill you.

Life seems to eat us alive. We may take on more and more until we are overloaded and incapacitated. Worse yet, we may start down a path only to find we are trapped by our wants, desires, or prejudices. Even though the trap is of our own making, we can refuse to stop or escape. The title, position, status, or persona we have built may be choking us, but we refuse to relinquish it. The techniques we are using may be ineffective, but we have devoted our lives to learning the style and refuse to change. We choose to choose no other path even though the one we are on is destroying us. By refusing to choose, we have chosen to continue.

The secret of living a better life is the patience and the foresight to see three steps ahead. If we cannot do this there will be many decisions we will make that will lead us to paths of destruction. Wrong associations, drugs, bad business deals, and many other choices may entrap us, making us a prisoner of our own keeping. It is a tiger that, once mounted, will become angry. We fear when we dismount we will face its fangs. Still, we must have the courage to dismount and flee bad decisions of the past. Some are miserable in their jobs. Some are miserable with lack of accomplishment. Some lack hope and feel trapped by life. Have courage. Face the tiger. Be free.

I shall be telling this with a sigh
Somewhere ages and ages hence
Two roads diverged in a wood
And I took the one less traveled by
And that has made all the difference
Robert Frost

Adapt or Die

One of the tests of intelligence and survival is the ability to adapt. Too many times in life we are held captive by the very training we depend on to keep us safe. Bruce Lee called this condition, "the classical mess". The mess is the inability to leap beyond the forms and formulas learned in our years of training. In life and in training, we choose to act and react in certain ways. We learn certain patterns and rules. If we become too entrenched or emotionally invested in "our " ways of doing things, we lose the ability to adapt. This is not a problem until we meet an enemy or a condition in life over which we cannot prevail. It is at this point we adapt or die.

Martial arts of today is an intricate game. It is a deadly serious game, but a game nonetheless. We, who are heads of systems, attempt with all we have in us to set before the students a set of rules in the form of techniques and philosophies that would be most useful in overcoming the greatest number of possible events. However, there are very few styles that are tested in the fires of real life and death scenarios. Even if the style we trained in was tested daily on the streets of major cities, it would simply point out and prove our inability to codify movements for the immense number of variables in an attack. We do not train in order to respond automatically with a perfect defense. We train that the art would become part of us and with it as our ally we would adapt, moment by moment to what life offers us; accepting it, deflecting it, and overcoming it with our will and mind and body.

The paradox is contained within the amount of time spent training in specific techniques in order to assimilate them internally so that, in some shattering moment of cognition, the realization of the underlying truth would occur to us. It is the truth, not the technique that is sought. If one must "run through" a list of techniques to decide how to react, he will die before coming to the right "filed away" technique. The key to life and death in an attack is reaction.

Like a master pianist who studies for decades, we can then play, having no thought of the rules, but from the heart, spontaneously and inspired.

The Natural Mind

A "natural mind" is a term used in the arts to describe a "frame of mind" the warrior must have to be able to react. A natural mind is relaxed, yet focused. Although life and death are decided in the moment, they are of no consequence. To be free of this concern is to be free of tension and angst. Thus, one moves strongly, quickly, freely. By not caring about death, one is more likely to live. The natural mind has no home. It is free to come and go as it wills.

There is Jackson standing like a stonewall. Let us determine to die here, and we will conquer. Brig. Gen. Barnard E. Bee at Bull Run 1866

When a fellow's time comes, down he goes. Every bullet has its billet. Pvt. E.E. Patterson, Confederate Army, 1862

The natural mind is clear and undisturbed. It is unaware of even itself, thus it sees everything clearly.

If he puts his mind in thoughts of his opponent's intention to strike him, his mind will be taken by thoughts of his opponent's intention to strike him. If he puts his mind in his own sword, his mind will be taken by his own sword. If he puts his mind in his own intention of not being struck, his mind will be taken by his intention of not being struck. Takuan Soho

This means there is no place to put the mind. When the mind is forgotten and we become unaware of self and of mind, there is no hindrance.

A certain person once said," No matter where I put my mind, my intentions are held in check in the place where my mind goes, and I lose to my opponent. Because of that, I place my mind just below my navel and do not let it wander. Thus am I able to change according to the actions of my opponent." ...This is a low level of understanding. It is at a level of discipline and training. It is at the level of seriousness. (It shows a lack of full understanding. If you place the mind in one place you will have no ability to move ahead and will be 'unfree'.) In what part of my body than should I put my mind? If you put it in your right hand, it will be taken by the right hand and your body will lack its functioning. If you put it in the eye it will be taken by the eye, and the body will lack its functioning. No matter where you put it, if you put the mind in one place, the rest of your body will lack it's functioning. Well, then, where does one put one's mind?

I answered, "If you don't put it anywhere, it will go to all parts of your body and extend throughout its entirety. In this way, when it enters your hand, it will realize the hand's function. When it enters your foot, it will realize the foot's function. If one thinks, he will be taken by his thoughts. Because this is so, leave aside thoughts and discrimination, throw the mind away from the entire body do not stop it here and there, and when it does visit these various places, it will realize function and act without error." The effort not to stop the body in just one place-this is discipline. ...When a person does not think, "Where shall I put it?" the mind will extend throughout the entire body and move to any place at all. ...

Put it nowhere and it will be everywhere. Even in moving the mind outside the body, if it is sent in one direction, it will be lacking in nine others. If the mind is not restricted to just one direction, it will be in all ten. ... Not stopping the mind is the object and the essence. No matter how much you try to figure or calculate by means of impressions or knowledge, it will not prove the least bit useful. You will be too slow and too often deceived.

Therefore, separate yourself from the discrimination of figuring things out. Departing from desire, realize a desire free mind -- this is the Way. When "No-Mind" has been developed, the mind does not stop with one thing. It is not captured and thus is free to be everywhere. One is not likely to achieve understanding from the explanation of another." Takuan Soho

The idea of "no-mindedness" is not a familiar one to the Westerner. To be aware of all things and nothing in particular is a difficult concept. To concentrate with a clear mind is not easily grasped. We are a goal-driven people. To be focused and alert, but not focused on or alert to any one thing seems on the surface a contradiction to our frenetic way of life. However, in the martial arts, to do otherwise wastes time, misdirects energy, and could cost our lives.

The mind is clear but not empty...it is focused but not on any one thing...concentrating but not bound.

There can be no preconceived ideas of how the opponent will attack. How will we know what weapon he will use or how it will be delivered? We cannot know. To look for something is to see it in everything, even if it is not there. This is true in relationships and in combat. Thus, to expect a kick is to think you see it coming even as the unexpected punch strikes you. So, let come whatever will come. If our minds are alert and open, we will not color sight with expectation. In that instant of time the attack is launched, we will see it for what it is.

To be ready to defend against a take down from the front makes you vulnerable to being slipped and suplexed from the back. To be ready to defend one side makes you vulnerable to a barzegar take down on the other. To watch your back means to give up defending your front. And finally men, to be ready everywhere makes you vulnerable to being taken down from anywhere. Pick your angles of defense wisely. - Mr. P.F. Stump.

To be swayed neither by the opponent nor by his sword is the essence of swordsmanship. - Miyamoto Musashi

Conquer the self and you will conquer the opponent. - Takuan Soho

Argue for your limitations, and sure enough, they are yours. - Richard Bach

The greatest cause of failure is lack of concentration.

Conquering evil, not the opponent, is the essence of swordsmanship. - Yagyu Munenori

Maturity and Courses of Action

Only the immature believe they are in control. Only the young and untried will use words such as "never" and "always". The mature person knows he is in control only of himself. Even that control is fleeting and thin. Anything can happen. We are not in control of life. Life is in control of us. All we can do is realize who is the author of life. The plan is His. As men, we set our minds to the task. Our will is directed at the goal. This is all we have to give. We will give it all to reach the goal.

In connection with military matters, one must never say what can absolutely not be done. By this, the limitations of one's heart will be exposed. - Asakura Norikage 1474 - 1555 A.D.

When one is mature, he controls his own spirit. He is not easily provoked. He realizes that war, like life, is a game, although a deadly serious one. To be provoked to anger, rashness, or wrath would cause the loss of judgment. Defeat would follow on its heels.

Though a warrior may be called a dog or a beast, what is basic to his nature is to win. Asakura Norikage 1474 - 1555 A.D.

If we have lived long enough we have seen the common man rise to be ruler and rulers brought low. Heaven has its own plan and God's timing cannot be foretold. It seems obvious even if we find ourselves in the position of leader or ruler, the common man should be treated with dignity and respect. One day the man sweeping your floors may be your master.

The fatal flaw of one promoted to a position of authority is to forget from where he came. If he remembers his previous low position he will see that every fool has a chance to advance. He will understand it is by grace and chance that he is there. If he understands there are many as good as he, the position will not seem so high and he will know all men are replaceable. This will keep him humble. In his humility he will treat others well and they will follow him willingly.

Men are your castles. Men are your walls. Sympathy is your ally. Enmity is your foe. - Takeda Shingen 1521-1573 A.D.

Although learning is as important as leaves to a tree, it is only half of what is needed to become a warrior. There must also be discernment. Learning is not only reading books and acquiring knowledge. The ultimate challenge is in knowing how to integrate knowledge into life. In this way and only in this way can learning become wisdom. Discernment is far-sightedness. It is the ability to see the outcome of a situation. To hone this skill one must be a student of human nature. He must allow knowledge, wisdom, and far-sightedness to gestate in his spirit. This patience will enable him to see the end of a matter from the beginning.

It is in the paradox of spontaneous reaction and careful planning life is lived to the fullest. We see the goal. We plan to journey. We begin the conquest. Yet, each step along the way brings unforeseen occurrences. With these unexpected deterrents we must spontaneously dance. We must keep pressing toward the goal through determination and intelligent planning. We must live moment by moment free of anticipation, dealing with what may come, never losing sight of the goal, and always moving toward victory.

One must not be negligent of learning. Lun Yu says, to study and not to think is darkness. To think without study is dangerous. Takeda Nobushige 1525-1561 A.D.

Keep your mouth shut – guard your senses and life is full. Open your mouth – Always be bustling and life is beyond hope. -Lao Tzu

Seeing the small (detail) is insight. Yielding to force is strength. Using the outer light returns us to insight. In this way we are saved from harm. This is learning constancy. Lao Tzu

There is no such thing as maturity. There is instead an ever-evolving process of maturity…. You continue to learn more and more about yourself every day. Bruce Lee

Remaining Fluid - Stretching

To change with change is the changeless state…Life is a constant process of relating. Bruce Lee

When it comes to stretching, martial artists should be the experts. Unfortunately, most are not. In fact, many stretching regimes are as dangerous as the supposed "masters" who use them. A good Sensei is a student of anatomy and physiology, and had learned how to properly work the body.

There is a proper, physiologically correct order to martial arts stretching. First, all of the joints should be rotated and loosened. Secondly, the muscles should be warmed up. Finally the muscles should be stretched. This is called "injury preventive stretching". This is stretching for safety, not flexibility.

After class, the same process should be repeated, but this time, greater emphasis should be placed on stretching for flexibility. If one works the muscles and never stretches, he may become strong, but he will become stiff. A man is born weak and gentle. He dies stiff and hard. Green plants are tender, pliable, and full of sap. They die withered and dry.

Stiff and unbending are the forerunners of death. Gentle and yielding are the disciples of life. An army without flexibility never wins. A tree that is unbending is easily broken. The hard and strong will fall. The soft and weak will overcome. Lao Tzu

To remain whole, allow yourself to be twisted. To be straight, let yourself be bent. Lao Tzu

The self is open, fluid, flexible, supple, and dynamic in body, mind, and spirit. Kisshomaru Ueshiba

As young children we are flexible and can normally take falls and bumps without being hurt to the point of broken bones and pulled muscles; but as we become older and don't stretch as easily, we become stiff and fragile. By stretching the body, we will stay more flexible and therefore resilient and childlike. Flexibility helps with health, balance and speed. In our lives we should always stretch our limits just as we stretch out to God in prayer to increase our spiritual balance.

A Balance of Power

One of the most important points in the training of the mind and spirit is balance. We should demand self-expression from any student, as well as from ourselves, in order to maintain balance between strength and compassion.

A true warrior should also practice some form of artistic expression. To freely express ourselves through mediums of music, poetry, writing, painting, sculpting, or drawing will serve to keep us from becoming brutes. In training hard and long in the martial arts, we harness and subjugate our feelings. We steel ourselves against pain, fatigue, and emotional distractions. In other areas of life, however, we must keep our hearts open. No one practicing any martial art, especially those who train hard to the point of pain and exhaustion, should be without some form of classical artistic expression in their arsenal.

I will go so far as to emphatically state that there are three components making up a complete and balanced martial artist. They are martial training, education, and artistic expression.

Artistic expression is one of individuality on the part of the martial artist whether it is the teacher or the student. Bruce Lee called it the "art of expressing the human body". We should be willing to tell our students:" Don't do things exactly as I do them because if you do, you will only be a copy of me". If the student puts his own personality into the techniques, the technique becomes the student's own. The student must be allowed the freedom to adapt the technique based on his individual strengths and weaknesses.

My strength comes from the abdomen. It's the center of gravity and the source of real power. Bruce Lee

He Who Flies, Falls

If you sit, sit. If you stand, stand but never wobble. Master Ummon

He who stands on tiptoes is not steady. He who strides cannot maintain the pace. He who makes a show is not enlightened. He who is self-righteous is not respected. He who boasts achieves nothing. He who brags will not endure. These are extra baggage. They do not bring happiness. They should be avoided. Lao Tzu

It never fails when new students begin training, they assume two things incorrectly. They believe they know how to breathe and they think they know how to walk. Neither is true.

When one examines the posture of an untrained person as he walks, the motion is one of falling forward off of one foot and catching with the other. In this action he pushes with the back foot and falls, stopping himself with the front foot, which is placed in front of the point of balance. Thus, as one is walking, he is off balance over and over again.

When a trained martial artist walks he does not push himself with the back leg. He pulls himself toward his front foot while keeping his leg under him and on balance. His posture is aligned and erect. He does not fall forward. He does not give away his balance so easily. He walks in balance, from foundation to foundation, rolling from heel to toe, keeping his feet on the ground as much as possible. This is the way he should attempt to live his life. The secret of living well is to have balance is all areas of life.

Nothing will succeed without a solid foundation. A foundation allows all of the strength we have in us to be delivered through us. That strength may represent learning, study, practice, strategy, or even corporate presence. Without a proper stance, we cannot use our foundation. Whatever we are delivering up, whatever techniques we use as an attack, cannot be delivered with power, precision, or speed unless our foundation is solid and our stance is firm. The stance is our way of tapping into the strength of the foundation.

Our stance goes further than our position. It is our posture, as well. Our posture delivers a message to the opponent. It is the conduit between our stance and the target. Whatever is thrown at the opponent will rise from our foundation and stance, travel through our posture, and be delivered to the target. If our posture is not correct, it will send the wrong message to the opponent or, worse yet, it will not allow delivery of power to the target. Our posture must be erect, as if suspended from the top of the head. The back is straight. The head is held up. The chin is slightly tucked. Eyes are fixed and firm. Our gaze takes in everything around us.

The stance should be light and quickly movable, able to adjust instantly to the changing competition. Yet, at the time of impact, the stance should settle and be rooted below the ground. The stance should push against the foundations and deliver that energy through the posture of the body to the attacking member. One should always punch in such a way as to hit the opponent with the solidness of the ground one is standing on.

Even though it may seem to be a small matter and of little consequence, I spend much of my time moving from stance to stance; simply stepping, walking, moving, standing, and watching my balance carefully. I do this during my daily life. Awareness and balance in daily life is the secret of the art.

If you must step back make it the preparation for leaping forward. Be resilient. Regarding stances, stability and mobility are mutually exclusive. The lower a stance the more stable it is, but the more difficult to move out of. The higher a stance is more mobile but less stable. The higher a person fights or stands the easier he is to tip over, however, it is easier to move out of a high stance. A compromise must be reached that will allow the stance to be stable yet mobile. The stance must allow the legs to adjust and kick quickly, yet it must deliver stability and power at the same time. For a faster delivery, the front foot is best to use for kicking. Meaning the weight must be predominantly on the back leg. For all of these conditions to be met, the "L" or "T" stance (Niunja Ja Sae) is preferable. One should place the weight equally on both legs or rock slightly back so that no more than 70% of the weight is on the back leg. This is called "loading" the front foot.

In our spiritual life, we depend on the foundation of our faith. *Upon Christ, the solid rock, I stand. All other ground is sinking sand; sinking sand.*

PSA 31:2-3 Bow down thine ear to me; deliver me speedily: be thou my strong rock, for a house of defense to save me. For thou art my rock and my fortress; therefore for thy name's sake lead me, and guide me.

PSA 61:2-4 From the end of the earth will I cry unto thee, when my heart is overwhelmed: lead me to the rock that is higher than I. For thou hast been a shelter for me, and a strong tower from the enemy. I will abide in thy tabernacle for ever: I will trust in the cover of thy wings.

PSA 62:6-8 He only is my rock and my salvation: he is my defense; I shall not be moved. In God is my salvation and my glory: the rock of my strength, and my refuge, is in God. Trust in him at all times; ye people, pour out your heart before him: God is a refuge for us.

1CO 10:1-4 Moreover, brethren, I would not that ye should be ignorant, how that all our fathers were under the cloud, and all passed through the sea; And were all baptized unto Moses in the cloud and in the sea; And did all eat the same spiritual meat; And did all drink the same spiritual drink: for they drank of that spiritual Rock that followed them: and that Rock was Christ.

Negotiations

There are two primary approaches to negotiation, a linear view, and a circular or ongoing view. Too often in the West, we see negotiations as a linear path ending with an agreement reached by all parties. We mistakenly believe the agreement ends the negotiations and the issues are settled. This is not the case for those who view negotiations as circular. Those who have a linear view are unprepared to continue. Why is there a need to continue if all parties have agreed? Because an agreement is a surface and transient state, changing with the posture of each question.

For example, most of us can agree we do not condone murder, but if I had to kill a man who was about to blow up a school bus filled with children, the state of agreement shifts with the situation. Now the posture of the question must change in light of this added detail. Murder must be redefined.

Knowing this is the way life works, many negotiators use such tactics simply to establish a new baseline in their strategies and safely continue from a firm position.

Combat is ongoing negotiations. It is a series of questions asked in a physical sense and answered in kind. A punch or kick is thrown. It is a question posed by the attacker. "Can I strike you if I do this?" The question is answered in a myriad of ways; yes, no, in a way, not very hard…even if there is agreement and the answer is yes, many times negotiations do not end. The question becomes yes one time and no the next as loopholes and exceptions are sought that will make the answer "no" and the reply painful.

If one were to look deeply into contracts and treaties between countries and individuals, there is often an ongoing change of status forced by one to the demise of the other. The treaties between the United States and the American Indians show this pattern in the full light of its destructive potential.

It is important that our blows and our words be firm. Our word must be our bond, but it is equally important to realize the opponent may be working on the theory of circular negotiations. What does this have to do with warfare? Don't turn your back on a man who is down. After treaties are signed, after the shaking of hands, after he has surrendered, continue to be watchful and aware. This is called, "perfect finish".

Awareness

Just because you are paranoid doesn't mean they are not out to get you.

The art of war teaches us to rely not on the likelihood of the enemy's not coming, but on our own readiness to receive him; not on the chance of his not attacking, but rather on the fact that we have made our position unassailable. **Sun Tzu**

Obviously, awareness is not paranoia. It is simply being cognizant of one's surroundings. To be aware of our terrain and positions allows us to stay clear of danger. Staying away from situations where we could be taken by surprise is half the fight to keep safe and alive.

When going to your car, office, or home, always have your keys in your hand. This is not only for speedy entry but also to be used as a weapon, if needed. Assume nothing. Even when approaching your car, look in the back seat before opening the door. Be aware of possible dangers and avoid them. Not to fight is the ultimate victory.

When entering a room, especially a public place such as a restroom, always glance in mirrors and behind doors. Scan the room quickly for anyone who looks suspicious. Never use a cordless phone when speaking of important personal information. Always assume the airways are not secure. When using access cards and numbers always shield yourself. Assume you are being watched.

When walking around corners, always do so in a wide arc, giving yourself plenty of space to stay out of reach of anyone who might want to accost you.

Officers trained in law enforcement are painfully aware of "the rule of twenty-one". The rule states even if an officer has a pistol on his belt, if the suspect has a weapon such as a knife or club in his hand and is twenty-one feet or closer to the officer, the policeman will not be able to draw, aim, and fire before being hit or cut. Given this general rule, it makes sense to keep a reasonable distance between yourself and those you do not know. Stay in lighted places. Evil men seek darkness to hide their deeds. If you find yourself in an alley, street, or unlit parking lot, be on your guard and correct the situation as quickly as possible.

PSA 74:19-20 O deliver not the soul of thy turtledove unto the multitude of the wicked: forget not the congregation of thy poor forever. Have respect unto the covenant: for the dark places of the earth are full of the habitations of cruelty.

ISA 29:15 Woe unto them that seek deep to hide their counsel from the LORD, and their works are in the dark, and they say, Who seeth us? and who knoweth us?

Whatever the situation, think self-defense. Imagine and prepare for various possibilities. See the situation in your mind before you are required to react. Be aware and prepared.

The Sphere of Influence

Stretch out your arms. From the point above the top of your head where your hands can touch outstretched, to the widest you can reach to your sides... this is your sphere of influence in combat. The body moves and the sphere moves with it, but wherever you go, your sphere will be the same approximate size. The sphere takes into account your legs and arms; feet and hands, and so, makes a bubble 360 degrees from the soles of your feet to the distance of your hands over your head. You may jump, spin, leap, or twist, but no matter what you do, this is all the space you can affect at any one time. The center of this sphere is located about two inches below the navel. This is the point of balance and the approximate halfway point in the body. The line of balance, like a plumb line, travels to the center of the feet. The hollow arch below the center of the ankle begins the connection with earth, foundation, and balance.

As one reaches further out from the center of the plumb line of his sphere, sphere speed increases but strength decreases. Distance allows acceleration and thus increased speed, but distance from the center, which is called the "Dan Jun", diminishes strength. A simple experiment will prove this true. Take a weight of 20 pounds and hold it up to your chest. Now slowly move the weight away from your chest to either side or straight out. As you move out from your chest you find the weight harder to support. Within the sphere there is a circle of optimum strength, which extends from chest to thighs with the Dan Jun as its hub.

Strength remains fairly constant within this area but drops off quickly outside the circle. For this reason, all movements such as wrist techniques and arm bars should be done from your center to offer your maximum strength. This means your hands should be kept close to your center and the opponent's hands should be kept as far from his center as possible. If strength is regarded as an army, then the Dan Jun is the camp. The further troops travel from the camp the weaker they are when they arrive. Join the battle fresh and strong. Fight from your center.

The sphere is comprised of three distinct but overlapping areas. Throws and grappling take place in the inner area. Punches and strikes are best used in the middle area. The outer area can be reached with kicks. Any area beyond the distance a kick can reach is outside the area of influence. Nothing happening beyond the outer area should cause a reaction.

In life, so many things are outside our areas of influence. If we are not careful we end up fighting the wind and wasting our energy. It is very important no battle be fought outside the sphere of influence. It will only lead to frustration. Friends and support are found within the camp. Rest, protection, and confidence are found in the camp. Do not wander far from your camp. To accept responsibility for anything happening outside your sphere of influence, is pointless and will lead to worry and confusion. One should never attempt to block a punch or kick that is not within the sphere.

Just as important is to have the area of the battle be within your circle of power. Maximize your strength with your sphere and fight on your personal "holy ground". In life, as well, we must identify our major areas of strength and fight within those limits, showing forth our strengths and keeping back our weaknesses.

Where there is an enemy value him, measure him, and never belittle him in your mind. When you are close enough to attack you are close enough to be attacked. Therefore, it is obvious, do not attack if you cannot defend.

Attack and Defend

Don't always think in a straight line. - The Way of the Spear

There is no difference between attack and defense. A hard block is an attack. If violence is imminent, a first attack is a defensive move. There is no difference in intent or posture. One must forget any distinction between attack and defense. As we "defend", we should be moving into a place of strength. To defend is nothing more than to evade harm while invading the opponent. If you were to ask the opponent his viewpoint he would have to say your defense was an attack, harming him and piercing his sphere. Attack and defense come down to points of view. After this is understood, the terms are little more than subjective perception.

Timing

A perfect technique, executed perfectly will fail without perfect timing. You must join at the
correct time and place. The moment of greatest strength and commitment is the moment of greatest weakness.

There is a tide in the affairs of men which, taken at the flood, leads on to fortune; omitted, all voyage in their life is bound in shallows and miseries. ... We must take the current when it serves, or lose our ventures. William Shakespeare, Julius Caesar, 1599

Sense the timing of your opponent and go counter to it. He will use rhythm to build power like a child on a swing. Without the ability to build on his movements with rhythm, he will have only his base strength at his disposal.

Break the rhythm between your own techniques. Give him no sense of your rhythm. Have no rhythm until it is time to attack, then your rhythm will appear illusive.

No greater words have been written for the warrior. One should not fight an enemy at the time of the enemy's choosing. One should be patient and ready to fight at the time when the enemy is weak. Though we may be drawn in, baited, and harassed, sit quietly and do nothing until it is time. Strike when his hand is out of place. Kick when his balance is faulty. Sweep when his step is in transition. Strike quickly when his mind has lost its focus. It is in that twinkling of an eye when his gaze fails him that he can be taken.

Evade, Invade, and Control. Timing is everything, but timing without action is nothing.

Training

Training is the education of instinct. Anonymous

Let your spirit precede your movement. Anonymous

Before movement there is intent. A strong intent will spontaneously drag movement behind it. The spirit of your movement is the strength of intent. Let each movement be done with full intent and strength of spirit. Commit fully to each movement with absolute intent and your spirit will precede you.

In the pursuit of learning, everyday something is gained. In the pursuit of the Way, everyday something is shed. Less and less is done until non-motion is achieved, then nothing is left undone. Lao Tzu

So often I have moved too often. So often I have made useless and unnecessary moves. Each movement brings the possibility of error and loss. Less is more.

It is said, "first the mind, then the body". If the abdomen relaxes, breath sinks deep, into the bones. If the spirit is relaxed, the body is calm. It is always in the mind. Being able to breath leads to agility. The softest will become the strongest. Chang San-feng

Storing up energy is like drawing a bow. Relax completely, sink, release, aim in one direction. When moving, there is no place that does not move. When being still, there is no place that is not still. If others are still, I am still. If others move, I move first. Let the postures be without breaks or holes, hollows or projections, or discontinuities or continuities of form. Motion should be rooted in the feet, released through the legs, controlled by the waist, and manifest through the fingers. The feet, legs, and waist must act simultaneously, so that while stepping forward or back the timing and position are correct. By alternating the force of pulling and pushing, the root is severed and the object is quickly toppled... Chang San-feng

The purpose of any style or technique is to transcend what is being practiced, without premeditation. An artist practices his music to strive to actually forget the notes and passages and transcend all things he has learned. Then the music flows from his instrument and he becomes the music. It pours from us, as Segar says, "Like sweat from our bodies". So does the martial artist train for decades in order to forget all training as he moves. Move as he may, he will manifest the ancient ways. Without thought, analysis, or intention, he will respond as a mirror to the image laid on him. There can be no thought of technique or posture. The reaction and response must simply happen. Music, art, and living must flow from us spontaneously. Only with years of training do we experience moments of complete freedom of technique without thought.

Training is the key to this art. Repetition is the key to training. Our journey through training changes us. In training we force ourselves onward. We sharpen and harden our will as we push toward the unattainable goals of perfection of form and ultimate speed. In this state of never ending dissatisfaction, we see all imperfections of skill and limitations of speed. As we push ourselves harder and harder, we reach a point of frustration as we hit the wall of our abilities. Frustration is good. It causes us to reach beyond our physical limitations and draw on strengths that would otherwise remain untapped. What results is a slow integration of body, mind, and spirit. This integration is the true purpose of training.

In this uncertain world, ours should be the path of discipline. Hiba Yoshimasa 1350-1410 A.D.

It is not wise to rush about. Controlling the breath causes strain. It takes too much energy and leads to exhaustion. This is not the Way. It is contrary to the Way and will not last long. Lao Tzu

Even if you have a partner who is vicious and determined to injure you, consider yourself fortunate. To know yourself, to know the opponent, to understand the relationship between the two: these are the true objectives of training. Shigeru Egami 1976

Hence the saying: If you know the enemy and know yourself, you need not fear the result of a hundred battles. If you know yourself but not the enemy, for every victory gained you will also suffer a defeat. If you know neither the enemy nor yourself, you will succumb in every battle. Sun Tzu

Simple practice is of minimal benefit. A fast kick or powerful punch is not the goal. Improvement to the point of proficiency is only the first step. This is why the achievement of the black belt is considered the beginning of the journey, not the goal, and certainly not the end.

I recommend you take care of the minutes, for the hours will take care of themselves. Lord Chesterfield

The Shodan, 1st Dan, or 1st Degree Black Belt is recognized as one who has mastered the basic techniques of the martial art he has studied, and has taken that first step toward a new beginning. Now a new path is illuminated toward a lifetime of knowledge. One must never stop learning, even until death. Obtaining the 1st Dan does not mean one has reached the end of the journey, it means he has proven himself worthy to be accepted as a student of the arts.

One should exert himself in martial arts absolutely. There are no weak soldiers under a strong general. Takeda Nobushige 1525 – 1561 A.D.

As we train in our field, our sport, or our art, we must push ourselves to the outer limits of our abilities, and then beyond. We must continue to beat against the walls and boundaries of our limitations until our frustrations set us free. Concentration, focus of will, full emotional content, and all of our strengths mix together to form a force previously unattainable.

There is no deadlier weapon than the will! The sharpest sword is not equal to it. There is no robber so dangerous as nature. Yet, it is not nature that does the damage: it is man's own will! Chuang Tzu Circa 300 B.C.

Beginning in the training of one's body, practice continues with the training of one's spirit. Finally one realizes that the body and spirit are not two but one. This is true practice. Shigeru Egami 1976

Turning your attention to anything should be tantamount to turning a spotlight on it. Your mind should focus on it, exposing all its aspects. Use uncontrollable change to our advantage.

Our actions, motions, and techniques must be practiced to the point of being reflexive. Thought takes time and divides the mind. Thought confuses the calm. Thought causes distraction. Distraction causes delay. Delay causes death.

A frog to a caterpillar said, "Pray, which legs comes after which?" This worked his mind to such a pitch; he lay distracted in a ditch considering how to run.   Japanese proverb

A fox and a cat met in the woods one day. The cat had been waiting for a chance to say, "I envy your mind, your skill, and cunning. So many tricks you have to my climbing or running." As the dogs closed in, the cat said farewell. She climbed the tree and wished the fox well. The fox was attempting to show off her skills. To give the cat a chill, a thrill, and was trying to decide which trick to do, but was still undecided as the dogs tore her in two. James Hiner

Seek to see and understand the philosophy behind techniques in your life. Seek to apply the philosophy more deeply in your soul.

The most important thing about goals is having one. -- Geoffrey F. Abert

One of the most important things we as instructors or teachers need to communicate to our students' is the difference between earning a black belt and being a black belt.

An instructor must communicate to the student that class will require his full, undivided attention, as well as his maximum energy level and focus. We are teachers, not private trainers or professional counters. We do not exist to assure each black belt candidate receives a good workout, or count Hana, Dul, Set, or Ich, Ne, San for them. We are there to illuminate the Way for them. We are not there to push them to be more committed about the black belt. We exist to shine the light over our shoulder; to illuminate the path we have taken. The student must make the journey alone. The student who must be pushed to earn a black belt should not receive one. A journey of 1,000 miles begins with the first step, but the student must continue the journey on his own.

You cannot push anyone up the ladder unless he is willing to climb a little. -- Andrew Carnegie

This approach opens the doors of opportunity for the student, creating an unprecedented future and limitless possibilities as a black belt for them, not only in martial arts, but also in life. We must remind the student that the large doors of opportunity swing on small hinges. To become a black belt requires one to live the martial way. The result of becoming a black belt should be an embodiment of the martial arts within the student, allowing life to be lived abundantly. He will not have just taken martial arts. The arts will have taken him.

It is strongly recommend instructors require students to keep a journal of their experiences and activities. Those who know how to write and don't write are no better than those who don't know how to write at all. Have them practice the art of writing.

The only thing worse than a man that cannot read and write is a man that can and won't. Bruce Lee

Students need to learn to practice by themselves. The student who does not practice on his own time isn't ready to be a black belt. Students must be shown how to let go of the past. One must always remember the past does not equal the future and the way you are "being" is the result of the way you have chosen to be. If habits are learned, they can also be unlearned. Nothing is hopeless. The future black belts should be encouraged not to waste their emotional energy on trivia. They must put their goals ahead of their moods. The man who wants to lead the orchestra must turn his back on the crowd.

If you want to succeed, you should strike out on new paths rather than travel the worn paths of accepted success. John D. Rockefeller, Jr.

Eagles don't flock -- You have to find them one at a time. -- H Ross Perot

"I will have no man work for me who has not the capacity to become a partner." J.C. Penney

Mr. Penney succinctly defines the way we should view our students. Future partners...future servants in the ministry...future black belts.

The Warrior Spirit and Christianity

We have already examined the concept of the warrior in chapters past. Now, let us turn our attention to the warrior spirit and Christianity.

Some believe by entering the ministry they must leave behind "the Warrior Way". The uninformed may think we merely teach people how to fight. In reality, we teach students so they do not have to fight. When one is confident in one's ability to defend oneself, one does not have to fight to prove it. We do not have to fight someone to prove our abilities. As martial artists, we have proven our abilities in the dojang, so those issues are long settled in our hearts. The first lesson of self-defense is to avoid confrontation all together. Without that as a foundation, the warrior becomes an uncivilized thug. There is a time to stand. An outstanding example of the "warrior spirit" from a Christian perspective can be found in the Bible in II Samuel 23:8-12 in the list of David's mighty men:

23:8-12 These be the names of the mighty men whom David had: The Tachmonite that sat in the seat, chief among the captains; the same was Adino the Eznite: he lift up his spear against eight hundred, whom he slew at one time. And after him was Eleazar the son of Dodo the Ahohite, one of the three mighty men with David, when they defied the Philistines that were there gathered together to battle, and the men of Israel were gone away: He arose, and smote the Philistines until his hand was weary, and his hand clave unto the sword: and the LORD wrought a great victory that day; and the people returned after him only to spoil. And after him was Shammah the son of Agee the Hararite. And the Philistines were gathered together into a troop, where was a piece of ground full of lentils: and the people fled from the Philistines. But he stood in the midst of the ground, and defended it, and slew the Philistines: and the LORD wrought a great victory.

The biblical warrior, Eleazar literally stood and fought until his muscles became so contracted his sword was literally stuck in his hand. Many Christian martial artists develop this spirit of Eleazar; they are willing to stand until the end and work for God with all their might and all of their heart. In this day and age when most people have the remote control frozen to their hand, we need men and women of God to develop the warrior spirit so exemplified in Eleazar, so we may shake the gates of hell for the glory of God.

We now turn our focus on the warrior spirit of Shammah, for he also exemplifies the Christian martial artist. The Bible tells us when the battle raged and everyone else ran away, Shammah stood "in the midst". The practice of martial arts instills the self-confidence needed to take a stand "in the midst". In today's world, many Christians try to straddle the fence and avoid difficult decisions. One who is taught to have confidence in God, through the Christian martial arts, is much more likely to take a stand in the midst of tragedy, when the non-warriors are long gone. Shammah stood during the crisis. It is easy to stand when all is well. Martial artists are taught to stand when all is not well. The principle of "mind like water" in Hapkido is a fine example of thinking rationally when circumstances are irrational.

Calm, But Not Passive

I believe every preacher ought to spend some time taking martial arts. If nothing else, it would give them some backbone. Dr. Peter S. Ruckman, pastor of the Bible Baptist Church in Pensacola, Florida

The mind must be calm, like an undisturbed body of water. Smooth water reflects accurately the image of all objects within its range; if the mind is kept calm, the comprehension of the opponent's movements, both psychological and physical, will be both immediate and accurate. If this can take place, then one's responses, both defensive and offensive, will be appropriate and adequate.

In the contrary state, if the surface of the water is disturbed, the images it reflects will be distorted. In other words, if the mind is preoccupied with thoughts of attack and defense it will not properly comprehend the opponent's intentions, creating an opportunity for the opponent to attack. When something enters our mind that disrupts its peace, it is like throwing a rock in a still lake. In order to use the "mind like water" principle to one's benefit, one must allow the lake to be free of ripples.

Water also has other properties. Water can be very destructive. In fact, over time, water is one of the most destructive forces on earth. Your mind must be like water. When necessary, be as destructive as you must, be no more than you need to be.

In a physical sense, the water principle is the essence of the Hapkido art. Water always follows the most efficient path to its end. It cannot do otherwise. As the water runs its course, barriers and obstacles arise, which at first, appear to impede or slow the flow of the water. Understanding this can lead to the better understanding of Hapkido.

On the surface water represents a gentle form of energy, always yielding, always conforming and soft, and yet, water is the great shaper of nature. It shapes the mountains and canyons, the shorelines of all the continents.

The first level of understanding the water principle in its Hapkido application, invokes the concept that even while yielding to our opponent's strength, we embrace him and his directed energy, which in the end dilutes his attack and focus. When an opponent thrusts, we recede, when he retreats, we fill in the space, just as water does. Though we are separate from our opponent, by applying the water principle, our movement becomes one with his. We are together; we are one; each apart, but each causing the movement of the other. In the end, there is only the flow, the one movement, the principle of water.

There is an even deeper level of the water principle. Water is free to flow in any direction, so the possibilities are limitless; but, in the end, water always chooses the most efficient path, regardless of terrain or obstacles. Water cannot make the wrong choice about its course of flow.

To properly apply the water principle in Hapkido, the martial artist must take an endless journey to flow to the sea. Success and failure become obstacles in the flow of the journey. In an advanced application of the water principle, we can become as water; totally fluid and integrated with all our surroundings. The martial artist does not learn the water principle; he experiences it, becomes it, and expresses it, thus entering the flow.

There has evolved a crisis of massive proportion in today's world: passive Christianity. There is a desperate need to instill the "warrior spirit" in the youth of our country so they will not bend, bow, or budge to the pressures of world and stand firmly on the principles of God's Word. This is precisely what Christian martial arts can be used to do...build the X-Generation into warriors for Christ.

Let us reflect upon the previous chapter and ask ourselves why did Shammah stand? He was not defending all of Israel, nor the Holy City, nor the Ark of the Covenant, nor the King. He was fighting to defend a "field full of lentils"; a pea patch! Fighting by himself, when the whole army fled, Shammah defeated the Philistine army, over a pea patch! And why you ask - because they were God's peas. We learn from Shammah the biblical concept of the warrior spirit, "If it is God's, it is worth fighting for". Don't the children belong to God? Aren't they worth fighting for? Without exception, the children reached through the martial arts ministry, have become radical Christians turning the community upside down for Jesus. Why? The warrior spirit. We must suit up for battle and shake the gates of hell.

PART SIX – A SIMPLE PHILOSOPHY – Evade, Invade, Control

Before Patience, There Must Be Peace

That which matters drives the mind. If the mind is driven, the spirit is in motion and we are not at peace. Soon, very soon, we will move too soon. If one is at peace, there is no need to move, no need to react, no need to fret. In the stillness of the mind, the spirit rests peacefully. All things are seen in perspective. There is no rush. There is peace.

PRO 16:32 He that is slow to anger is better than the mighty; and he that ruleth his spirit than he that taketh a city.

Courage is the price that life exacts for granting peace. Amelia Earhart

No drives. No compulsions. No needs. No attractions. Then your affairs are under your control. You are a free man. Chuang Tzu Circa 300 B.C.

When an archer is shooting for nothing he has all of his skill. If he shoots for a brass buckle he is already nervous. If he shoots for a prize of gold he goes blind... his skill has not changed, but the prize divides him. Chuang Tzu Circa 300 B.C.

To have life and death held in balance and be free from care; this is the inner peace that brings consistency. It is not that I wish to die. It is not that I wish to live. It is that I am not, and it does not matter.

Patience, Then Enter At One Stroke

Who can wait quietly while the mud settles? Who can remain still until the moment of action? Lao Tzu

Patience is not passive. Patience is concentrated strength. Bruce Lee

Although the tiger is hungry, it will not eat bad meat. Chinese proverb

The quality of decision is like the well-timed swoop of a falcon, which enables it to strike and destroy its victim. Therefore the good fighter will be terrible in his onset, and prompt in his decision.

Energy may be likened to the bending of a crossbow; decision, to the releasing of a trigger. Sun Tzu

LUK 21:19 In your patience possess ye your souls.

For timing to work, patience must be practiced. For timing to work, you must be aware of his intentions. For timing to work, you must act without indecision. In your patience, you possess your very life.

Distance – Taking the Initiative – Determining to Move An Interval of Grace

What sense does it make to meet force with force? This strategy pits attacker and defender against each other with equal opportunity for damage. Whether we are going to attack or be forced to defend, we must consider other means.

Remaining patient, detached, and unencumbered by fear or ego, we watch as the play unfolds and wait until it is time to play our part. We join the battle in one moment. In a blink of an eye, we enter with one stroke, our force against the enemy's weakness. We evade the attack, invade the defenses, and there we control and conquer the enemy. The essential question then becomes…how do we evade?

Instead of attempting to block, it is best to evade the attack. The rule of thumb is, "don't be there." However, to simply step back is to allow a continued attack. Thus, something must be done to take us out of danger and stop the cycle of his attack and our evasion. If there is a change of direction the attacker must stop and renegotiate his attack. We must force him to change direction to continue his attack. We must step out of the way at an angle that demands the attacker renegotiate his direction and distance if he is to attack. This is one of the critical strategies within the martial art of Hapkido. If one goes no further than to evade, there could be another attack, and another, until the attacker exhausts him, or worse, the defender makes a mistake and is killed. To simply evade is not sufficient. We must place ourselves where his next attack would be difficult, and ours would be easy and swift. If our attack is swift, it will be done before his next negotiation of direction. To accomplish this, we must invade his sphere and assume a position in his Dead Zone.

How we invade is extremely important. How we enter his sphere is a matter of life and death. We enter at such a time and at such an angle as to prohibit defense. We enter at his point of weakness with all of our strength. To evade the attack we step in toward the opponent at an angle that will evade the punch or kick and place us behind the attack. We invade his sphere and pass behind his lines of defense. At most, a soft, deflective defense is offered. The block is not obtrusive and is meant to move the punch or kick just enough to avoid contact. This is nothing that would draw attention; nothing to cause him to break off his attack. We want him to continue his attack. As he reaches out from his sphere, we can step in. Wait for him to extend himself, then, attack.

It could be said that Shinsei is spiritual and philosophical training made physical. In the pressure of battle, logic and philosophy are forgotten. All that is left is reality. It is at this point your art must be a part of you. It must be real.

Waiting Within Your Sphere

For many years, the Japanese had the upper hand when it came to negotiations. In business they would very frequently emerge victorious. In most martial events they were also victorious. Many said it was their superior training and intelligence. In fact, it was simply patience. They knew how to wait.

Their patience was for a reason. When the first offer is made the position is given away. Negotiations will only erode the first proposal and never enhance it. Similarly, in the first attack, position is given away. Tactics are disclosed. Weaknesses are shown. In most cases the first attack tells volumes about the opponent. Does he fight from a distance or in close? Does he kick, punch, or grapple? Is he left or right handed? All may be disclosed with the first move. It is always best to play the guest. Let the host make the first move. As the guest, always take what is offered. Allow the opponent to graciously show his hand, his weakness, his tactics, and his strategy. Wait within your sphere. Watch and wait.

To paraphrase Lao Tzu: "Soldiers have a saying: 'I dare not make the first move. I had rather play the guest. I dare not advance an inch. I had rather withdraw a foot'. This is called marching without appearing to move, rolling up you sleeve without showing your arm, capturing without an attack. Never underestimate the enemy. There is no greater catastrophe. By doing so you will lose what you value. Therefore, when the battle is joined, the underdog will win."

Never assume the opponent thinks the way you think. That would be a deadly mistake.

A functional military does not rely on the enemy not coming, but relies on the fact that he himself is waiting; one does not rely on the enemy not attacking, but relies on the fact that he himself is unassailable. Sun Tzu

When the world is peaceful, a gentleman keeps his sword by his side. Wu Tzu

The non-action of the wise man is not inaction... The sage is quiet because he is not moved... The heart of the wise man is tranquil... emptiness, stillness, tranquility, tastelessness, silence, and non-action are the root of all things. Chuang Tzu Circa 300 B.C.

The good fighters of old first put themselves beyond the possibility of defeat, and then waited for an opportunity of defeating the enemy. Sun Tzu

Disciplined and calm, to await the appearance of disorder and hubbub amongst the enemy - this is the art of retaining self-possession. Sun Tzu

Set yourself in place and watch.

Whoever is first in the field and awaits the coming of the enemy, will be fresh for the fight; whoever is second in the field and has to hasten to battle will arrive exhausted. Sun Tzu

Every move creates an opening. By seizing the opening while the technique is being performed, an instant victory is possible. As the first move gives you information on the strength of the opponent, it also opens him up. If a punch is extended, the response should be directed at an opening the punch has created. Openings come in ways seen and unseen. We can easily see how a punch may create a lack of defense for the ribs, stomach, or head since there is no block available for that side. Sometimes openings are there but they have to be inferred. An attack depletes balance, resources, and defense. Generally speaking, while a hand is attacking, it cannot defend. When energy is being expended it can be led. When one attacks for too long, he will tire. If the attack reaches too far, balance is sacrificed. These opportunities may be unseen but are there nonetheless. Openings created by the act of attacking are often so great that the contest ends with the first move.

Perseverance is a sign of willpower. He who stays where he is will endure. Lao Tzu

Opening the Door – Avoiding the Attack

A force of four ounces deflects a thousand pounds. We know that the technique is not accomplished by strength. Chang San-feng

If he is secure at all points, be prepared for him. If he is in superior strength, evade him. Sun Tzu

The best defense against an attack is not to be where he attacks. To step out of the way of an attack is called, "opening the door". This type of defense relies totally on timing. The opponent must be allowed to commit himself to the attack and to the technique before we dare to move. To move one instant before he is fully committed to his attack will allow him to adjust the angle of his attack to our new position and strike us.

As his punch or kick proceeds out from him, and his balance shifts forward, he will be committed. The punch or kick will enter your sphere of influence. Do not move until the attack enters your sphere. Timing is everything!

Assuming you are standing face on to the opponent, to open the gate, step back with one foot forming a balanced "T" stance so one foot is behind the other by about one to two feet (Niunja Ja Sae). This allows the punch or kick aimed at the center of your body to pass safely through the opening you have created. Your body has hinged like a gate or door. When you move, the door is opened for the attack to pass through. This is the most basic defense.

Thirty spokes together make one hub. Where the nothing is, lies the cart's use. Throwing clay to make a vessel; Where the nothing is lies the vessel's use. Cutting windows and doors for a room; Where the nothing is, lies the room's use. So what we deem useful is in the void. Lao Tzu

In the physical world, nothing is softer or more pliant than water. And yet when it attacks hard, inflexible things, none of them is able to win. This is due to their lacking the ability to yield to it. That which yields wins over the rigid, the soft wins over the hard... Taking the lower status one is deemed the king. Lao Tzu

In another translation:

Under heaven there is nothing softer or more yielding than water. Yet for attacking things solid and strong there is nothing better. It has no equal. The weak overcomes the strong. The pliable overcomes the stiff. Everyone knows this but no one puts it into practice. If you can take upon yourself the humiliation of the people you are fit to rule them. If you can take upon yourself the county's disasters you deserve to be its king. Truth often sounds paradoxical. Lao Tzu

Ways to Invade

You may advance and be absolutely irresistible, if you make for the enemy's weak points; you may retire and be safe from pursuit if your movements are more rapid than those of the enemy. Sun Tzu

Speed is of the essence if you are to avoid his attack and enter his defenses, as he leaves himself open. In the moment of commitment, move at an angle in and beside him. You have avoided the blow, closed the distance, and stand beside him ready to control. You have both evaded and invaded him completely.

Military tactics are like unto water; for water in its natural course runs away from high places and hastens downwards. So in war, the way is to avoid what is strong and to strike at what is weak. Sun Tzu

Strike those places that cannot be tightened or strengthened; knees, shins, feet, groin, throat, eyes, nose, all of the places training will not toughen.

Rapidity is the essence of war: take advantage of the enemy's unreadiness, make your way by unexpected routes, and attack unguarded spots. Sun Tzu

As the sword is raised or the punch chambered, there is a moment, a twinkling of time as up becomes down and in becomes out, that one, with sufficient courage, may attack. The speed and determination must be great, but victory can be won by this action. You must be ready. You must be prepared. Have no hesitation from fear. Be fearless and rush in like a mighty wind. Strike as he prepares to strike.

And if we are able thus to attack an inferior force with a superior one, our opponents will be in dire straits. Sun Tzu

When invading hostile territory, the general principle is, that penetrating deeply brings cohesion; penetrating but a short way means dispersion. Sun Tzu

That the impact of your army may be like a grindstone dashed against an egg--this is effected by the science of weak points and strong. In all fighting, the direct method may be used for joining battle, but indirect methods will be needed in order to secure victory. Sun Tzu

The essence of Shinsei philosophy is comprised of indirect methods employed directly. When facing a direct, face-to-face attack, move toward the opponent while avoiding his oncoming offenses, so leaving you beside or behind him, and able to attack him in an indirect angle. When he attacks, don't be there. Another way of turning defense into offense to is "check" his punch or kick, to stop it before it becomes a mature force.

Trouble is easily overcome before it starts...Deal with it before it happens. Set things in order before there is confusion. Lao Tzu

It is easier to block a chambered punch or kick before the attack is launched. Before it accelerates or develops power, one can halt it with a touch. Stop the punch; check the kick, as you move in. Kick the leg as it starts to kick. Strike the arm as it begins to punch. Jam and invade.

The transition from the defensive to the offensive is one of the most delicate operations of war. - Napoleon

Working Toward the Center of His Sphere

In keeping with the philosophy and practices of Shinsei, one must seek to invade. There are only two places to go. Step to the Dead Zone or to the very center of the opponent. Starting with long-range weapons, such as kicks and punches, we assault the enemy while closing the gap between us. We use our kicks and punches to establish openings we can move into. Moreover, we must be able to defend against his attacks while seeing the openings in his defense caused by his attacks. As we block his attack we move into those places. It is, therefore, necessary to defend and attack at the same time as we move. This is not as difficult as it sounds if we understand the two basic theories or evasion and invasion.

Evasion is accomplished by stepping out or back at a distance great enough for the attack to miss its target. Invasion is accomplished by stepping into the opponent. Both can be done at once if we were to step toward the opponent and out to his side at about thirty degrees.

This angle allows the strike to miss us as we step to the opponent's side.

Invasion is accomplished by closing the distance between you and the opponent. We do this, in part, by transitioning from long-range attacks and defenses to short-range attacks and defenses. Any weapon in our arsenal can be broken down into two primary categories. They will either be long-range weapons or short-range weapons. In life, this may be regarded in the concept of time versus space, wherein our activities will be effective over the long term or short term. Long-range weapons in the martial context are punches, strikes, and kicks, used to batter the opponent, whose sphere is touching the outside of ours.

As the spheres merge and we are closer to the opponent our weapons will begin to change. There will be a distance at which long-range weapons will not be effective. At that point you must begin to use shorter ranged weapons such as knee strikes, elbow strikes, and projections. At this point or slightly closer, wrist techniques, locks, and bars become effective. At the opponent's center we may add throws, grappling, and chokes to the list of possible weapons. The mix of weapons and the space at which each becomes effective is dependant on your body type and fighting style as well as those of the opponent.

In life it is best to make three kinds of plans and have three kinds of goals:

1.   Short term plans. Goals to be accomplished within the next few weeks
2.   Mid-range plans. Goals to be accomplished in the next few months
3.   Long range plans. Goals to be accomplished in the next few years

Short and mid-range goals should always be part of the eventual long-term goal. They are pieces and sections of the larger puzzle. By planning in this way one will not tire or be diverted easily.

Collapsing the Circle and Closing the Gap

In theory, penetrating the opponent's defenses is simple. Wait for an attack. Defend by blocking his attack toward his opposite side, and slip in and behind the attacking member by stepping toward him and to his side. If he is punching you should block and slip your hand to his elbow. This keeps him from attempting a backfist or elbow as you enter his Dead Zone. If he has kicked and you have blocked, be watchful of his hands. He should try to answer with a punch. If you have blocked his punch first, he may attempt to kick as you enter. You must be able to block while moving.

The perfect setup is to block a punch thrown off the side of the front leg. In this scenario, you can slip passed his punch and he cannot kick, as his weight is on his front foot. As you step in, place your leg close to his knee or calf to restrict his movement. The block and sliding step should place you beside the opponent in his Dead Zone just behind his shoulder.

Creating a Sense of Space

To draw your opponent into you, pull your body in and become small. Create the illusion that your reach is less than it is. It will seem your sphere is smaller than it actually is. He will step within your sphere and still feel safe. Explode!

In life, promise little but deliver much. In life, set expectations of long term and humble goals. Deliver quickly and exceed expectations.

The Dead Zone

You can be sure of succeeding in your attacks if you only attack places, which are left open and not defended. You can ensure the safety of your defense if you only hold positions that cannot be attacked. Sun Tzu

Attack him where he is unprepared, appear where you are not expected. Sun Tzu

We must try to position ourselves within the "Dead Zone" of our opponent – that is the place from where he cannot easily attack or defend without having to turn his body.

It is the place where his foundation is weakest. It is both a place and direction. The Dead Zone can be seen if an opponent would stand prepared and you were to push against him. If you push in line with his feet, he is strong. But, there is a place where a line bisects his stance at 90 degrees. Here, even a child could topple him. This is the direction to attack when in the Dead Zone.

The Dead Zone is also a place. It is normally located just behind the shoulder. It extends in to an area around the shoulder blade and out about 1 to 2 feet from the center of the back to about two feet behind the shoulder and slightly outside. From here he is too close to kick and his fists and elbows will not reach. Even his elbows will deliver no power. You can stop him from turning by placing your hand on his shoulder or triceps. You will then have your other hand and both legs free and aimed into him.

All things and situations have a Dead Zone. It is located somewhere behind the direction of their attack. Psychologically, the Dead Zone is the opposite of how they are laying their plans and conducting their attacks. If they are rational, you must arouse their emotions to a fever pitch. If they are emotional, you must remain calm, watch, and plan. Wait for their internal storm to overtake them.

Most martial artists have a general knowledge of the Dead Zone, as it pertains to "stand up" fighting. Slipping into the Dead Zone is a major factor in the Shinsei strategy. Many martial arts lack knowledge of the Dead Zone in ground fighting. There are "holes" in the opponent's defense there, just as there are when he is standing. The objective in grappling is not to "tap out" an opponent, it is to move into an area to avoid his attacks and defenses, and destroy what he leaves vulnerable.

Ways to Control – Checking the Kick

Entering his Dead Zone may take more than one step. It is possible to enter his sphere right away by checking and deflecting a punch or a kick. This may not put you in his Dead Zone but you will be able to jam him and prevent his attack while you work to the zone or to his center.

As his foot comes up to kick, extend yours low and straight, with your foot turned to the side. He will kick his shin against the blade of your foot. Keep your foot aimed at his shin. As he releases his kick, he will catch your shin and his kick will rebound causing pain and lack of balance. With your leg still extended, step in toward him checking or blocking his hand. You are now eye-to-eye and close enough to enter his center. Most throws take place when the opponent's center is controlled and his balance is taken.

There are only two places we can win - in his Dead Zone or in his center. Yet, when at his center, you must be watchful of his attacks.

A leader will not foresee all problems. He is not omniscient. A good leader will fix a problem so it is not likely to occur again. When an attack occurs, whenever possible, tie up, incapacitate, or damage the attacking limb, person, or organization so as not to allow a repeat attack.

Position of Power

All techniques have within them elements of timing, balance, and position. Movement hinges on them. When to move, how to move, and where to move make up the outcome of an attack or a defense. We must move from a position of power to a position of power. If a technique does not accomplish this, it should be discarded. Most techniques are enhanced when the foot with which we step out is on the same side as the hand we use to push or turn. This does not allow the body to turn as it would were we were to step with the left side and use the right hand. Because of this, angles remain intact and power is delivered directly. The symmetry of your body must be guarded. Getting off on the correct foot is essential.

In the simplest terms, the power of most techniques comes from having your hands as close to your center as possible, and having the opponent's hands as far from his center as possible. All things being equal and on balance, the further from the Hara you are, the weaker you become. In using this rule, you will find most throws are done by either having the hands at your center or bringing them back toward the center.

Power Versus Speed

Principles of physics teach us if an object increases in mass, the power delivered when it impacts is increased accordingly. However, as the velocity of the object increases, the power delivered increases by the square of the speed. Thus, velocity becomes critical. It seems counter intuitive that it takes more time to produce velocity on impact. Inertia of an object at rest must be broken and the object must accelerate in order to be traveling at higher speeds when contact is made. In this way of thinking, it becomes a trade-off between the velocity of an attack at delivery and the speed with which it can be delivered.

To further demonstrate this theory, let us examine a simple punch. If the punch is thrown from the right side with the right foot stepping forward, it has less distance to travel than a punch delivered off of the back leg. The punch thrown off of the front foot will contact the opponent in less time but may not be traveling at maximum speed, since it has not had time to reach the greatest possible velocity. If the punch is thrown with the right hand while stepping with the left foot it has further to travel. Even though it may take more time to reach the target, the fist will be traveling at a higher velocity when contact is made.

When speaking of delivery speed, there is always a trade-off between speed and power. Life is full of trade offs. There are many things that cannot exist simultaneously. Hate and loyalty, respect and loathing, leadership and disrespect, power and speed are but a few.

Leaving a Way Out

To fight and conquer in one hundred battles is not the highest skill.
To subdue the enemy with no fight at all, that's the highest skill. -
Sun Tsu

What is the objective, to win the battle or argument, or to save the
person from destruction?

Let the battle be for conversion and not for destruction. Too many
souls are lost in wars of words and battles that could have been won
with the conversion of the person. There is nothing so fierce as an
opponent who is fighting for his life. It is usually unnecessary to kill.
Victory is won not in destruction but in the minds, heart, and spirits.
When taking a position and fighting the battle over land or market
share, it is best not to leave dead bodies to trip over. Conversion,
affiliation, persuasion, and influence are preferable to destruction.

If faced with a violation within your ranks that is not disastrous,
confront the person assertively. Announce his error to him alone.
Always leave the person a way to explain and correct his actions. He
is more likely to apologize when able to gracefully retreat. Having
been caught and allowed mercy, he will attempt to avoid the
situation again. In this way you will win his gratitude and respect.

A man will fight if his pride or life is threatened. When approaching
an opponent with whom you wish peace, you must always give him
the path and circumstances for an honorable withdrawal. Even if an
army has an open road for retreat the men will stay, fight, and die if
honor is at stake.

When you surround an army, leave an outlet free. Do not press a
desperate foe too hard. Sun Tzu

Even if the enemy must be completely destroyed, it is still best to
leave the appearance of a way to escape. Be there on the other side,
waiting. When they relax in safety; strike.

Show him there is a road to safety, and so create in his mind the idea that there is an alternative to death. Then, Strike! - Tu Mu quoted in the Art of War by Sun Tzu, 490 B.C.

Thus, there are two reasons for allowing a path of escape. One is to keep peace by permitting the opponent an honorable retreat. The other is to induce a sense of security to destroy them in a time and manner of your choosing. We should always be aware of this strategy when we ourselves are retreating lest we be entrapped as well.

Allowing bravado and the venting of verbal aggression without responding prevents wars. It takes discipline to listen and not react, but as emotions are emptied, common sense may prevail.

Likewise, if it is obvious the enemy is bent on war and will not relent, it is best to act quietly, intently, and deliberately in preparation for battle. Say little, plan for victory, attack without warning when the enemy is in mid-sentence. Talking is not conducive with action.

Do not allow the enemy to escape too soon. They will regroup and attack again. Do not seek to destroy or you will arouse their fear of death and the strength and speed that it entails. Sun Tzu

There are ways to make the opponent understand you have willingly allowed him a way to justify or explain his actions, or that you left a way open to retreat. He must see himself trapped, then you may open the escape.

When the opponent's spirit is broken, and he knows he will be defeated, it is the time of mercy. Mercy is the tool of choice. It is the salvation for many and the velvet hammer of the wise.

If men are not afraid to die, threatening them with death gains you nothing. If men live in fear of dying and breaking the law means death, who will dare break the law? Lao Tzu

If a man is willing to give his life for a cause, he cannot be stopped by anything short of death, that is, unless you remove his cause.

A student and master walked in a field. From behind them came a rabbit followed closely by a dog. "Who do you think will win?" asked the master. "The dog", said the student, "he has superior strength and speed." The master replied, " the dog runs for his lunch, but the rabbit runs for his life."

Moving Without Appearing to Move

Stand your ground. Be strong of spirit. Do not retreat. The enemy will begin to move, waver, and test you. The enemy will rush forward to try you and then fall back. As he falls back, close the gap by staying just outside of his sphere of influence. Move with him like a shadow. Never retreat, but advance as he retreats.

When it was to their advantage, they made a forward move; when otherwise, they stopped still. Sun Tzu

By moving in this way, one lures the opponent into a trap or leads him into a weaker position.

Control of Terrain

If the enemy has any measure of fear, you may back him into the position or direction you wish. By breaching his sphere just enough to make him withdraw into a defensive posture, the opponent will give space. He will stay in the defensive posture for a time and then relax into a waiting posture. Again, breach his sphere and wait for the cycle of drawing back and waiting. By entering his sphere at angles in line with the destination you seek, you can push him into a corner, force him to fight up hill, or place him looking into the sun.

You must be aware of the terrain, sun, and all other obstacles and use them to your advantage. Throw him into mud and water. Make him step through oil to reach you. Each time he blinks, slips, or trips you should be there to take advantage.

Fighting the Opposite Way – An Opponent's Strength Becomes His
Weakness

**Your opponent's style is his strength. Assuming no one defends
using a weak stance or technique, we will know the style of the
opponent by his posture. Is he a boxer or kicker? Does he fight
high or low? Does he want to grapple or stay distant? All of these
things will tell you what he is before the battle begins. In addition
to the physical posture of the man, is his mental posture. Is he
aggressive or defensive? Does he act or react?**

Never do battle on the opponent's terms. If his posture tells you he
wants to fight from a distance using kicks and punches, you must
attack to gain the inside advantage using grappling, Judo, and
wrestling. If the opponent fights low and inside as if he wants to
grapple, you must attempt to keep him distant and fight him with
long-range techniques such as kicks and punches.

To be capable of competing against various styles we must train hard
and long in all major strategies and styles. We must understand and
be competent in kicking, punching, throwing, and grappling, then
we must chose a way opposite of theirs.

We must be attune and observant. With each word or move, a story
of personality and fighting style is told. From the time he assumes a
fighting position you will know his ways. Is his front foot light? Then
he may kick. Does he crouch? He wants to take you to the ground.
Does he dance with his fists up? He wants to box. Whatever style he
chooses, fight him in the opposite manner.

*When the dragon is in shallow water even the frogs laugh. Chinese
Proverb*

In some Chinese folklore dragons were water beasts. The proverb
above shows the weakness of the most fearsome beast when it is out
of its element.

Most of those who train, defend against the style they practice. Most of their time has been spent studying their own movements and the counters to them. We spar with those in our own camps. The weakness in defense comes from inexperience with other styles of combat. For this reason it is absolutely necessary to have a wide range of knowledge exposure to opponents of varying styles. Your opponent practices the way he knows best. Attack him with the opposite style. Soft against hard – hard against soft – close against long and long against close.

Just Breathe

JOB 27:3 All the while my breath is in me, and the spirit of God is in my nostrils.

JOB 33:4 The spirit of God hath made me, and the breath of the Almighty hath given me life.

The mind and breath are king, the bones and muscles are the court. Chang San-feng

In Motion, all parts of the body must be light, nimble, and strung together. The breath should be excited, and the spirit should be internally gathered. Chang San-feng

Control your breathing but do not force its control! Breathe before you need to. What general would wait until battle to amass his troops? Breathing is the only essential bodily function that can be controlled both consciously and unconsciously. Our control of it must be both.

If the Dan Jun, (the center of our body and balance) represents the camp of our internal troops, the rhythm of breathing would be the cycle of day and night, sleep and work, of the troops. The comings and goings of breath are small cycles of strength and weakness. We know this intuitively. No one ever tries to lift a heavy object by breathing in. We always breathe out. This cycle goes deeper than the generation of power. It goes toward attack and openings.

If we watch the opponent closely we can see patterns in his actions. Attack and defense will have a style based on training and personality, but the rhythm will be very individual as if it was hard-wired. Attack will always occur when breathing out. This is an extremely crucial observation. Watching the opponent's breath will give insight to the timing of his attack. Even though breathing patterns can be changed very quickly, it does not happen immediately. If we attack at the bottom of the opponent's exhale or the beginning of the inhale cycle the opponent will have no breath to exhale and will be weak and open for attack. His defense will be slightly weaker and he will not be able to withstand a forceful blow. Part of the armor of defense against a strike to the body is to breathe out and tense strongly. When breath is gone, armor is lacking. Your attack must happen in concert with your breathing. To attack strongly, breathe out strongly.

Approach with a spirit of bouncing the enemy away, striking as strongly as possible in time with your breathing. If you achieve this method of closing with the enemy, you will be able to knock him ten or twenty feet away. It is possible to strike the enemy until he is dead. Train well... From A Book of Five Rings

The way we breathe is important and controls our calmness, balance, and strength. Our breath must be unlabored and deep. The breath should be taken into our chest just as water is poured into a bottle. The breath is centered low in the abdomen and fills the lungs from the lowest point upward. The chest never rises but the abdomen expands. Breathe out in long, controlled cycles. Pain and tension are breathed out. Strength and preparation are breathed in. Guard your breath. If you lose your breath you lose your life.

Rhythms

All people have rhythms of weakness and strength. They try, attack, and tire. They retreat, rest, and observe the damage to the opponent as a result of their attack. They pause, taking time to judge the enemy's defenses. We see this with clarity when in combat. Most people will attack with a set series of combinations of techniques. The combinations will usually have a number of movements. In general, the series will be three or four movements in combination. After each series the opponent will retreat, recover, and observe. The number of movements represents the timing the opponent is also expecting. The time directly after the attack in which the opponent begins to fall back and retreat is called the moment of relaxation.

The Way of strategy is the Way of nature. When you appreciate the power of nature, knowing the rhythm of any situation, you will be able to hit the enemy naturally and strike naturally... From A Book of Five Rings

To use this knowledge to your advantage you must lead the opponent into the moment of relaxation. Using the same timing and number of moves in combination the opponent attempted, attack him fiercely. Stop as he did. Begin your retreat by an obvious relaxing of your body and begin to pull back but keep your mind centered and concentrated. You must never allow yourself a moment of relaxation. As you see the opponent begin to respond to your feigned retreat by relaxing also, attack quickly and with full force.

In large-scale strategy it is important to cause loss of balance. Attack without warning where the enemy is not expecting it, and while his spirit is undecided, follow up your advantage and, having the lead, defeat him. Or, in single combat, start by making a show of being slow, then suddenly attack strongly. Without allowing him space for breath to recover from the fluctuation of spirit, you must grasp the opportunity to win... From A Book of Five Rings

Continue your attack by doubling the number of movements the opponent used in his series plus at least two more movements. For example, if the opponent uses a 3 or 4 movement series you would use a ten movement series in your attack. This will allow you to force the opponent through two complete periods of relaxation and will overwhelm the opponent. It will give you two periods of force in his rhythm of weakness. In business, life, and war all movements of attack, acquisition, take over, or progress are always followed by periods of recuperation and observation. This is a time of weakness and relaxation when the person, company, or opponent is vulnerable.

... when you have closed with the enemy, hit him as quickly and directly as possible, without moving your body or settling your spirit, while you see that he is still undecided. The timing of hitting before the enemy decides to withdraw, break or hit... From A Book of Five Rings

When entering into battle or a physically demanding situation, one should begin breathing deeply. No general would go into battle before his army is gathered, amassed, and ready. All beings have a rhythm of movement and rhythm between movements. All beings have their own natural timing. It does not take long to feel and understand the rhythm of the opponent. It will not take him long to determine yours. In combat, it is important to vary your rhythm. Strike - - - Strike. Strike, Strike, Strike!

Habits, Compulsions, Patterns and Responses

To telegraph your intention is to give permission for your own defeat. Do not move in spirit or body before you attack.

Know yourself, find your habits, and eliminate them. Do not deal with one-time mistakes. Correct habitual mistakes first.

The ability to understand and affect your own behavior is proof of being self-aware or sentient. Even though we have the ability to change, new habits take time to instill. Because of the time needed to establish new patterns of behavior or eliminate old one, in the short term it is best to view the brain as a hardwired, preprogrammed machine. In this light we can see that habits, die-hard or not at all. Add to this, the way the brain and body interact and you have the stage set for habits. There are hundreds of small habits happening within each moment of life. We fight how we train. In training, if there is any preparation, shift of stance, or movement repeated during training it will be programmed in with the techniques as a habit.

Habits can even be formed while helping others train. For example, a person might develop the habit of touching or tapping the leg they are about to kick with, while helping to sharpen a partner's blocks.

This may originally have helped the partner anticipate the incoming kick and prepare to block, but now it has been programmed into his kick as a habit. Everyone has habits, though they may be subtle. Watch carefully and you will see what he will do before he even knows it himself.

The same can be said for patterns of attack. As we train, we develop sets of patterns used for attacks. These are called combinations. They are also habits. We hope the pattern will be effective enough to devastate the opponent before the pattern can be detected and used against us. Our job is to stay alive and observant long enough to detect the pattern of the combination and use the knowledge against the opponent.

Moving to the outside of the opponent's arm can stop most combinations. One does not have to be in the Dead Zone; however, that is always best. One must only force the opponent into a major change of direction. Combination attacks are programmed to cover an area of space or a direction of movement. The direction is usually backwards, in a linear fashion. The opponent counts on forcing you straight back so he may continue the direction of attack with the programmed combination of techniques. To defeat this, one must not retreat. Avoid the attack by evading the combination as a whole. Do this by stepping in to the attack and to the outside of the arm. Avoid by invading. This causes a 90-degree change of direction for the opponent and results in the combination attack being broken off short. Before he can adjust, you can attack.

Remember, evade, invade, control.

All Movement and Power Begins in the Hips

The waist is like the axle and the breath is like the wheel. Chang San-feng

Larger muscles take longer to activate. One should activate them first. The generation of maximum power depends on the larger muscles. They should be used to drive the smaller muscles into the target. In this way, the power of the legs being pushed and the waist twisted produce the power used to force the fist into the target as the shoulder and arm compel it forward. Push with the back leg, twist the waist, launch the fist, and strike the target as all of the actions culminate in one laser sharp moment of time.

It is not only the simple punch that pulls its awesome power from the ground, through the legs, into the waist, and out to the arm. All movement should be produced in the same way and coordinated with the breath. In this simple rule is the true power of the art.

Circle of Defense and Attack

The skillful tactician may be likened to the shuai-jan. Now, the shuai-jan is a snake that is found in the Chung Mountains. Strike at its head, and you will be attacked by its tail; strike at its tail, and you will be attacked by its head; strike at its middle, and you will be attacked by head and tail both. Sun Tzu

It flows far away. Having gone far away, it returns. Lao Tzu

Returning is the motion of the Way. Yielding is the way of the path Lao Tzu.

Yield the side on which you are being attacked, while advancing the other side to attack the opponent without warning. Step forward with the left side if the right is being attacked. This opens the left so the attack will pass while the right advances into him. If the attack is in the Dead Zone, you must spin and backfist him with the other arm. Turn toward him, withdrawing the side closest to him first.

To Orbit Around the Great Center

Take a child's rubber ball. With a pen, mark dots in various areas on the ball. Now turn the dots so, one by one, you are looking straight over them. At the moment each comes directly under your eyes, it looks as if that dot is the center of the ball. In much the same way, during a throw, we become the center of everything happening. We capture the center, balance, and orbit of our opponent. We become his center. Our center becomes the pole he orbits around. At that instant, we have his movement and his balance. It is focused within our center. Our Hara (center) is still and quiet and he orbits around the great center in us. This is how an Aikidoist can see himself as the center of all things.

The Accepting Mind

If you want to understand the truth in martial arts, to see any opponent clearly, you must throw away the notion of styles or schools, prejudices, likes and dislikes and use what works for you. Bruce Lee

One of the most difficult tasks in combat is to evaluate the opponent and still keep an open mind. If the person is obviously one who likes to use kicks and he has kicked 49 times, it is difficult not to look for that 50th kick. But, if you look for it and you are wrong it could mean your life. You must be prepared to respond to whatever he throws at you and not be misled by patterns.

If you are looking for a dragon every cloud in the sky will look like one. Chinese proverb

Your mind must be open to any possibility. This includes his attack and defense as well as your own. Let the movements flow spontaneously according to the opening and opportunity. Use the patterns you see in his movements against him, but be ready for changes. An open mind does not ignore or overlook information. It sees and uses all information. The open mind makes no assumptions. It is ready for all possibilities. Face your enemy with an open mind; expect nothing; see all.

Big Circle, Small Circle, No Circle

The difference between a master and a black belt is the master's ability to apply the principle of conservation of movement. The beginning student tends to exaggerate movements due to a lack of body awareness. He does not know where his body is in space at any specific time. The advanced student understands the mechanics. He knows where his body is and its relationship to the opponent, but he has not mastered the interactions. The master has eliminated all useless movement, including the movement of mind as well as body. What is left can be seen as a simple and determined movement. There is no waste of space, movement, or time.

The beginner moves in large circles. The advanced student moves in small circles. The master has eliminated all excess movement and the circle has diminished to nothing. When Michelangelo was asked how he carved the statue of David, he remarked that David was already in the block of stone. He simply chipped away what did not belong to reveal the image. In this art we must polish the mirror, chip away at the block, see clearly, and reveal what is inside of us, waiting to come out.

Straight Lines and Circles

There are few advantages to widely circular techniques. Circular techniques may develop more power over a shorter distance, but they always take longer to land. Spinning kicks are the slowest to deliver and must rely on the element of surprise in order to land. The use of roundhouse and spinning kicks opens up an avenue of direct and straight counter attacks. By stepping slightly into the circle formed by the approaching kick, one can easily block while simultaneously issuing a front kick to the opponent. The block and counter-attacking kick should contact the opponent at the same time.

Hard and Soft

Defend using soft blocks against hard attacks and circles against straight lines. Hard can be defined as a straight or linear approach; soft techniques as circular paths. Hard techniques develop speed and power using muscle expansion and contraction. In soft techniques the body remains relaxed and never tenses. Power comes from body movements such as stepping or turning the waist.

The area of two intersecting circles is a single point. The point will have the sum of energy developed by the momentum of both circles, if the circles meet at a phase of 180 degrees. In other words, when using a soft or circular block, damage and pain can be inflected on the opponent if his circular technique is met at the proper angle.

For example, if the opponent kicks with a roundhouse kick, your block should meet his kick going out and toward the kick's angle of approach. Leave your arm relaxed but controlled. Breathe out and turn your waist into the path, holding your arm out from the shoulder and up at the elbow.

The turn of the waist, not the muscles, will produce the power. Use the soft block to crush the attack. In this way, a circular or soft block against a circular attack will yield a hard result.

To divert an attack without impeding his momentum, use a soft block against a hard attack. When blocking a hard, fast, and straight technique, a slight push sideways will divert the attack sufficiently to miss its target. Meet the side of the attacking fist at the wrist with a soft touch from the palm of your hand. Drag it in towards you while pushing out toward your shoulder's edge. In this way, a mere four ounces will cause a killing blow to completely miss its target.

To defend against a hard attack with a hard block is possible but it puts attacker and defender equally at risk of harm. The most conditioned body will win. It is wise to choose better odds.

The Ellipse, a Circle in Motion

From before the time of PTOLEMY (about A.D. 100–70), man believed the circle to be the perfect form. All things in heaven, it was said, traveled in circles. Ptolemy, an astronomer and mathematician, whose synthesis of the geocentric theory (that the earth is the center of the universe), dominated astronomical thought until the 17th century.

Ptolemy proposed a geometric theory to account mathematically for the apparent motions and positions of the planets, sun, and moon against the background of fixed stars. He began by accepting the generally held theory that the earth did not move but was at the center of the system. The planets and stars, moving eternally, were considered (for philosophical reasons) to move in perfectly circular orbits. This geocentric theory held until the time of Copernicus.

Nicolaus Copernicus (1473-1543), was a Polish astronomer, best known for his astronomical theory that the sun is at rest near the center of the universe, and that the earth, spinning on its axis once daily, revolves annually around the sun. This is called the heliocentric, or sun-centered, system. Sometime between 1507 and 1515, he completed a short astronomical treatise, De Hypothesibus Motuum Coelestium a se Constitutis Commentariolus ("A Commentary on the Theories of the Motions of Heavenly Objects from Their Arrangements") (known as the Commentariolus). Even in this great work, the circle was considered the pattern God used for the orbit of all planets.

The circle was considered the holy template of heaven until Kepler found the ellipse in the orbit of Mars. Johannes Kepler, (1571-1630), was a German astronomer and philosopher, noted for formulating the three laws of planetary motion now known as Kepler's laws.

 Kepler accepted Copernican theory immediately, believing the simplicity of Copernican planetary ordering must have been God's plan. In 1594, Kepler worked out a complex geometric hypothesis to account for distances between the planetary orbits — orbits that he mistakenly assumed were circular. Kepler later deduced that planetary orbits are elliptical; nevertheless, the preliminary calculations agreed with observations to within 5 percent. Kepler then proposed that the sun emits a force, which diminishes inversely with distance and pushes the planets around in their orbits. Kepler published his account in a treatise entitled Mysterium Cosmographicum (Cosmographic Mystery) in 1596. The first law states that the planets move in elliptic orbits with the sun at one focus; the second, or "area rule," states that a hypothetical line from the sun to a planet sweeps out equal areas of an ellipse during equal intervals of time; in other words, the closer a planet comes to the sun, the more rapidly it moves. In 1612 Kepler became mathematician to the states of Oberösterreich (Upper Austria).

While living in Linz, he published his Harmonice Mundi (Harmony of the World, 1619), the final section of which contained another discovery about planetary motion: The ratio of the cube of a planet's distance from the sun and the square of the planet's orbital period is a constant and is the same for all planets. The ellipse had been established as the form of heaven.

An ellipse is but one of a group of shapes called conic sections. By slicing a cone in various ways, one can see all of the conic sections. There is the circle, ellipse, parabola, and hyperbola all visible in the cutting of a cone. In these forms, and especially in the ellipse, are found the patterns to all throws and projections. By using and combining these conic sections, the patterns of all throws and projections are traced in space.

It is said, "as above, so below". But even after we accepted the place of the ellipse in the heavens, we did not update our view within the martial arts. In some ways, our thoughts about throws and projections in martial arts are still viewed from the Ptolemaic framework. The circular view of throws and projections are outdated and were never true to begin with. Life is never stagnant or stationary. A perfect stationary throw with a single center is the only way a true circular projection can exist and it is not likely to happen since inertia and thus movement must exist before a projection can occur. There is no dynamic sphere of Aikido.

Aikido, Hapkido, Judo, and Jujitsu are spheres in motion and spheres amidst the transference of focus from person to person. **Therefore, a throw or projection has a beginning center within the person attacking and an ending center within the person defending. These two foci constitute the shape and definition of an ellipse.**

As energy is aimed at the person being attacked, it is redirected back toward the direction of origin. If this redirection is unnoticed, the throw will proceed. If the transition from defense to offense is transparent to the attacker, the technique will be flawless. The pattern of the transition is the ellipse. The shape starts with the center of energy being within the attacker, but as the attacks progresses he projects his energy outside his center in the form of a punch, a swing, or a kick. The energy is received by the defending party and is redirected in order to control the attacker. When this is done properly, the defending person becomes the center of all balance for both persons. Now he or she is the center of movement. This transference from one focus to another forms the ellipse. It is one of the forms that is conducive to a 180 degree reversal of energy.

The major axis of an ellipse is a straight line that passes through the two foci lengthwise along the ellipse and extends to meet the curve at the farthest point of each end. Any ellipse is symmetrical with respect to its major axis — the portion on one side of the major axis is a mirror image of the portion on the other side. It is along this major axis that most throws take place.

When training for throws or projections, find the ellipse within the technique and work to perfect the transference of energy from direction to direction.

Strategies

Strategies and tactics arise with the use of selected weapons. When using a long sword, there will be certain tactics, which cannot be employed favorably when using a short sword. Tactics and strategies must always follow the instruments and abilities at hand.

In the practical art of war, the best thing of all is to take the enemy's country whole and intact; to shatter and destroy it is not as good. So, too, it is better to capture an army entirely than to destroy it, to capture a regiment, a detachment or a company entirely than to destroy them. Hence to fight and conquer in all your battles is not supreme excellence; supreme excellence consists in breaking the enemy's resistance without fighting. Sun Tzu

Avoid the enemy's strength, strike at his weakness. Sun Tzu

If the enemy leaves a door open, you must rush in. Sun Tzu

Methods of Attack

RELAXING AND TURNING: THE POWER OF SOFT
A soft answer may turn away wrath, but a turning soft block causes great pain. The most forceful block is soft, thus making the block an attack. The arm is soft, yet overwhelming power is produced by turning the waist into the attack.

DIRECT ATTACK: THE POWER OF HARD
Strike the opponent, quickly and efficiently. Some of the most common forms of the direct attack are the backfist, the jab, and even the lead shin/knee kick.

COMBINATION ATTACK
Attack by throwing a combination of strikes. Example combos are the 1-2-3 (jab-cross-hook), lead shin/knee kick to round kick, etc.

ATTACK BY DRAWING
Open a target for the opponent. Await his attack. Be prepared to defend and attack.

ATTACK BY TRAPPING
Trap one or more of your opponent's possible defenses, such as an arm. Using one hand to pull down the defending arm, attack the face his face with your other hand. With your front hand open, grab the opponent's front hand and pull it down slightly. This will make an opening through your opponent's defenses. Execute a rolling backfist to the face with your free hand.

ATTACK BY CONTINUING
Feint at one or more targets on your opponent in order to make your opponent defend, then attack the intended target without withdrawing your attacking weapon. Examples are: round punch into an elbow strike or elbow strike into a backfist.

ATTACK BY CIRCLING
This is best done with a spinning backfist. As the opponent attacks, block with a soft block spinning down the opponents arm. Landing behind the opponent in his Dead Zone execute a backfist, elbow, or punch.

ATTACK BY MISDIRECTION
Attack high to draw the block, then quickly counter with a low strike or kick.

ATTACK HIGH-LOW, OR LOW-HIGH
This type of attack draws the opponent's attention and defenses to an area so an attack can be launched in the opposite region. An example would be a snap kick to the groin followed by a roundhouse to the face. Draw the defenses high then attack low. Draw the defenses low then attack high.

That which shrinks must first expand. That which fails must first be strong. That which is thrown down must first be raised up. Before receiving there must first be giving. It is the perception of natural things. Soft and weak will overcome hard and strong. Lao Tzu

Evade, Invade and Control Using Angles

By stepping in toward the attack and out to the side of the attack, the kick or punch will miss you as you step solidly into his Dead Zone (that area beside and behind his shoulder where he cannot defend). Angles of avoiding must be closely controlled. Not enough and you will get hit; too much and you will be out of his Dead Zone, vulnerable for another attack such as his kick. Step from danger into safety and attack with speed.

Evading Angles

Within the Shinsei System, the evading angles correspond with the approaching angles, in that they are the same basic eight angles in relationship to one's position. An attack may be directed from several angles, and these same angles can be used in reverse to evade. The eight attacking angles become the eight evading angles. The eight angles are visible by imagining you are standing in the center of a great compass facing north. The angles correspond to North, Northeast, East, Southeast, South, Southwest, West, and Northwest. There is a secondary way of looking at eight basic directions within a three dimensional space. They are North, South, East, West, up, down, in, and out. All combinations of these make up our possibilities in life.

If someone yelled "fore", while you were on a golf course, and you looked back to see a white ball zooming for your head, you would automatically know which evading angle to use to avoid being smacked by the ball. If you are entering an elevator as someone is exiting, you turn your body slightly to your right or left to avoid contact, thus again using an evading angle.

While the above examples may seem overly simplistic, these evading angles and invading angles are used in combat and self-defense. It is a theory many fail to teach or understand, but is one that cannot be ignored, no matter from which walk of life you come. Whether jumping out of the way of a golf ball, maneuvering a Bradley fighting vehicle through oncoming Iraqi attacks, parrying a bayonet thrust on the battlefield, dodging a crash on the NASCAR track, or "bobbing and weaving" in a boxing ring, evading angles are being used in relationship to attacking angles. To properly understand and utilize the evading angles and attacking angles, one must be able to determine the direction of force.

Evading angles permit one to avoid direct contact with the enemy's attack. In Shinsei Hapkido, they do much more. One can evade all day, but never win the battle. By properly using the evading angle, one is able to lead the attacker into a technique of neutralization by drawing him into the sphere of defense. Once he is drawn into your defensive sphere, he is in a position of physical instability, and strategically located for you to invade and control. You thus neutralize his immediate attack and implement measures to avoid any secondary attacks.

The eight basic evading angles are the basic movements that which comprise the foundation for the extremely mobile footwork of

Shinsei Hapkido. Movements along these evading angles may be linear, angular or straight-lined, circular, or a combination of these movements, but they will always fit into the basic eight angles. They are the key ingredients of evading the attack.

Evasion must be accompanied by a simultaneous attack to the weakest points in the opponent's stance or defense. This is the where evasion transitions to invasion. When invading, one should direct the attack along angles of the opponent's weakness in order to penetrate his defense and establish control. The understanding of these angles and their practical use is important when defending against several opponents attacking from different directions. At this moment, positioning becomes crucial.

Coupled with the use of evading and invading angles is the concept of strategic positioning. This concept allows one to enter the opponent's defensive sphere and apply a technique of control or neutralization. One must position oneself so one of the opponent's angles can be utilized to engage him. In doing this, one can off-balance an opponent and step into his Dead Zone. This limits the opponent's ability to defend, and makes it nearly impossible for him to initiate a secondary attack. When properly used, this concept may be implemented in a multi-directional defense, using the angles to penetrate the attacker's defense, thus evading, invading, and controlling.

Projecting, Throwing and Tripping

It is not physics or anatomy but human psychology which causes many throws to be effective. Desire and attachment can put us in a position to be defeated. Many projections and throws are successful only because the opponent tries too hard, holds on too long, steps too far, or resists too strongly. Due to his greed he will off-balance himself, first emotionally, then physically. As he reaches too far to grab you he gives away his balance. By holding on too long he sets himself up to be thrown. By resisting your motion strongly he loses his ability to move. We must remain balanced in mind, body, and spirit. If you do not reach further than your center will allow, you will be very difficult to defeat.

Effective Self-Defense

If you're really interested in self-defense, then you need to train specifically for that. Because when you get hit you can't stop to think. ALL you can do is react, and you will react however you trained. -- Wally Jay

Whatever you teach, teach it well, but understand there are differences between martial sports and martial arts. Never mistake a sport for self-defense. Too often the Olympic stylists call what they teach, "self-defense", but have no idea how to defend taken to the ground. On the other hand, the practitioner whose techniques are all ground or grappling-based, cannot defend well in a typical boxing match. Most arts and sports do not train for techniques that are outside of their sphere of study. If someone were to bite your leg while you are applying the infamous arm bar, would you be prepared? Arts taught for the purpose of self-defense should be well rounded. The teacher should have examined the techniques and responses from a street fight perspective. In sport karate, such as point sparring, there are rules against groin strikes, kidney strikes and eye gouges, to name a few. Because of such rules, the students do not train to defend against those assaults. On the street there are no rules. An art of self-defense should be designed with this critical issue in mind.

There is nothing wrong with martial sports. It can be very enjoyable "rolling" with well trained grapplers, however, as true martial artists, we must think outside the box if we are going to teach self-defense. We must teach our students how to defend when there are no rules, no mats, no judges. When the need to defend occurs in the student's life, he will react exactly the way his instructor has trained him. Unfortunately, most will throw one strike, then pause, as if the mystical self-defense genie is going to appear and say, "Point!"

One of the most serious issues facing modern martial arts systems is the failure to address the reality of multiple attackers. Thugs like to travel in gangs. In the earliest years of the martial arts, even the thugs had a sense of honor, and it was considered "dishonorable" for a group to attack a single man. Therefore, many of the old masters did not train for that occurrence. Viable self-defense must be taught utilizing multiple-opponent training.

In the Shinsei system, we teach true self-defense must be "stacking". This movement puts one in a position to cause the would-be attackers to get in each other's way. The positioning of the body in line with the attackers, in such a way that sets them up one behind the other, makes it difficult for the attackers to reach the defender en masse. They are instead in a queue, approaching, one attacker at a time. Now the odds are even.

What if He Has a Weapon?

When the attacker possesses or is wielding a weapon, he has two distinct advantages over you. One, he has a psychological advantage. In his mind, regardless of your abilities, he is better equipped for the fight by having possession of the weapon.

Secondly, regardless of the weapon type, he will be able to attack while remaining outside your defensive sphere. A bat, club, flashlight, or any other object can be a deadly weapon if wielded by someone with mal intent, even in the hands of an untrained attacker.

When a weapon is involved, you must act without hesitation to either draw the attacker into your defensive sphere, or you must move along one of the evading angles. In either case, you should evade into the Dead Zone in order to gain control of the weapon and neutralize your attacker's ability to use it. Controlling the attacker may be accomplished by disarming him of the weapon and in turn, using his own weapon against him. Projections and throws may also be incorporated at this point, as well.

When defending against an opponent who has a knife, controlling the weapon is of the utmost importance. Soft blocks, and circular movements should be used when subduing him. Re-direct or evade the weapon while following up with a technique of neutralization.

There are many systems that teach weapons defenses that are just not practical. Many tapes and seminars teach knife defenses that are, for the most part, useless. If attacked by a knife-wielding opponent, the situation is completely different than when being attacked by a blunt weapon. When an attacker is using a blunt weapon, one is able to grab the weapon. Needless to say, grabbing an edged weapon is not a feasible or intelligent maneuver. The attacker will be gripping the only area of the edged weapon which is safe to handle.

The Battle Is Joined

Defend, Stick, Attack. Once you touch him do not let go. Don't allow the fight to restart after you begin the attack.

In nature, things move violently to their place and calmly in their place. Sir Francis Bacon

The law of entropy affects all things in the universe, living and not. It also affects all situations and states of existence. Stated briefly, the law of entropy is:

In a closed thermodynamic system, a quantitative measure of the amount of thermal energy not available to do work...a measure of the disorder or randomness in a closed system...a measure of the loss of information in a transmitted message...the tendency for all matter and energy in the universe to evolve toward a state of inert uniformity. The inevitable and steady deterioration of a system or society.

What is entropy? It is the idea that all things evolve to a state of random uniformity. All things deteriorate. Hot becomes cool, cold heats up, mountains flatten, valleys fill in, and whatever is standing will eventually fall. In other words, all fights, if they last long enough, will end up on the ground. If for no other reason than this, one must be prepared.

On Evolution, Entropy, and Organizations

In high school and college, through courses of physics and earth science, we are taught entropy is the law of the universe. Entropy is the law of "disorder". It says all things will wind down. The sun will burn out – systems break down – mountains wear down – valleys fill up – batteries discharge – things decay – energy is used up and is dissipated as heat, which spreads out into the universe until its effects are randomized into nothingness. All systems move toward disorder or decay. There is only one arm of science that dares to run counter to this argument. It is the theory of evolution. Evolution implies in the very premise of its theory that systems will become more complex and ordered as time goes on.

It is an *a priori* that is contrary to science itself, except for one possibility. If there is an intelligent creation, there must be an intelligent creator, far superior to the creation itself. The only way a system can increase in complexity and defy the laws of nature is for it to be guided by force that is itself rational, ordered, and more complex. Scientific theory and law disproves the theory of evolution, and proves the existence of God. To quote the first axiom of biology, only life can produce life. One might add, only intelligence can produce order.

From one-celled creatures to the human mind, we see a mixture of order and entropy. We see a cycle of order in newness and birth and the entropy that brings aging, death, and decay. We see all stages of grow, maturity, and chaos between birth and death. As the complexity of the organism increases so does the necessity of order and the order of stages. As a one celled organism reproduces it divided in a violent splitting of itself. It tears itself apart and becomes two independent creatures. Both are small, limited, and simple life forms.

As we move up the scale of complexity, more and more order is established until we reach a mammalian stage. At this stage there is nurturing and caring. There is devotion and loyalty and without these things life could not continue. The mother gives birth, nurtures, feeds, and, if all is as it should be, the child and parent are devoted and loyal to one another. If either were not devoted to the others, life would cease. Assuming there was no other person to take over the chore, without the devotion of the parent the child would die. If the child were to leave the parent or in some way destroy the parent, the child would die. Both must be attached and in some way devoted to the other. As the child grows it begins to claim its independence. It walks, talks, and views the world as an individual. The parent continues to teach the child because of the child's limited experiences and viewpoint. This is how we transmit wisdom and knowledge. Through the stages of feeding, growth, experience, and independence there should remain devotion, loyalty, and love. And, so it is until the next generation. When any of these steps are violated there is disaster.

This simple story is mimicked in every aspect of human life. In the family, in business, in organizations, and among countries, familial piety is the key to healthy growth. However, it seems the same rebellious and arrogant growth cycle of adolescence occurs in these systems as well. Just as the angry pubescent flings himself in the face of the parent and rushes out on his own into failure of conflict, so do those in business, churches, and martial arts systems. The parent organization could have nurtured and helped, but the rebellious branch breaks away from the tree before roots are even considered. They find themselves on their own without nourishment or the wisdom to gain it. Barring being held captive in the parent body, there should be a mutual respect and loyalty that allows a transmission to needs from parent to child. It is an entropy of integrity that brings on failure or suffering.

The growth cycle of all systems inhabited by man follows the same laws of maturity as man. Take religion for example. In the beginning of most religious movements there is discovery and joy, just as a child. Then there is growth and independence. But, when the time of puberty comes there is at times a violent rush toward acquisition of territory, souls, and self-righteousness. It happened in Christianity with the crusades, and in their current stage of religious maturity, the Muslims are killing now. It is a stage of growth gauged by a simple rule: the larger the movement the slower the growth pattern. Thus, in business, local churches, and martial arts systems, there will be splits that should not occur, led by rebellious people who should never have been given power leading away the immature ones into suffering; a condition the parent system could have lessened.

To diverge brings multiplied opportunities to survive and discover. To split, as a one-celled organism would, is to destroy or damage a higher form, as we and our society functions. What keeps this from happening is mutual respect, devotion, and loyalty. What we have in the world today is entropy of integrity. Pieces of society split and splinter into smaller and smaller groups and disperse into the world with less and less influence, much like energy dissipates into the cosmos as randomized molecular movement. Those who show lack of loyalty or devotion to the parent group should not be expected to have these traits in their lives or organizations. They should be avoided at all cost. In families children must be educated in these areas. In the adult world, souls must be converted or removed. One cannot wait too long to remove them since they will split the group if allowed to take root. To prevent splintering from occurring in our organizations, integrity, loyalty, and devotion must be encouraged. The traits we want to produce can only be reaped if they are first sown, shown, and taught. We need to act in our various environments of business, church, groups, clubs, and countries in the same way God would have us act toward the family. The parent should nurture and mature each member. The higher wisdom, intelligence, and integrity from God, which keeps entropy at bay, must be learned by the parent, passed down to the child, and carried forth to the next generation. Those who remain loyal will stay longer, learn more, and be rewarded by roles of leadership. They will begin to mentor the younger members. When it is time for the leaders to assume the parent roll for themselves, their parent group will help them begin an independent unit, separate but united with the parent group. In the same way birth and nurturing occurred in life, it should also occur in society. In this way integrity, love, and devotion will overcome the entropy of integrity.

There is a huge problem in the martial art's world today. There is an explosion of new systems and false masters. It is the slitting and splintering which begins the entropy and downfall of any organization. The young practitioners are untested and unwise. Their art is ill defined. They start systems and splinter away from the parent group too early, too fast, and in the wrong way. They are rebellious children who, as adolescents are wont to do, have decided they have all of the information, wisdom, and knowledge needed to strike out on their own. They claim ranks unearned and exhibit techniques without insight. Insight in the martial arts takes a lifetime to gain and only after you have acquired it do you know it was missing. The earned wisdom of the parent group is not in them. The transmission was aborted. Integrity, devotion, and loyalty are missing from their understanding. They are not nurtured by the parent group and have no resources on which to draw. Entropy is overtaking the martial arts world today. It is an epidemic of arrogance caused by an infection of pride. The symptom is the entropy of integrity.

How do we avoid these groups? Ask their parent groups about them. Find out who taught them. Ask about their attitude and their ability. Find out if they are young adults or angry adolescents. Find out if you are stepping into a system of chaos or one that has been transmitted through the care and wisdom of elders. In martial arts we seek the knowledge of those who have come before. This knowledge, when added to your own, will raise you up to be a better person. Seek those who have themselves learned the full measure of knowledge and wisdom from their teachers. Make sure your teacher was not in his spiritual diapers and crawling on the path. His folly will begin your entropy if you are not careful.

A Theory of Asymmetry

The body functions well as long as there is symmetry of movement. Without symmetry one part of the body will overextend or retard movement of the entire body. Balance is essential for living a happy life, and for fighting a good fight. Asymmetry destroys balance.

There are two conditions in which balance may be violated and a man may be brought down. One can be extended beyond his capacity to keep balance, or one can be forced into a direction where movement is not possible. Forced asymmetry is putting a person in a position where he cannot function.

When over-extended, the person will be led to take a step larger than is possible to maintain balance. Either through greed, avarice, or being pulled, pushed, or twisted, the opponent will fall if part of his body goes too far away from his base of balance.

By limiting his movement in one direction and forcing his step in that direction he will trip and fall. In both of these situations, the symmetry of the opponent is violated. His body and mind are brought out of balance. He is stretched or stopped short. He is distorted and asymmetrical. This should be a lesson to us. We should never go further than our balance will allow. We should never take on more than we can handle. We should never be so prideful as to go unprepared into our areas of weakness.

If, by locking the spine, the enemy is prevented from moving, then any direction he is taken which does not unlock the spine will result in him falling.

Locking the Spine

The body is made to function, move, and generate power only in a symmetrical fashion from a certain range of distance. To defeat movement, thwart balance, or diminish strength one has only to make the opponent's body asymmetrical.

Generally speaking, this can be done by dropping a shoulder and twisting the backbone. By pushing a shoulder down and back, the spine locks and motion stalls. By bringing a shoulder out and down the balance is stolen. By turning the neck so the shoulders turn and drop, all perception of space is distorted.

There are three basic methods of stealing balance: lifting, leading, and tripping. The opponent can be lifted off of his base. The opponent can be lead or extended beyond his base. The opponent's base can be restricted as his momentum continues.

When leading him into weakness – keep your hands close to your center and his hands as far as possible away from his center.

Our ability to function in life is based on where we see ourselves in relationship to our immediate surroundings. Even if we see our relationship in the macrocosm, we judge our relationship and path based on our posture and position in the present. When posture or position change enough, our balance, strength, and perspective is lost temporarily. This is why in life and business it is best to make dramatic changes at once and then wait for adjustment. To do it in slow steps leads to continual annoyance and resistance by the masses.

Taking Him Off Line

If the body is not kept in symmetry, it does not have balance or strength. Strength comes from an alignment from ground to target. If this alignment is thwarted, the opponent loses the ability to strike with any power. To perturb balance and position, "take him off line". One of the most effective ways of doing this is to cause a change in direction that is impossible for him to make without an intermediate step. If he cannot change direction he will fall. Another way to "take him off line" is to change the perspective between him as attacker and you as target, so that to re-acquire his target he must stop, redirect his balance, and start again.

As an example, from a kneeling position when one is sitting so the feet are at the tailbone, push the opponent back and to the side. He will fall. From a standing attack, as he steps in toward you, step toward him and to his side. Pull his shoulder back, to the side, and down. The direction should be about 30 degrees off. Stepping into his Dead Zone is not enough. You must attack from that point or move him when entering. To move him in a direction away from his intended direction of attack is to take him off line. Taking him off line steals his balance, his direction, and his power. It stops the attack.

When the Battle Goes to the Ground

Over ninety percent of all fights end up on the ground, so one must be well prepared to do battle there. While the Shinsei Hapkido system is well rounded, to include and extensive array of grappling techniques, we do not advise diving headlong into a grappling match. Ground grappling is simply not effective against multiple attackers. You may be able to lie on your back and apply a triangle choke, but what do you do about Bubba and his pals?

Joseph Lumpkin

The Bubba Factor

Idiots tend to flock together, like vultures or buzzards. When
something is putrid, the whole flock comes. The Bubba factor is a
simple concept...if you get yourself out of position while engaging
one thug, another one may be lurking in the shadows, waiting to
pounce. Be the best ground fighter you can be, but never depend on
grappling alone.

A system of martial arts that does not teach ground fighting is weak,
but the opposite is also true. Many Jujitsu styles, to their own
detriment, discredit kicking. What is the best way to fight a grappler?
Stay out of grappling range. However, ground fighting is still a
necessity. Most martial artists are taught to throw, breakfall, and roll,
but never learn defense and offense from the ground. Shinsei brings
this to light.

Breaking the fall is the single most important aspect of ground
fighting self-defense. If you are out cold, or stunned from being taken
down, you will not be able to launch a defense and counter attack
from the ground. Breakfall training should be stressed in all martial
arts. Sparring in a Shinsei Hapkido dojang includes strikes, locks,
throws, and grapping, just as attacks occur in the dark alleys of the
real world.

To teach ground fighting one must always consider position.
Training to fight from one's back, side, and even face down is
essential. The position of the attacker also plays into the reaction of
the defender in ground fighting. One should practice defending with
the aggressor kneeling, mounting, standing, and applying locks. Do
not assume you will always end up in the same configuration when
being taken to the ground. It could be a fatal assumption. Prepare for
the unexpected.

The environment also plays a key role in self-defense and especially
ground fighting. Why use an arm bar when you can pick up a
wrench and "clock" the attacker. Is there room to move? What if
there is not room for Judo throw number seven? Contrary to the
some martial schools of thought, most attacks don't happen on a

football field. If you need one hundred yards to accomplish the technique, toss the technique, it is useless.

The Weapons of War

Within the Shinsei Hapkido Kwan exists an extensive array of weaponry. The Shinsei weapons system acts as an extension of the techniques within the system, including fan, rope, belt, stick (Than Bong, Jung Bong, Jang Bong), a ballpoint pen, and an everyday walking cane.

The Cane

What good does a weapon do one if one is unable to use it? In the heightened security of this post 9-11 society, the law does not look well upon one carrying a weapon. Mark Shuey says the cane is "most practical weapon a martial artist can learn. What other weapon can you carry on a plane or in a casino? Try getting nunchaku past airport security. And if you think about it, what good is learning a weapon if you can't bring it anywhere?"

The cane may very well be one of the last "permissible" weapons available. The cane functions as the base weapon of the Shinsei Hapkido system. It is practical, can be used as an extension of the existing techniques of Hapkido, is easy to learn, and is a highly versatile and an extremely effective weapon for self-defense.

Hapkidoists, as well as the Hwa Rang of old, have used the ji pang e (Cane) for self-defense, and in war for hundreds of years. Canes were used for hiking, and for defense against invaders, rebels, and fending off wild animals. The cane techniques of Shinsei Hapkido are designed with self-defense in mind. Even the elderly can use the cane for self-defense with a small amount of training. Using the cane as an extension for the locking techniques of Shinsei creates one of the most practical tools of self-defense today.

In using the cane against an armed attacker, it may be used to re-direct or block an attacker's weapon. Strike the arm holding the weapon, re-direct the arm holding the weapon, strike other targets, disarm the attacker, attack pressure points, apply locks, disarm, and control the attacker.

The Sticks

Shinsei also incorporates the use of the short stick, middle staff, and long staff (Than Bong, Jung Bong, Jang Bong). One may not carry along a six-foot staff, but a pool cue could be available. One may not have a short club, but perhaps a flashlight, or a bat is nearby.

Tahn bong techniques can be found in martial arts styles in almost every country in the world, as the martial artist requires a variety of "weaponry". A stick, a bat, pencil, ruler, or a piece of metal may be adapted to practically any self-defense situation. Usually the "short stick" is anything under 3 feet in length and is used as an extension of the Shinsei Hapkido open hand techniques, including the blocks, strikes, traps, joint locks, and grappling locks. By holding the weapon loosely and striking in a slapping motion, tightening the grip at impact, the extended technique becomes devastating. Blocking is done in the same manner.

Weapons of preference are those, which can be found anywhere. The Sai is a beautiful weapon, but where is one to find a small pitchfork of steel when one needs it? Instead, we focus on those weapons that can easily accessible. It is not difficult to find a three-foot long piece of wood. Any mop handle of tree limb will do. We must continue the process of adapting to our surrounding conditions.

New May Not Be Better

If a man knows not what harbor he seeks, any wind is the right wind. Seneca

That we henceforth be no more children, tossed to and fro, and carried about with every wind of doctrine, by the sleight of men, and cunning craftiness, whereby they lie in wait to deceive. Ephesians 4:14

Some philosophies are simply not compatible. If students are learning the philosophies and linear techniques of traditional Shotokan Karate, then imposing a regime of Judo with its circular approach, creates havoc in the students' learning process.

Shinsei Hapkido, is a blended art. It borrows from differing styles. Every single technique, philosophy, and concept fits within the philosophy of Shinsei Hapkido itself: Evade, Invade, Control. If the movement does not balance in the philosophy of Shinsei, it is discarded.

Teachers must be extremely careful about teaching mixed arts. There is nothing wrong with teaching a soft style but do not confuse the students by arbitrarily throwing in hard techniques that are in opposition to the philosophy of the basic art being taught. Likewise, if teaching a hard style, do not confuse the novice with circular techniques. Remember, the philosophy of the style will seep into their souls. If only for this reason, we should teach a well-defined philosophy within the art. If you were coaching basketball, you wouldn't have your students practicing slap shots.

Respect the art and accept the traditions and wisdom of teaching with a grateful and humble heart. The teachings are an ancient legacy transmitted by many teachers who worked trained hard to learn and pass on an important body of knowledge. You, as a teacher, will have others follow you. The responsibility to teach something of value is yours. The students will follow.

While many self-proclaimed "experts" of the art will probably disagree and continue to teach contradictory styles together, it is to the detriment of their students. Most of the modern "eclectic" styles today come from the instructors' unwillingness to explore the depths of their own system. Your students will reward you in full one day by teaching what they are taught. If it is useless, you will have an empty body of work. Individual teachers will, understandably, modify the original material, adding their own personal preferences. However, we should not change things simply for the sake of novelty. There is nothing new under the sun. Do not try to make the art more precious or secretive under a "traditional" blanket. Rather we must study all aspects of the art, its history and its philosophy before passing it on.

Examine what is being taught. Observe the reactions and responses of the students to what and how you teach. Never forget your students are your only reason for teaching. Be willing to admit when you don't know the answer. "I do not know" is a legitimate answer to your student, but is should be followed by, "But I will make it a point to find out".

In some martial arts circles the old methods and traditions are ignored and teaching is done in a "modern" way, supposedly far superior to the old ways. Some even claim to have moved beyond Kata. Kata is merely a way to practice by repeating movements. Are these practitioners of the arts who claim they do not need Kata stating they do not need practice? You certainly do not need to practice, if you want to end up dead.

The Mind Factor

You can prevent your opponent from defeating you through defense, but you cannot defeat him without taking the offensive. Sun Tzu

There is much talk these days about "leading the mind". This seemingly mystical practice of Aikido is based solidly on human greed and the propensity of men to over-extend the reach. By keeping just within reach of the opponent and just far enough away that the opponent cannot grasp us, he will continue to extend his reach. Knowing he is touching us and thinking he will be able to grasp or strike us any moment, he will pursue until he is over extended, off balance, and ready to be thrown. It is not a mystical force. It is an understanding of human nature.

Yet, there is an easier way to lead the mind. The mind will always go where pain resides. By inflicting a sharp pain, even a pinch, the mind leaves its first concern and flies to the area of pain, leaving the body unprotected and unaware. Now your assault can be carried out in the midst of his distraction. Due to the demands of exact timing, leading the mind is difficult, but the mind will go to where pain drags it.

Confusion Stops the Mind

We have already discussed the function of breath and its relationship to strength and attack. An attack usually occurs on the exhalation. Strength is manifested when exhaling. An unexpected event causing a startle reflex is needed to stop the breath or cause an involuntary inhalation. This will temporarily freeze an opponent for a fraction of a second. It is all the time needed for an attack.

For this a KiHap or Kiai is used. KiHap or Kiai means, "shout of spirit". It is an explosive outburst forced up from the lowest parts of the abdomen, pushed onward by the stomach muscles, diaphragm, chest, and finally escaping through the throat, which emits a loud and explosive yell. The strength of a properly executed KiHap is awful. The enemy will flinch and halt, taking a sharp breath in or forgetting to breathe all together. It is in this window you must attack. Soon after his body relaxes there will be a momentary rush of adrenaline, during which he may resume his attack.

There will be many times when a simple distraction will turn the conflict to victory. In those times when we must buy a split second, it is best to use a great commotion requiring little energy. Do not let the distraction of the enemy pull you out of your place or timing in the battle.

If your opponent is as clever as you are, you must steel yourself to ignore his distractions.

The Moving Mind

One of the greatest mistakes many instructors make in teaching is in failing to teach the importance and value of the mind in the arts. Mental training, mental preparation, and dealing with proper thought processes can make up for great deficit in the students' physical abilities. On the flip side of that coin, the reverse is not true. Great physical ability can never take the place of proper mental training in the arts.

While training outdoors, two of Master Vu's students were arguing about the school flag flapping in the wind. One, wanting to sound enlightened, said, "You know, it's the wind that is really moving, not the flag." The other student responded, "No, the flag is moving, but it is the wind that moves it." Master Vu turned to look at them and said, "You are both wrong, it is neither the flag nor the wind that is moving, it is the mind that moves". We must never forget the basic philosophical principal of Shinsei, "The Moving Mind".

Observe the Enemy

To be near the goal while the enemy is still far from it, to wait at ease while the enemy is toiling and struggling, to be well fed while the enemy is famished - this is the art of husbanding one's strength. To refrain from intercepting an enemy whose banners are in perfect order, to refrain from attacking an army drawn up in calm and confident array - this is the art of studying circumstances. It is a military axiom not to advance uphill against the enemy, nor to oppose him when he comes downhill. Do not pursue an enemy who simulates flight; do not attack soldiers whose temper is keen. When the soldiers stand leaning on their spears, they are weak from want of food. Do not swallow bait offered by the enemy. Do not interfere with an army that is returning home. Sun Tzu

In a basic application, one should allow the enemy to move, test, and try while we remain calm and watchful. One should never move needlessly. Let the enemy exhaust himself, while we wait for weakness to appear. This tactic is especially evident in Judo matches. One opponent will begin to move and manipulate the other into a place where balance is lost. If the man being moved remains calm and centered, the one attempting to manhandle the opponent will become tired and weak. His aggression will cost him the fight.

Be in "your place". Remain calm and watchful. Don't fight against balance or gravity. Do not be drawn in if he offers you an opening. When he becomes weak, slow, or unbalanced, attack quickly and decisively.

Hold out baits to entice the enemy. Feign disorder, and crush him.
Sun Tzu

You intentionally leave an opening in your defenses or leave some
limb vulnerable to attack in order to draw your opponent into
making a predictable attack which you can in turn counter. Like
baiting a fish hook. Bruce Lee

Pretend to be weak, that he may grow arrogant. Sun Tzu

Simulated disorder postulates perfect discipline, simulated fear
postulates courage; simulated weakness postulates strength. Hiding
order beneath the cloak of disorder is simply a question of
subdivision; concealing courage under a show of timidity
presupposes a fund of latent energy; masking strength with weakness
is to be effected by tactical dispositions. Sun Tzu

If we wish to fight, the enemy can be forced to an engagement even
though he be sheltered behind a high rampart and a deep ditch. All
we need do is attack some other place that he will be obliged to
relieve. Sun Tzu

Draw your enemy forward, on and on in his attack, by keeping just
out of range. In this way you can lead him into exhaustion, then
overwhelm him. Draw the enemy forward and you control his
timing. Draw the enemy in to meet your attack.

Never Prolong the Battle

Somewhere in life, we will get old. Our strength will fade quickly.
Our breath will diminish, but our spirit to fight will remain strong.
We will demand a quick end to the battle. Victory resides in the
demand of engagement without hesitation.

When you engage in actual fighting, if victory is long in coming, then
men's weapons will grow dull and their ardor will be damped. If you
lay siege to a town, you will exhaust your strength. Again, if the
campaign is protracted, the resources of the State will not be equal
to the strain. Thus, though we have heard of stupid haste in war,

Joseph Lumpkin

cleverness has never been seen associated with long delays. There is no instance of a country having benefited from prolonged warfare. In war, then, let your great object be victory, not lengthy campaigns. Thus it may be known that the leader of armies is the arbiter of the people's fate, the man on whom it depends whether the nation shall be in peace or in peril. Sun Tzu

Hence, when able to attack, we must seem unable; when using our forces, we must seem inactive; when we are near, we must make the enemy believe we are far away; when far away, we must make him believe we are near... If he is taking his ease, give him no rest. If his forces are united, separate them. Sun Tzu

The Eyes Lie

Watch the triangle of Dan Jun (belt buckle area) and chest. See all. Watch balance. Do not move in anticipation of his eyes. Move as a mirror. Respond to him as if you were one with him. You cannot lie to a mirror. It is a dance of life and death.

Draw by Angering

The general, unable to control his irritation, will launch his men to the assault like swarming ants, with the result that one-third of his men are slain, while the town still remains untaken. Such are the disastrous effects of a siege. Sun Tzu

To lose one's temper is to lose the stillness needed to be at peace. To lose peace is to lose control. Be angry and move too soon, be too rash, become thoughtless. Die.

PRO 16:32 He that is slow to anger is better than the mighty; and he that ruleth his spirit than he that taketh a city.

Ki and Emotional Content

The power of commitment is wondrous and can transcend all other forces. Tak Kubota, The Art of Karate

Emotional content is not anger or fear. It is not colored by any particular feeling. It is emotion itself, raw and clear as water. Emotional content is determination on an emotional level. It is the glue and fuel that holds and controls Ki. Ki is not some metaphysical force, as some would suppose. It is the synergy of strength, speed, technique, timing, and willpower coming together in harmony of purpose. Ki appears to be a force greater than the sum of all its parts. Humans often perform acts that seem far beyond normal capability. These acts always come in times of complete focus of mind and determination. This is Ki.

Ki is not a mystical or magical force. It is the normal result of an ongoing and intense practice. Ki is a power that occurs when body, concentration and timing come together. It is the potential of the human body realized. As we practice the movements of our martial art, we hone the timing and coordination those movements require. We practice the rhythm, timing, concentration and muscle toning needed for the movements until we have smooth, powerful, focused techniques.

At this point the technique becomes more than the sum of its parts. This is Ki. When a mother finds her son trapped under a car and lifts it to get him out, this is Ki. When a man runs into a burning building and carries out those larger than himself, this is Ki. When a martial artist puts a hand of flesh and blood through layers of brick, concrete, or wood, this is Ki.

Christianity is not made up of just prayer, or just giving, or just service, or just faith. It is centered on Jesus, and he demands our faith, our prayer, our service, and our actions. It is in the combination that we come fully into a Christian life. Practice! Practice to build strength. Practice to form your body. Practice to improve timing, concentration, and coordination. Practice enough and you will transcend these parts of the technique and experience the power of the whole. Ki is nothing more than the power of the whole.

A Problem of Good and Evil

All warfare is based on deception. Sun Tzu

Deception only works once. The same deception used twice becomes a pattern. Once detected, the opponent will catch you in your own trap.

Humble words and increased preparations are signs that the enemy is about to advance. Violent language and driving forward as if to the attack are signs that he will retreat. When the light chariots come out first and take up a position on the wings, it is a sign that the enemy is forming for battle. Peace proposals unaccompanied by a sworn covenant indicate a plot. When there is much running about and the soldiers fall into rank, it means that the critical moment has come. When some are seen advancing and some retreating, it is a lure. Sun Tzu

The Western concept of good and evil tends to view evil as a weaker force, which will always lose in the end. Although it is true in the end times when Christ comes again, good will be victorious, however, it is not necessarily true in day-to-day life. Too often we see the rats keep winning the rat race.

For the purpose of simplicity we will define evil in two ways: complete selfishness, or a lack of conscience. In this concept of evil, good must be very careful. The code on which Shinsei is built is one of honor and respect. It is also one of mercy and grace. With this in mind one can easily see how, in a matter of life and death, a good person would hesitate to harm or kill. Our love of God and our fellow man will stay our hand and force us to think twice about our actions. This may be twice too long. Evil does not hesitate. It feels no remorse. It wants only its own gain. It cares for no one else. Good must be very careful not to turn its back nor to put itself in the hands of an evil person. Good may lose the battle by its very nature. Know yourself and know your opponent.

Moral Force

Conflict may be avoided, and men may be led by moral force. This does not mean the leader is more righteous or better than the followers. Moral force is a term given to the "strength of presence" that affects and sways others. It is present naturally with some but can be developed by practice of martial training; demanding concentration, strength of will, and physical prowess. Moral force develops less because of physical strength and more because of the force of will produced as a person pushes himself beyond his limits time and time again. It is not charisma. It is an inner strength sensed by others. Moral force comes from having known and overcome fears and limitations through determination and patience of prolonged practice. It is a steeling and strengthening of character. Moral force can be used to lead others, but the warrior will do what is right and allow others to follow.

It takes courage and vision to lead, but determination and courage are demanded from those who follow. It is very important never to ask anyone to do that which you would find difficult to do yourself. Moreover, age notwithstanding, a leader should perform all he asked of his men before he asked it. Do it. Demonstrate it is possible, then, ask others to follow you.

Moral force is a self-confidence and self-assurance others can sense. People are drawn to it and respect it. Enemies respect and fear it. Moral force truncates attacks before they happen. It is one of the primary qualities of leadership. People feel secure under a wise leader who has moral force. In all areas of life, it is better to lead by example and moral force than by any other means. People tend to follow the person and not the goal or vision. If they trust the leader they will trust his vision. Fear and tyranny will fail. These only compel people to do what is necessary to save their lives or positions, but respect and loyalty raises followers and students who will follow you to the gates of hell and give their all when they arrive.

Order and direction result from good leadership. There is a mutual vision and singleness of mind within the community, resulting in the abolition of conflict. Once authority is established, order is kept with a single word.

The consummate leader cultivates the moral law, and strictly adheres to method and discipline; thus it is in his power to control success. Sun Tzu

All men come to he who keeps unity. For there lie rest, happiness, and peace. Lao Tzu

The humble is the root of nobility. Low is the foundation of high. Princes and lords consider themselves orphaned, widowed, and worthless. Do they not depend on being humble? Too much success is not an advantage. Do not tinkle like jade or clatter like a stone chime. Lao Tzu

God's presence is everywhere. This is the concept of the omnipresence of God. But, God only manifests himself in two places. One is in heaven; the other is with the humble man. Isaiah 57:15 says, "For thus saith the high and lofty One that inhabiteth eternity, whose name is Holy; I dwell in the high and holy place, with him also that is of a contrite and humble spirit, to revive the spirit of the humble, and to revive the heart of the contrite ones". There are only two places to experience revival, the full manifest presence of God, one is in heaven, the other is here on earth, but only if you are humble and contrite.

Gratitude and Respect

Gratitude is the key to humility. We owe all things to God and to Him goes all gratitude. We owe our knowledge and training to our teachers and fellow students. We train, holding the lives of those with whom we train in our hands. We trust them because we know in their hands are our lives. Martial arts training brings a bond of trust that goes far beyond friendship. It is a matter of life and death.

The same can be said of our mentors in business and in life. For a man to share his wisdom and transmit his knowledge is a matter not to be taken lightly. Should disagreements occur, respect must be maintained.

A story was once told about Cory Teinboon, a famous heroine of the Christian faith. She was asked how she made survived the torturous events of the holocaust. Her reply was, "When you achieve something great, turn it into a rose and lay it at the feet of Jesus Christ". When asked what that had to do with suffering she replied, "Then, when you experience the trials, you can turn them into a rose and lay them at the feet of Jesus. If you lay down the glory at his feet, you can lay the trials there as well, but if you want to keep the glory, you'll have to deal with your own trials".

Teacher-Student Relationships

You, who are on the road, must have a Code that you can live by,
And so, become yourself, because the past is just a good bye.
Teach your children well, their father's hell did slowly go by,
And feed them on your dreams, the one they pick, the one you'll know by.
Don't you ever ask them why, if they told you, you would cry,
So just look at them and sigh, and know they love you.
Graham Nash

1CO 12:28 And God hath set some in the church, first apostles, secondarily prophets, thirdly teachers, after that miracles, then gifts of healings, helps, governments, diversities of tongues.

ISA 30:20 - 21 And though the Lord gave you the bread of adversity, and the water of affliction, yet shall not thy teachers be removed into a corner any more, but thine eyes shall see thy teachers: And thine ears shall hear a word behind thee, saying, This is the way, walk ye in it, when ye turn to the right hand, and when ye turn to the left.

The relationship of teacher and student should be one of trust and respect. Devotion and deference should be given to the teacher. Students are expected to their complete and undivided to the instructor. The student's life may depend upon getting the next techniques exactly right. That is not to say there should be blind obedience, but in class all things must be done in order. Only in this environment can the Way be passed on, straight and true.

The purpose of a teacher is to produce students who surpass him at all levels. The secret is to protect your students from yourself. You must shelter them from your weaknesses and shortcomings. Give them only your strengths while nurturing theirs. Limit the effects of their weaknesses by teaching them balance, fan the fires of their strengths, enable them to accomplish their goals, demand from them success, and above all, encourage them to exceed their limits, as well as yours.

Throughout all of this, the teacher must also learn from the student and correct mistakes he, himself, has made; therefore, each learns from and strengthens the other.

If the teacher is not respected and the students not cared for, confusion will arise, however clever one is. This is the crux of mystery. Lao Tzu

Crestfallen

There is nothing more heart wrenching than a child with a broken spirit. These would-be warriors come to us without hope or self-confidence, and they don't even know it. As instructors, it falls to us to rebuild these young ones into confident and assured individuals. But, how do we proceed?

Spiritual growth is akin to learning to walk. As we begin, our steps are small and unsure. Only by trial and success can we build the confidence to take larger steps and venture further afield. Teaching the dejected and crestfallen takes special care. For every small accomplishment there should be celebration. The dejected soul has not been properly celebrated as a person, much less for its accomplishments. It is up to us to correct this. By pushing the student to perform and then rewarding each accomplishment, we establish a bond. When love and trust is established, the teacher's ability to lead is increased. It is through their confidence in us that we motivate the student. This circular yet vital detail is important to understand and crucial for a successful relationship. The student will develop a strong confidence in the teacher, and the confidence the teacher has in the student gives them confidence in themselves.

The key in maintaining this circular growth is in never assigning the student a task he cannot accomplish. It falls to the teacher to know the limitations of each student and to assign tasks that test the extremes of the student's ability without inviting failure. Success must become a habit. With each technique the student performs, his confidence increases. This is especially true when the student himself did not think he could accomplish the goal, yet still succeeds.

Self-confidence takes time. We do not know the history or duration of any emotional abuse the student may have experienced. We may not know if it is continuing, even now. But we do know we fight against its insidious ravages. The teacher becomes the good parent image the child will carry with him for the rest of his life. If the roots of our confidence and caring can be sunk deeply enough into the heart of the child it can nourish him for the rest of his life. This is success in teaching. Crestfallen and dejected they will come. Warriors they will leave. Just give us time.

The Space Between Brown and Black

...with no other light or guide than the one that burned in my heart. The Dark Night by John of the Cross

For years the student has struggled to sharpen his skills. For years he has pushed himself to all limits of strength, endurance, and precision. Now, he has reached his limits and can go no further; or so it seems. This is the time of the brown belt, a time of waiting and of maturing. It is a time of frustration and abandonment.

For years the teacher poured himself into the student. He has invested sweat and blood over time. For the past four or five years, the instructor has communicated the system and philosophy in its entirety to the student. Now both must wait.

I will never forget the day my instructor bowed to me and said, "I have nothing more to teach you." Looking down at the pitiful unfulfilled color of my belt, I pondered. If this were true why was I not a black belt? My teacher seemed to pull away as I struggled to perfect my knowledge. My limits had been reached. My strength was tapped. There was nothing left to do but practice what I was becoming.

Over the next year a slow transformation took place. It was imperceptible to most. Certainly it was to me. But he saw it. He was watching for it. He was waiting for that time when knowledge became truth and truth became reality. This is the place between brown and black.

We, as Christians and as warriors, must be aware of the tourniquet around our necks. It is not easy for knowledge to become truth. It must drip slowly from head to heart. It can be a painful and laborious process. It is a journey most do not complete. There in the dark, without a teacher, and without a guide we become what we have learned, if we continue to practice with patience. "…with no other light or guide than the one that burned in my heart."

Grafting in Leaders

As a dear friend said to me this morning, "things that happen in the physical world have their counterparts in the spiritual world." This statement seemed very pregnant with wisdom to me since I had seen things happen around me in church and with friends in the ministry of late that had me questioning man's ability to rise above his natural sinful nature. Certainly, I question my own ability constantly and see I fall sort of "doing the greater good". It is difficult to see beyond what one wants to do into what is best for the whole. If we are Christians then why is this such a struggle? I believe the following verse gives us insight:

ROM 11:16 For if the first fruit be holy, the lump is also holy: and if the root be holy, so are the branches. 17 And if some of the branches be broken off, and thou, being a wild olive tree, wert graffed in among them, and with them partakest of the root and fatness of the olive tree; 18 Boast not against the branches. But if thou boast, thou bearest not the root, but the root thee. 19 Thou wilt say then, the

branches were broken off, that I might be graffed in. 20 Well; because
of unbelief they were broken off, and thou standest by faith. Be not
high minded, but fear: 21 For if God spared not the natural branches,
take heed lest he also spare not thee.

You see, we are grafted (graffed) in. Today, this is seen as a good
thing because we take a hardy root from a wild shrub and graft in a
branch from a beautiful but less robust rose. What we get is a good
root system supporting a magnificent flower, but that is not what this
verse says. It is the other way around for Christians. Our root is
Christ Jesus. He is the supply of our spiritual nourishment, and we
are grafted in to him. We are grafted in to the root of Jessie and the
linage of Abraham. Our roots are those of the rose of Sharon and the
lily of the valley, but we are still a thorn bush.

So, now, having a deep and hardy roots and being planted in high
places, we draw from our spiritual roots, hoping to be changed by
the substance from which we feed, but still producing thorns until
that day our nature of shortsightedness, pride, and selfishness is
changed slowly by the sweetness of the root. We will never be a rose,
but in time, we may produce fewer barbs.

We cannot change our own nature. If we could, Jesus would not have
had reason to die to redeem us. It is because we cannot change our
own nature that so much confusion arises. Whether we are invited
into a club, group, style, or church, we are expected to follow the
rules and bylaws of that organization. If that type of conduct is not
already part of your nature you will fail to adhere to their standards
because you will always revert to your true nature. So, groups seek
like-minded members. This is because they have no real power to
change themselves or others.

This knowledge should serve as a warning for those in leadership
positions. As we begin a ministry we will seek others to help. We will
seek capable, driven souls who can assist us in the creation or
advancement of the ministry, church, or project God has given us.
Many times those we choose and trust so deeply will disappoint us
in ways we cannot imagine. They will fail by selfishness or by lack of
vision. Most of all, they will fail by simply doing nothing. They fail
because they were grafted in by you to do a job they had no

intentions of doing. They simply wanted the place of power, or honor, or title. In short, they were grafted into the vine and they are expected to produce the fruit that the vine produces. They will not. They will do what any graft will do. They will draw off of the vine or the root and continue to produce their own fruit. We will be surprised and hurt by this, but it will be our own fault for not examining their fruits before we did the graft.

It is just as important to examine why they produce the fruits they do. You cannot take a desert plant and expect it to produce in a jungle. Some people only produce fruit when they are acting alone or for themselves. They are movers and shakers but they have no sense of serving others. This type is like a vine that will try to amass rank and power but will add nothing to the ministry. They are dead weight and will strangle you. There are "suckers". These are small shoots that take energy from the root but stand separate from it as a beginning of a new tree. These seek their own path and will attempt to create a separate ministry covertly off of the hard work and foundation you have planted. There are those that simply are a separate species and cannot adapt to your environment. They are like the Morning Glories. They will look beautiful, full of color and life, but they will not endure the first trial of heat or dryness. They will fall way.

The only hope of selecting good men to assist you is to choose those with a vision like your own, who have produced the fruit you seek in the same conditions you are in.

Never depend on someone to change. You cannot change them, and they can only alter their actions for short periods at a time. But He who made us can change us. If we endure the flames – we can be reformed. In the church I now attend, there are drug addicts and white collar professionals. All sinners, all nourished from the same roots, all being changed into the likeness of Christ – slowly – ever so slowly – becoming more like Him. All with thorns – All with hope of being less like themselves and more like the root.

We are a thorn bush cared for by a lily. – ugly, hurtful, harmful, but with a sweet and pure substance coursing through our veins. It is the Blood of Christ.

A Unique Relationship

Of all the relationships formed within the dojang, or martial arts community, the most unique is that of the Judo concept of Tori and Uke. Tori – the defender or demonstrator of a technique, and Uke – the person receiving a technique, have an amazing and complex relationship of trust and embattlement.

To practice martial arts there must be conflict. As techniques are perfected, they must be practiced in a real-time scenario. This means there must be a person attacking with intent and speed. This person is a trusted friend, a training partner, a fellow seeker. On first blush, it may seem that is it Tori who holds Uke's life in his hands, after all, he is the one performing the technique to be practiced, however, this is not the case. It is true as training progresses the speed, power, accuracy, and lethal potential of the techniques increase. It is possible Uke will be subjected to more and more devastating techniques, but it is not so. Tori is maturing in his control. His precision is acquiring equal heights to his ability.

By allowing Tori to use him for his practice, Uke will save Tori's life if the time ever comes to use his technique. Uke knows Tori holds his life in his hands, but this trust will enable Uke to attack with more and more ferocity even as Tori's defense becomes more and more deadly. Tori knows if Uke is harmed he could die also for lack of training. Uke is Tori's lifeline. It is Uke who allows Tori to learn the techniques he must perfect for use in a crisis situation. They are friends, combatants, and brothers in the art.

Leadership

Leadership is a combination of strategy and character. If you must be without one, be without the strategy. General H. Norman Schwarzkopf

The principle on which to manage an army is to set up one standard of courage which all must reach. Sun Tzu

It is the business of a general to be quiet and thus ensure secrecy; and to be upright
and just, and thus maintain order. Sun Tzu

One matures into leadership. It overtakes him as he learns and expands. Those who seek leadership usually do so prematurely. When people see those things in you they desire in themselves, they will follow. Only then are you a leader.

Those who desire to govern their states should first put their families in order. And those who desire to put their families in order would first discipline themselves. Confucius

A leader will never have understanding and compassion until he fails at least one great time. It is necessary to have failed before one can be considered a mature leader. Both the leader and his followers must know he can and will continue after failure, overcoming defeat, and refusing to surrender when things are at their worst. People do not expect perfection, but they do expect wisdom and strength of will.

It should be stated emphatically, a warrior NEVER stops learning. Even the highest ranked teacher should also be a student. New information and acquisition of other viewpoints should be implemented to adjust and re-define one's philosophy. The warrior must be humble enough to follow before he is fit to lead. Moreover, he must be open to any possibility, unencumbered by the possession and attachment of his own ideas and philosophy.

The philosophy is detached from the person with the exceptions of honor and faith. Only our sense of honor and our faith remain constant in our philosophy. As mortal men, our honor may flag and our faith may fail, but the philosophy remains the same. To make sure our philosophy and actions remain constant, we should come under the protective eye of another. For this purpose we all need teachers, particularly if we are teachers.

After I, a man of little rank, unexpectedly took control of the province, I have put forth great effort both day and night, at one time to gather together famous men of all kinds, listened to what they and to say, and have continued in such a way up to this time. Asakura Toshikage 1428 – 1481 A.D.

Likewise, we seek to uplift ourselves through the company of better men. Although we should never turn away from those in need, we should never seek to befriend those of lower character in any deep way. Treat all people with equal respect but seek to befriend better men than yourself.

It is the nature of man that the good is difficult to learn, while the bad is easily taken to, and thus one naturally becomes gradually like those with whom he is familiar. Shiba Yoshimasa 1350 – 1410 A.D

Just as water conforms to the shape of the vessel that holds it, so will a man follow the good and evil of his companions. Imagawa Sadayo 1325-1420 A.D.

Humility and the learning, searching, and keeping the open mind of a child are traits of a true warrior. Insecurity is its own wisdom. This may seem contradictory to the idea of "moral force", which, upon first glance may look as if it is leadership by unbending will. Yet, even if the will is unbending, the mind and options remain open. We are not swept to and fro by any means. We set our course and command our will to go before us. Yet each time we learn and observe, the path may shift and tactics alter. For this to happen, we must remain teachable. To remain teachable we must be positive of one thing and that is our lack of knowledge. If the leader remains teachable and humane, his followers will continue to learn his ways.

A leader must be most careful. The lives of his students and their welfare depend upon him. If he is arrogant or incorrect, all will perish.

ISA 9:16 For the leaders of this people cause them to err; and they that are led of them are destroyed.

If one family has humanity, the entire state will become humane. If one family has courtesy, the entire state will be courteous. But, if one man (the leader) becomes grasping and perverse, the entire state will be brought into rebellion. Confucius

If one will fix his heart on the way of assisting the world and its people, he will have the devotion of the men who see and hear of him. Gokurakuji

The general who advances without coveting fame and who retreats without fear of disgrace, whose only thought is to do good for his country and to his sovereign is the jewel of the kingdom. Sun Tzu 490 B.C.

It is very regrettable that a person will treat a man who is valuable to him well, and a man who is worthless to him poorly. Gokurakuji

Encourage and listen well to the words of your subordinates. It is well known the gold lies hidden underground. Nabeshima Naoshige 1538-1618 A.D.

In caring for others and serving heaven, there is nothing like using restraint. Restraint begins with giving up one's own ideas. This depends on virtue gathered in the past. Lao Tzu

No ruler should put troops into the field merely to gratify his own spleen; no general should fight a battle simply out of pique. Sun Tzu

It is truly beneficial in the understanding of what makes an excellent and respected leader, to look at the chinks in that leader's armor. The following are eloquent writings of Sun Pin, exposing his concerns and observations of the weaknesses of leaders. Sun Pin was the great-grandson of Sun Tzu and an accomplished military strategist in his own right. His insights are as applicable today as they were in ancient China.

If he is incapable but believes himself to be capable, if he is arrogant, greedy for position, greedy for wealth, light (flippant), obtuse, cowardly, weak, not credible, not decisive, slow, indolent, oppressive, brutal, selfish, confusing, he is defective. When defects are numerous losses will be many.

If he loses mobility, if he lack resources but acts anyway, if he is argumentative, if he wrangles over right and wrong, if he does not see to it his commands are carried out, if his subordinates are not submissive, if his forces have stirred those around them to bitterness, if his troops are not fresh, if the troops are suffering and wanting to go home, if the soldiers are in disorder, if he is eager but unprepared for battle, if the troops are afraid, if the army does not view the leaders as capable, if the army has been victorious and has become lax, if he is brutish and relies on deceit and ambush, or if he is worried too much about strengthening one part of the army and is not watchful over the whole, he can be defeated. Sun Pin

The good company has no place for the officer who would rather be right than be loved, for the time will quickly come when he walks alone, and in battle no man may succeed in solitude. Brig. Gen S. L. A. Marshall 1964

It is important to have those of weaker character closely tied to and positioned between those who are loyal and strong. The weak will be goaded to good deeds by the fellowship of great men. The host thus forming a single united body, makes it impossible either for the brave to advance alone, or for the cowardly to retreat alone. This is the art of handling large masses of men. Sun Tzu

Rank and position are not to be sought. They will come in the proper time and by nature of the person. Those who seek rank and position for themselves are poor judges. How can a man judge himself objectively? A man who seeks rank and position will do so over the bodies or reputations of others. They are not to be trusted. Instead, one should always seek to improve oneself and then test those improvements in the fires of life. It is never the rank that makes the man. It is always the man who makes the rank. Take away the certificates and titles and we are still only what we are. Here, in the person, is the only thing that counts.

The clever combatant looks to the effect of combined energy, and does not require too much from individuals. Hence his ability to pick out the right men and utilize combined energy. Sun Tzu

The control of a large force is the same principle as the control of a few men: it is merely a question of dividing up their numbers. Fighting with a large army under your command is nowise different from fighting with a small one: it is merely a question of instituting signs and signals. Sun Tzu

Any organization that relies on one central person is bound to fall when the figurehead falls. There should be structure and redundancies within the organization. President and vice presidents should be in agreement. Men in authority should have their assignments and also be trained in ways to equip them with overlapping knowledge. A leader should lead more through his captains than directly. In this way the troops and thus the entire organization, can continue should the leader fall.

Lead by teaching, nurturing, and planning. Lead in humility, calmness, and a teachable spirit. Be direct, honest, and kind. Share your vision with others.

Lead by communicating. Write your thoughts and plans down for those who are carrying out your wishes, Write clearly. The written word lasts forever.

If you cry "FORWARD!" you must make sure the direction... If you fail to do so, the monk and the revolutionary will go in exactly opposite directions. Anton Chevok

Cultivate virtue in yourself and it will be real. Cultivate virtue within the family and it will spread and abound. Cultivate virtue in a village and it will propagate. Cultivate virtue in a nation and it will abound. Cultivate it everywhere and it will be everywhere. Lao Tzu

I have three treasures which I hold and keep; mercy, economy, and humility. From mercy comes courage. From economy comes generosity, and from humility comes leadership. Nowadays men reject mercy, and try to be brave. They abandon economy but still try to be generous. They do not believe in humility, but always attempt to be first. This is certain death (failure). Lao Tzu

To defend your faults is to covet your own weakness. One should be always ready to acknowledge a fault and correct it.

To justify a fault is to argue for your own downfall.

No leader should trust a person who will not take correction. Those who will not take correction cannot be led. They will turn on you one day.

He who will not apply new remedies must expect old evils. Sir Francis Bacon

Obtaining victory may be easier than preserving the results.

Generally speaking, people do not care who is in charge as long as things run well. The charisma and expertise of the leader should be used to build the organization, but things should not be tied so closely to the leader that his downfall would adversely impact the organization.

Humility must always be the portion of any man who receives acclaim earned in the blood of his fellows and the sacrifice of his friends. Gen Dwight D. Eisenhower, 1945

Comrades, you have lost a good captain to make a bad general. Saturninus 100 B.C.

I think with the Romans, that the general of today should be the soldier of tomorrow if necessary. Thomas Jefferson, 1797

Never forget the fact that all leaders have strengths. In addition to that, all leaders have weaknesses. The problem for the leader is to avoid pride. If not, a leader will see his strengths become his weaknesses.

Herding Cats

Leading people in the martial arts community has its special challenges. Martial artists are normally people of strong will and determination. Most have their own aims, goals, and agendas. Those in positions of authority tend to be reserved and even somewhat aloof. In many ways they remind me of cats. They can be hard to organize and difficult to direct. If pushed, they scatter or strike. They disperse easily and quickly go in various directions. Each is capable of making it alone and most think they need no one else. Because they are self sufficient, they have no individual need to stay together.

Leading strong individuals is like herding cats. To herd cats one must make the direction desirable for all. There must be goals that do not inspire competition outside the ring. The goals must allow all to grow and benefit from the journey. In other words, to lead you must serve each one.

To lead, one must have the strength to remain invisible. It is best for a project to be completed and the people to exclaim, "Look what we have done!" Never should a leader hear, "Look what he has done!" In attempting to gain recognition, men look weak. A leader should allow this to happen in heaven's time, if at all. Instead, focus on those with strength and ability within your organization. Raise them up. Delegate responsibility. Do not fear a loss of control.

To gather strong and capable men, there must be a clarion call to a higher goal agreed upon by all. It must be a goal that benefits all equally. To herd cats you feed them all and love them all. Leading men is much the same.

The Purpose of Kata

Kata is a term given to a routine exercise comprised of movements performed exactly the same each time. Kata means "form". In traditional Hapkido there was not Kata in the modern sense. However, each time the students were told to practice a set of techniques, the result was a spontaneous formal exercise. Each time the instructor gave a command to practice a set of movements over and over, it could be viewed as Kata.

The routines that make up a Kata are put together by those with many years of experience and are designed to bring out certain strengths during practice. Each time you practice a set of movements you are doing a Kata of your own making. Kata functions because of one's mental state during practice and not because of the routine.

To this day there are ongoing debates regarding the usefulness of Kata in training. The problem arises in part due to a misunderstanding of how to use Kata. The Kata should be useful and direct. There should be a complete understanding of each move and its purpose. This may not be for the reason you suspect. For Kata to be effective, you must put your opponent in the Kata as if your life depended on each movement. You must see him as you do each technique. In this way, you visualize the opponent and you are in combat while doing the Kata. What is more, if your techniques are done as if life hangs on each one, they will seem weak and slow. With each move you will try harder and harder to get it right. You never will. Always, as if life depends on your techniques they will never seem good or fast or right enough. In this way the ego is killed with every kick and punch.

For the first twenty years of my training, I had little use for Kata. For the next ten years, I respected Kata and trained. For the last five years, I have known nothing. I simply try to get just one or two Kata right. I am yet to succeed.

You must be deadly serious in Training. When I say that, I do not mean that you should be reasonably diligent or moderately in earnest. I mean that your opponent must always be present in your mind, whether you sit, stand, or walk or raise your arms. Gechin Funakoshi 1868-1957

Knowing is not enough; we must apply. Willing is not enough; we must do. - Bruce Lee

Viewing Life

There is no standard in total combat, and expression must be free. This liberating truth is a reality only in so far as it is 'experienced and lived' by the individual himself; it is a truth that transcends styles or disciplines. Bruce Lee

Great Knowledge sees all in one. Small knowledge breaks down into the many. Chuang Tzu Circa 300 B.C.

Where the foundations of passion lie deep the heavenly springs are soon dry. Chuang Tzu Circa 300 B.C.

Minds free, thoughts gone – clear brows, serene faces. Were they cool? Only as cool as autumn. Were they hot? No hotter than spring. Chuang Tzu Circa 300 B.C.

There are no fixed limits. Time does not stand still. Nothing endures, nothing is final. Chuang Tzu Circa 300 B.C.

Let each one understand the meaning of sincerity and guard against display. Chuang Tzu Circa 300 B.C.

Lee explained, "reality is the truth." With this in mind, your mission becomes one of seeking reality in combat. To make things easier to understand, you can substitute the word "reality" whenever you see the word "truth."

Reality is a perception. What you perceive to be reality may not be precisely what your neighbor or your opponent perceives to be reality. Lee took this into account when he said your truth is not my truth and my truth is not your truth. We are all unique in how we perceive the world around us, and this includes combat or self-defense. This subjectivity does not extend to the spiritual realm, however. In this realm we are guided by an objective truth set before us as scripture. But, in the corporeal world there are variances in perceptions. One man may be colorblind. To him there is only black, white, and gray. This does not mean the colors do not exist, but it does mean his truth is divergent from others'. The same can be said with all martial artists. Since the art breaks down truth into minuscule gradients of timing and distance, each body and each art will have its truth. We must each seek fully after our truth. It is bound to take us outside our art and ourselves.

Know what you are looking for and do not be in denial when you discover it. Martial artists who have devoted years to training in a traditional system and have trained according to what they have been taught is truth, sometimes have difficulty accepting they might have spent years studying a lie. Not only might they have studied a lie, but they may have spent years training according to that lie.

At best styles are merely parts dissected from a unitary whole. All styles require adjustment, partiality, denials, condemnation and a lot of self- justification. The solutions they purport to provide are the very cause of the problem, because they limit and interfere with our natural growth and obstruct the way to genuine understanding. Divisive by nature, styles keep men 'apart' from each other rather than 'unite' them. Bruce Lee

The important thing is to not dwell on the lie. Be thankful that you have become aware of it and adjust your training to what you now know is real. Bruce Lee

When the Hard Times Come

Here is the secret of endurance in hard times; this too shall pass.

My life is simple, my food is plain, and my quarters are uncluttered. In all things, I have sought clarity. I face the troubles and problems of life and death willingly. Virtue, integrity and courage are my priorities. I can be approached, but never pushed; befriended but never coerced; killed but never shamed. - Yi Sunshin, Last letter to an old friend.

In crossing salt marshes, your sole concern should be to get over them quickly, without any delay. Sun Tzu

There are patterns and seasons in life. One should realize this and prepare accordingly. Set your spirit to endure the bad and excel in the good. Strong winds and heavy rains do not last long. In the darkest times of life we must see the sun and know that joy will come in the morning.

Nurture strength of spirit to shield you in sudden misfortune. But do not distress yourself with imaginings. Many fears are born of fatigue and loneliness. Beyond wholesome discipline, be gentle with yourself. Desiderata

Joseph Lumpkin

There will be times when we think hard work and training are of no use. We will blame our lot in life on fate alone. In our despair we may say:

Valor is of no service, chance rules all, and the bravest often fall by the hands of cowards. Tacitus A.D. 54-119

It is not true. Chance will play its hand. Fate will take its shot. It is up to us to do our best with what we are given. Our strength of spirit will determine the outcome.

When valor preys on reason, it eats the sword it fights with. "Anthony and Cleopatra" by William Shakespeare, 1697

Strength Made Perfect In Weakness

There is a story of a 10-year-old boy who decided to study Judo despite the fact that he had lost his left arm in a devastating car accident.

The boy began lessons with an old Japanese Judo master. The boy was doing well, so he couldn't understand why, after three months of training, the master had taught him only one move. "Sensei," the boy finally said, "Shouldn't I be learning more moves?" "This is the only move you know, but this is the only move you'll ever need to know," the sensei replied. Not quite understanding, but believing in his teacher, the boy kept training.

Several months later, the sensei took the boy to his first tournament. Surprising himself, the boy easily won his first two matches. The third match proved to be more difficult, but after sometime, his opponent became impatient and charged; the boy deftly used his one move to win the match. Still amazed by his success, the boy was now in the finals. This time, his opponent was bigger, stronger, and more experienced. For a while, the boy appeared to be overmatched. Concerned that the boy might get hurt, the referee called a time-out. He was about to stop the match when the sensei intervened. "No," the sensei insisted, "Let him continue."

Soon after the match resumed, his opponent made a critical mistake. He dropped his guard. Instantly, the boy used his move to pin him. The boy had won the match and the tournament. He was the champion. On the way home, the boy and the sensei reviewed every move in each and every match. Then the boy summoned the courage to ask what was really on his mind: "Sensei, how did I win the tournament with only one move?"

"You won for two reasons," the sensei answered. "First, you've almost mastered one of the most difficult throws in all of Judo. And second, the only known defense for that move is for your opponent to grab your left arm."

The boy's greatest weakness became his greatest strength. We don't often view our weaknesses in the same way, but we should. I am reminded of the time Paul prayed fervently for God to remove some affliction, which he called a "thorn in the flesh."

Refusing to remove it, God said to Paul, "My grace is sufficient for you, for my strength is made perfect in weakness." (2 Cor. 12:9).

It is nearly incomprehensible to us, and yet we see throughout the Bible how God is able to work despite the weaknesses of men and women, showing forth his power -- David with his small stature against Goliath the giant; Gideon a man of no significant background leading a greatly outnumbered band of men; Jesus taking on humanity in the form of a helpless baby. In fact, the greatest demonstrations of God's power have come when men and women have felt the weakest. Remember that the next time you feel inadequate.

"Therefore most gladly I will rather boast in my infirmities, that the power of Christ may rest upon me. For when I am weak, then I am strong." (2 Cor. 12:9b-10)

Imperfect Strength

Be careful of whom you make an enemy. Everyone has a family and friends. When David met the brother of Goliath, whom he killed, David was almost slain.

1SA 17:4 9 And there went out a champion out of the camp of the Philistines, named Goliath, of Gath, whose height was six cubits and a span. 5 And he had a helmet of brass upon his head, and he was armed with a coat of mail; and the weight of the coat was five thousand shekels of brass. And he had greaves of brass upon his legs, and a target of brass between his shoulders. And the staff of his spear was like a weaver's beam; and his spear's head weighed six hundred shekels of iron: and one bearing a shield went before him. And he stood and cried unto the armies of Israel, and said unto them, Why are ye come out to set your battle in array? am not I a Philistine, and ye servants to Saul? choose you a man for you, and let him come down to me. If he be able to fight with me, and to kill me, then will we be your servants: but if I prevail against him, and kill him, then shall ye be our servants, and serve us. And the Philistine said, I defy the armies of Israel this day; give me a man that we may fight together.

1SA 17:24 And all the men of Israel, when they saw the man, fled from him, and were sore afraid.

 1SA 17:32 - 37 And David said to Saul, Let no man's heart fail because of him; thy servant will go and fight with this Philistine. And Saul said to David, Thou art not able to go against this Philistine to fight with him: for thou art but a youth, and he a man of war from his youth. And David said unto Saul, Thy servant kept his father's sheep, and there came a lion, and a bear, and took a lamb out of the flock: And I went out after him, and smote him, and delivered it out of his mouth: and when he arose against me, I caught him by his beard, and smote him, and slew him. Thy servant slew both the lion and the bear: and this uncircumcised Philistine shall be as one of them, seeing he hath defied the armies of the living God. David said moreover, The LORD that delivered me out of the paw of the lion, and out of the paw of the bear, he will deliver me out of the hand of this Philistine. And Saul said unto David, Go, and the LORD be with thee.

1SA 17:42 - 46 And when the Philistine looked about, and saw David, he disdained him: for he was but a youth, and ruddy, and of a fair countenance. And the Philistine said unto David, Am I a dog that thou comest to me with staves? And the Philistine cursed David by his gods. And the Philistine said to David, Come to me, and I will give thy flesh unto the fowls of the air, and to the beasts of the field. Then said David to the Philistine, Thou comest to me with a sword, and with a spear, and with a shield: but I come to thee in the name of the LORD of hosts, the God of the armies of Israel, whom thou hast defied. This day will the LORD deliver thee into mine hand; and I will smite thee, and take thine head from thee; and I will give the carcasses of the host of the Philistines this day unto the fowls of the air, and to the wild beasts of the earth; that all the earth may know that there is a God in Israel.

1SA 17:49 -51 And David put his hand in his bag, and took thence a stone, and slang it, and smote the Philistine in his forehead, that the stone sunk into his forehead; and he fell upon his face to the earth. So David prevailed over the Philistine with a sling and with a stone, and smote the Philistine, and slew him; but there was no sword in the hand of David. Therefore David ran, and stood upon the Philistine, and took his sword, and drew it out of the sheath thereof, and slew him, and cut off his head therewith. And when the Philistines saw their champion was dead, they fled.

2SA 21:15 - 22 Moreover the Philistines had yet war again with Israel; and David went down, and his servants with him, and fought against the Philistines: and David waxed faint. And Ishbibenob, which was of the sons of the giant, the weight of whose spear weighed three hundred shekels of brass in weight, he being girded with a new sword, thought to have slain David. But Abishai the son of Zeruiah succored him, and smote the Philistine, and killed him. Then the men of David sware unto him, saying, Thou shalt go no more out with us to battle, that thou quench not the light of Israel. And it came to pass after this, that there was again a battle with the Philistines at Gob: then Sibbechai the Hushathite slew Saph, which was of the sons of the giant. And there was again a battle in Gob with the Philistines, where Elhanan the son of Jaareoregim, a Bethlehemite, slew the brother of Goliath the Gittite, the staff of whose spear was like a weaver's beam. And there was yet a battle in Gath, where was a man of great stature, that had on every hand six fingers, and on every foot six toes, four and twenty in number; and he also was born to the giant. And when he defied Israel, Jonathan the son of Shimeah the brother of David slew him. These four were born to the giant in Gath, and fell by the hand of David, and by the hand of his servants.

To Choose a Few Good Men

The place of a leader is to empower others. Allow their advancement, and they will lift you up as well. Thus, one of the most difficult things to successfully accomplish is the choosing of friends, students, and employees. If one has chosen well, all else will fall into place. If one has chosen well, he and his organization will prosper.

There is a saying that goes, "Even though one associates with many people, he should never cause discord." In all things one should support others. Hojo Nagauji 1432 – 1519 A.D.

Leaders should select and train two or three trusted men or women younger than themselves. These students should be groomed with every good and strong trait from the leader and all strengths the students themselves possess. They will perpetuate the organization after the leader is gone.

There will be those you value in your position of leadership. They are your inner circle of advisors. Your advisors should be those with differing and varied ideas from your own. Other viewpoints must be sought to balance your perspective. Seek strong and wise people who may not agree with you. Respect their advice above others. Yet, in public, there should be no difference in the way respect is shown to advisors, friends, and strangers. In public, respect all men equally.

Respect is not the same as reward. If one who serves you well is rewarded the same as one who does not, the sacrifice will lose its value. His efforts will fail and you will have lost a loyal servant. Be watchful, quick, and fair to reward loyalty.

Select people of quality and you will succeed. The men selected to be warriors in the story below were those who kept their heads up and stayed alert.

JDG 7:2 - 9 And the LORD said unto Gideon, The people that are with thee are too many for me to give the Midianites into their hands, lest Israel vaunt themselves against me, saying, Mine own hand hath saved me. Now therefore go to, proclaim in the ears of the people, saying, Whosoever is fearful and afraid, let him return and depart early from mount Gilead. And there returned of the people twenty and two thousand; and there remained ten thousand. And the LORD said unto Gideon, The people are yet too many; bring them down unto the water, and I will try them for thee there: and it shall be, that of whom I say unto thee, This shall go with thee, the same shall go with thee; and of whomsoever I say unto thee, This shall not go with thee, the same shall not go. So he brought down the people unto the water: and the LORD said unto Gideon, Every one that lappeth of the water with his tongue, as a dog lappeth, him shalt thou set by himself; likewise every one that boweth down upon his knees to drink. And the number of them that lapped, putting their hand to their mouth, were three hundred men: but all the rest of the people bowed down upon their knees to drink water. And the LORD said unto Gideon, By the three hundred men that lapped will I save you, and deliver the Midianites into thine hand: and let all the other people go every man unto his place. So the people took victuals in their hand, and their trumpets: and he sent all the rest of Israel every man unto his tent, and retained those three hundred men: and the host of Midian was beneath him in the valley. And it came to pass the same night, that the LORD said unto him, Arise, get thee down unto the host; for I have delivered it into thine hand.

**When Two Tigers Fight, One is Sure to Be Injured,
The Other Killed**

Nothing except a battle lost can be half so melancholy as a battle won. Duke of Wellington, 1815

It is a sad state and a failure to endure when we are forced to fight. Even if we win there will be pain, at least, and death, per chance. It is the end of peace in our world. The most we can hope for is to find peace in having done all we could to avoid conflict. Now we stand firm.

In the final choice, a soldier's pack is not so heavy a burden as a prisoner's chains. Dwight D. Eisenhower, 1953

Stand your ground. Don't fire unless fired upon, but if they mean to have war let it begin here! Capt. John Parker of the Minute Men of Massachusetts, 1775

It is an unfortunate fact that we can secure peace only by preparing for war. John F. Kennedy, 1960

Cry, "HAVOC!" and let slip the dogs of war. Julius Caesar by William Shakespeare, 1599

They shall not pass. Marshall Henri P. Petain, referring to the attacking Germans and the defense of France at Verdun, 1916

This aggression will not stand. President George H. W. Bush referring to Saddam Hussein's invasion of Kuwait, 1990

Uncommon valor was a common virtue. Adm. Chester Nimitz 1945 Iwo Jima.

All right, men, we can die but once. This is the time and place. Let us charge. Brig. Gen William Haines Lytle, Union Army, Chickamauga, 1863

The warrior feels and never reasons, and thus is always right. His mind and body have been trained all of his life. His reason has become his natural way. Now, move as he will, he will move in the "Way".

Adversity reveals the genius of a general; good fortune conceals it. Horace in Satires, 25 B.C.

It is well that war is so terrible – we should grow too fond of it. Gen Robert E. Lee, 1862

I love combat. I hate war. I don't understand it, but that's the way it is. Gen Chuck Horner, Commander in the Gulf War, 1999

When It Is All Over

When all is said and done, it was not war we wanted, but peace we sought wholeheartedly. It was not the training of the body we pursued, but we moved to find stillness and battled our own souls to find the peace within. In the final analysis, each kick and every punch was directed against our egos, our defective natures, and ourselves. This fighting art is an instrument in the war for peace. It is peace, which we must take by storm.

Hear me, my chiefs! I am tired; my heart is sick and sad. From where the sun now stands I will fight no more forever. Chief Joseph of The Nez Perce, 1877

Joseph Lumpkin

SO SPEAK THE MASTERS

What we are and what we may become is determined by who we are and what wisdom we glean from others. In this work, wisdom was captured. Through association, friendship, and instruction these great men have influenced us and added to our understanding of the art and our understanding of ourselves. They are masters and teachers in their styles. We wish to thank them for their contributions.

Joseph Lumpkin

The Old Way
by Mark Barlow

My sensei often said that a Shodan (1st degree black belt) merely signified the student could tie his obi correctly and walk across the tatami without falling on his face. He first said this to me when I was a recently promoted Shodan and I resented what I saw as an attempt to take the wind out of my sails. Now I understand he was trying to remind me that training should be an ongoing endeavor.

Alex Marshall was my Judo and Jujitsu sensei for almost twenty years. Already in his 60s when I became his student, his vitality and humor buoyed his hard-core students both in their training and as we faced life outside the Dojo. While I've been fortunate to have had the opportunity to train with many talented and well-respected instructors, I feel truly blessed that I was able to spend so many years with Mr. Marshall. While my style of teaching evolved quite differently from Mr. Marshall's, his attitude and outlook continues to exert a strong influence on me.

Dough Boys returning from France offered Mr. Marshall his first exposure to Jujitsu. As a student at Alabama Polytechnic, now Auburn University, he excelled at fencing, boxing and the lance. After graduation, he relocated to the Chicago area and became a factory manager. Upon learning of a nearby Judo class, Mr. Marshall was waiting on the Dojo doorstep when the teacher arrived. However, the Japanese instructor informed Mr. Marshall that the class was only for Japanese. Mr. Marshall said he understood but still wanted to join. The sensei refused and ordered him to leave. For the next several weeks, this scene was repeated every night the Judo class met. Mr. Marshall would arrive early and stand by the door. The sensei would attempt to hurry past him as Mr. Marshall asked if he could join and the instructor would say, No, only Japanese!" as he hurried past. Thinking that watching a class might scare him off, the instructor finally agreed to let Mr. Marshall observe a class.

In retrospect, Mr. Marshall said he realized that the sensei was much rougher than usual that night, hoping the very sincere screams of pain from the students would scare the crazy gaijin away. After class, the sensei told Mr. Marshall never to return. Mr. Marshall was waiting at the door the next class and whether out of exhaustion or the hope that he wouldn't last long, the instructor allowed Mr. Marshall to join.

Jumping ahead forty years we find Mr. Marshall teaching his own class. When I first met him, Mr. Marshall was teaching a Judo class and was the Defensive Tactics Instructor for the Birmingham Police Academy. He was the quintessential little old man, late 60's, 5'3" and, unless a camera was pointed his way, wearing a perpetual smile. Upon joining his Judo class, I quickly learned to not be deceived by his size or age. As one of his larger students, he frequently used me to demonstrate techniques. While this was a wonderful training experience, it also encouraged me to spend a lot of time learning to fall correctly. While I can't stress enough how even tempered and unfailingly polite Mr. Marshall was, I have to admit he was often brutal in his teaching. His classes were notable for the high drop-out rate and injuries.

The majority of Mr. Marshall's sensei were pre-World War II Judoka and Jujitsuka. His training was geared for self-defense and was harsh and punishing. The more aggressive and promising the student, the more personal attention the instructor gave. Since this personal attention usually consisted of extensive cranking, stretching, twisting, throwing and striking, it often seemed to be a gift of questionable value. Mr. Marshall took this method of instruction to heart and his students suffered accordingly.

On the other hand, Mr. Marshall also followed the example of his Japanese instructors in utterly ignoring those he thought were insincere or inattentive. A student Mr. Marshall perceived as having potential would be pushed to his physical limit, night after night. No pats on the back, no encouragement, just correction and training to the point of physical exhaustion. On the other hand, the student who was content to play and chat on the mat was never kicked out of the Dojo. He was permitted to pay tuition and take up space on the mat but was never given rank or responsibility. If one of these students asked Mr. Marshall how they were doing, he'd say "Great!" and keep walking. I once asked him what I needed to work on for an upcoming rank exam and he simply said, "Everything."

Mr. Marshall promoted only two of his students to Black Belt. I'm honored to have been one of those students. When I began teaching in 1980, Mr. Marshall frequently observed my classes and could be very harsh in his corrections. While no one enjoys being corrected, it was reassuring to know that he thought I was worthwhile enough to chastise.

Until shortly before his death, Mr. Marshall was constantly training, exploring and asking questions. He never quit being a student. This is what I want to pass on to my students. Mental and physical training should never stop. As hokey as it may sound, the journey never ends.

My years with Mr. Marshall provided me with an amazing opportunity to not only train in a highly effective martial art but to catch a glimpse of training methods that have fallen by the wayside and to experience methods little changed from the Golden Age of Jujitsu. At Mr. Marshall's urging, I also trained with some of the best Judo, Jujitsu and Aikido instructors in the world.

Eventually, as any instructor will, I developed my own style of teaching. I know that I have retained the practicality, which was so important to Mr. Marshall, and I hope I have made the training a bit more accessible. If I had to put my teaching style in a nutshell, it would be train the students hard, push them physically and mentally, pat them on the back as needed and never accept less of them then they are capable of giving. I'd like to think Mr. Marshall would be pleased.

Debt and Duty
by Dr. Chris Dewey

While thinking about what to write in this chapter, I looked back over more than three decades of learning and teaching martial arts and gave thought of what was important to me in the process of teaching both when I am the student and when I am the teacher. Specifically, I was interested in the nature of this journey of self-discovery and the things that have kept me engaged and motivated since I started martial arts back in the late sixties. I came to the conclusion that many of the major factors have little to do with the martial arts, *per se*. I suspect many of the reasons also keep other people in the martial arts, especially those of us who eventually become teachers.

Learning

First and foremost, in my mind, the teacher is a student, and the student is a teacher. When I was a colored belt, I didn't realize this. I don't think I really came to terms with the full impact of this statement until I had been teaching for several years. There is always something new to learn. When we think there is no more to learn, then our minds have closed and we are of little use as teachers or students. It never ceases to amaze me how much I learn from watching my newest students execute some drill in a way that I hadn't previously considered possible. To me, learning is a journey, not a destination. It is an ongoing process that defines everything I do in my life. You might say the desire to learn is a defining characteristic of my entire life process. As a lifelong student, I sometimes look at the journey and think how little I have truly learned compared to what is possible. Such thoughts are exciting to me. It is like taking a bucket of water out of the ocean. I have a piece of the ocean, but I have not even begun to drain the ocean, yet. Martial arts represent the ultimate mental and physical challenge in this regard. I can never learn it all. I can swim in it, I can learn some of it, but I'll never really make a dent in what is possible.

There are several things resulting from being a perennial student. One of them is that I do not see style differences in the martial arts. Someone once described the martial arts to me as a pie, and each art as a slice of the pie. Well, call me greedy; I want to eat the whole pie! As a student of the martial arts and as a teacher thereafter, I have always seen similarities where others might see differences. I see unity of principle rather than difference in technique. Think about a simple wrist turn out throw, for instance. How many ways are there to do the thing? How many black belts does it take to change a light bulb? Answer: All of them. One to change the light bulb and all the others to say, Yes, but that's not the way we do it in *my* dojo. So, back to the wrist turn out throw, there are countless ways to perform the throw, but they all rely on the same principles of body mechanics. You still have to remove the old light bulb, get a new light bulb into the socket and switch the light on…the similarities in mechanical principles outweigh the differences in technique.

Another aspect of being a perennial student is every time a new student takes his or her first steps onto my deck, I wonder what he or she will teach me. And another thing: Age is not a factor. I have been taught some great lessons by some of my youngest students. Such teaching is not always about the technique either; often the teaching can take the form of a life lesson. All it takes is a willingness to see the lesson and have an open mind to receive the teaching. Perception is everything: You can either see a child doing something wrong or you can choose to see the child discovering a new path. Sometimes the paths work, sometimes they don't; either way there is a lesson to be learned.

You see, I believe we are all teachers in one form or another. We all teach through our words and our deeds. We either teach good lessons or bad lessons. As Brian Tracy, a highly respected and renowned motivational speaker, says: "Everything counts, nothing is neutral." As a martial arts teacher, I teach not just through my technical skill (although that is critical), I teach through the standards and values I hold to be important. If martial arts are to be propagated as a tool for personal growth, then it is critical that we live the martial way and that we do so with ethical conduct. Gichin Funikoshi (founder of modern Shotokan) and Jigoru Kano (founder of Judo) both talk about the martial arts being used for the perfection of the human condition and for mutual benefit and welfare. It is significant to me that in neither case do these great leaders talk about technical skill. Technical ability may attract students to our dojos, but it is our ability to educate, motivate and inspire those students that keeps them coming back week after week and year after year.

Sharing

From learning, I come to sharing. Teachers give of themselves and they do it in every class. They do it out of love of their craft and because they love their students. Teachers give because to do otherwise would be wrong; it would be incongruent with the nature of what they feel drawn to do. Teachers are not, however, the source of knowledge, they are merely a conduit for knowledge. There is a huge difference. To teach is to share. We share in all manner of ways. We share with the victories and the defeats of our students; knowing the road to success is walked with the tears of disappointment and of failure. Teachers know the road is hard and often unforgiving. We know that sometimes the student's greatest obstacle lies within the student, rather than in the lesson material. Teachers understand each lesson has a purpose and each lesson has a price tag. True learning does not come cheaply. Knowing these things, teachers in the martial arts are mindful of their students' potentials, needs and goals. Teachers know some lessons will be easy for some students and difficult for others and they offer a helping hand at some times and a compassionate word at others. Sometimes also, the teachers must stand back and wait, but this does not mean that they do not care. Teachers who are truly invested in the process, share in the struggles of their students because struggling is an important part of the process. Those same teachers will also fade into the background, when their students stand on the pinnacle of their own successes.

Some years ago I was told as a brand-new national level Judo referee, that the best referees are invisible. All you should see on the Judo mat during the match is the players. Ideally, all you should see is the Judo…two players locked in their strategic ballet of Judo. In much the same way the best teachers are invisible. All you should see are the students learning, achieving and growing.

Some of my greatest teachers in life were not necessarily those who helped me out by giving me answers, but those who showed me a path and expected me to walk it alone. The great teachers believed in me far more than I did in myself, they could see in me potential of which I was blissfully unaware. The mentors in my life today fill much the same role. They see things in me that I have difficulty seeing, they have great expectations of what I can achieve, but they also know that only I can achieve these things. No one else can do it for me.

So paradoxically, sharing as a teacher is as much about standing back and letting your students walk, stumble, fall, get up, walk, climb and eventually achieve, as it is about being a bastion of strength and an unflagging believer in their ability. Sharing and giving: These two qualities are so inextricably bound together they almost define the life of a teacher. Sharing and giving imply bonding and caring. In every student and teacher relationship is a bond of trust and a bond of communication. It is possible for either party to break this bond, at which point the learning stops. Truly invested teachers grieve at such moments because they know they have lost something precious, but equally wise teachers also know that it is not for them to reach all students.

On and Giri

"On" and "Giri" are two Japanese words meaning debt (on) and responsibility or duty (giri). On and giri are the yin and yang of teaching. On and giri are the unavoidable and inseparable pair that arise as a matter of consequence, the instant we become a student. Each of us acquires a debt when we put on a white belt for the first time and set our feet on the hard wood floor of the Karate dojo or the tatami of the Judo dojo.

We owe a debt to our teachers and their teachers for what they have given us and what they will give us. This is a debt that accrues over time. The knowledge we have gained comes from the centuries of study that preceded the day we took our first steps onto the deck. It is therefore important for us to see ourselves as a link in the chain of learning. The knowledge we each carry is a gift; it is not ours to keep, nor was it ours first.

We hear a lot of talk about "blended" or "non-traditional" or "eclectic" martial arts programs these days, but they are really no different than "traditional" programs. Each martial arts teacher was taught by someone, or perhaps by several instructors during his lifetime. In thirty-five years of training, my direct instructors have come from Judo, TaeKwonDo, Hapkido, Jujitsu, Aikido and Kendo. I do not have a single sensei in the strict sense of the term, but I do have a debt owed to each of those who taught me and to their instructors before them. I can never repay that debt, just like I can never repay my parents for the gift of life they gave to me, or the freedom they gave me to find my own way in life. By teaching, we give back that which we have been given. We honor those who taught us and we respect those who genuinely wish to learn what we have to offer. On and giri in teaching, therefore, are about loyalty and honor. In its simplest state obligation resides at the root of honor; we can never repay the debt to the past or those who taught us, but we can honor their actions by passing on the trust we have been given.

Unfortunately, students sometimes think they have no obligation to their teachers. Too often these days, students have a mentality of expectation; that something should be given to them at no cost, because *they* are owed it. Nothing could be further from the truth. Students have an obligation to learn and teachers have an obligation to teach, but when students blame their teachers for their lack of performance, it is not always the teacher who is to blame. I was one of those students. I blamed my poor teachers for my lack of performance, when all along I knew I was responsible for myself if any learning was to occur. I have had great teachers and lousy teachers over the years, but in each case I gained the most when I took personal responsibility for what I learned, rather than expecting the teacher to tell me what I was supposed to learn.

Now, here's the rub: it doesn't matter that I have had good or bad teachers, good or bad parents, good or bad friends, to each of them I owe a debt I can never repay. It is, once again, a question of perspective. In every situation and every experience is the opportunity for learning. I either choose to see the opportunity to learn or I complain about how the system failed me or how rotten my childhood was, or how my friends let me down. One road leads to success and happiness in life, the other leads to failure and regret.

If we lived our lives with a little more awareness of "on" that we carry, perhaps we might be more willing to see the opportunities that meet us every moment of every day. We might also be willing to do something with those opportunities, which leads us neatly into "giri".

Giri, therefore, is about debt service. Giri is concerned with meeting our responsibilities, which means it carries the obligation of becoming the best possible teacher, to honor what was given to us and in so doing, provide the best learning environment and the greatest number of opportunities for our students. To illustrate this point, I want to use another example from more than twenty years of being a Judo referee. I have been fortunate enough to be mentored in my refereeing by three IJF-A (Olympic) referees. One of these mentors once told me the reason he refereed was that the players would have a better chance of the getting the correct outcome than if he was not on the mat. This is what being a good teacher is all about. When giri is operating at its highest level and I am doing my very best job teaching the martial arts, I am giving my students something they would not be able to acquire if I were not instructing.

I guess that my final thought in this regard is that giri isn't simply about paying back a debt because we are expected to do so; it is about doing more than is expected. To truly honor our teachers, we achieve our potential and will settle for nothing less. To truly honor those who taught us, we always go the extra mile and seek to improve ourselves.

What is Physical is Also Mental

Judo teaches us so much about life. If I am resistant, I am merely a wall to be overcome by a superior force. If I am off-balance I can be overcome by a weak force. If I am stable and mobile, I am flexible and adaptable. I can roll with the punches, avoid the traps and move more efficiently and effectively. Needless to say, these things also apply to all aspects of my life.

As a martial arts teacher I am increasingly aware what I do physically also applies mentally and spiritually. I believe that this is what Funikoshi and Kano saw in their arts. They saw the direct links between physics, psychology, physiology and philosophy. Another aspect of yin and yang emerges here: it is virtually impossible for us to learn the martial arts on a physical plane without being affected on the psychological plane. The transition of the physical to the mental and spiritual development of the student is almost a guaranteed outcome of a great leaning environment. Martial arts instructors are full of stories of how a student has gone from being a poor student in school to becoming an honor-roll student, simply by applying the skills they learned on the deck. In order to apply a punch with skill or a throw with precision, our students must learn self-control and the self-control will have application beyond the training deck. Likewise, overcoming the frustrations, disappointments and obstacles that will naturally arise during the training process, will lead to the development of determination and persistence, both of which have effects outside the dojo.

The physical path of the martial arts is completely ambivalent to the student. The skills of each martial art are simply what they are, they must be faced, learned and assimilated by each student as he or she progresses through the ranks. The techniques do not care about age, size, shape or natural talent, ability or grace of the student; we all start out the same. How we face the training determines to a large degree how we progress. I have watched young, talented students give up while older, out-of-shape students work through their tears of frustration and refuse to give up. The latter is the road to courage.

The physical lessons become the lessons which teach us how to live and to grow as people. The martial arts simply become a conduit for that process to occur.

Personal Growth

Finally, teaching martial arts is all about personal growth. As my body ages my strength will fail me, my tournament skills will fail me, my metabolism will slow down and my strength will slowly dwindle. I expect these things to happen very slowly over the greater part of a century. I have been competing since I was fourteen years old, and I am still competing in Judo to this day. I have no intention of stopping. But when I stand on that line and face my opponent, it's me on the other side of the referee, not some stranger I must beat. My only enemy is my own self, my own weaknesses and my own inadequacies. When I compete I am competing against myself, I am testing my own skills against myself. It's not about winning or losing; it's about being and doing. It's about living.

Martial arts are a lifetime furnace that forges each of us and burns away the weaknesses and draws out our strengths. As a teacher in the martial arts, I want each of my students to experience that same precious gift of self-growth. You see, in the final analysis it is not about how sharp our kicks look or how well we throw our partners (although technical excellence is a mirror to the internal drive process), it is about using the martial arts as a vehicle through which we become better people.

Ultimately, being a martial arts teacher is about congruency. Brian Tracy says something akin to the notion that we become what we think about most. In essence, what you are on the outside is who you are on the inside. As martial artist instructors we cannot escape the responsibility or the fact that we are icons to those we teach. We are expected to live by example and live with integrity and ethics. We are expected to demonstrate superior skill and control. But if we do not also do the same thing in our personal lives, then the martial skills are merely a sham. It is incumbent upon each martial arts instructor to commit himself to the process of personal growth. Only in that way will we each achieve congruence between what we do, what we say and what we think. Perhaps this is the greatest lesson of all.

Mind in Internal Gung-fu
by Erle Montaigue

In order to understand internal Gung-fu, it is important to also understand at a basic level how the mind works. In knowing this, we are able to understand why we have to perform forms or Katas and why they are so important.

Try this. I would like you to do this right now. As you read this chapter, imagine that you are holding a big yellow juicy lemon in your hand. You must SEE the lemon in your mind's eye; you must feel the waxy texture and that little lump at the end. Take a big knife and cut the lemon holding one half up to your mouth and squeeze the juice into your open mouth.

What happened? Your mouth produced saliva, didn't it. You really didn't have a lemon, you were imagining it. Your mind however, still caused your body to do what it would have done had you a real lemon!

It is the same in the martial arts. When we practice our forms or Katas, we imagine the opponent in front of us. Provided you have a good imagination, your sub-conscious mind will be doing all the self-defense applications as you go through your forms. The good thing is, you do not have to imagine the applications every time you practice. Only once or twice do you have to be told what the applications are and only once or twice do you have to go through the whole form imagining you are performing those applications. After that, those movements go into your 'long term' memory and you no longer have to think about them, they will just happen sub-consciously. It does not take long for a 'short term' memory to become long term. I have a way of remembering my pin numbers where I make up a story using those numbers. Every time I have to use my pin, I remember the story rather than the pin number. But the amazing thing is, after a short time, I no longer even remember the story, as it has become sub-conscious or 'long term memory'. So the pin number just comes without any conscious thought. Doing it this way, there is no chance that you will forget the number after some time of not using it, as it has become long term, like your name. No one forgets their name unless they have a memory disease. It is the same with the martial arts, before long, the self-defense applications go into your long term memory and you no longer have to think about them, they will just happen if you are ever attacked.

This is the beauty of doing form or Kata correctly, you do not actually have to go out and fight people in order to learn about self-defense. You have to know how to strike hard objects, and you have to have some form of interaction with other people in a self-defense area such as in push hands or in what we in our schools call 'offense/defensive' sparring. I use the word 'sparring' with caution, however, as most people nowadays know my thoughts on this silly pastime. What most people call sparring is the most damaging area of one's martial arts/self-defense training. Of course, if you are only doing a martial art as opposed to using a martial art as a self-defense tool, then you should do sparring.

Sparring is a sport and if you are doing martial arts for sport, then you will have to do sparring in order to win trophies. But do not expect to be able to ever defend yourself using the same movements you use in the ring against a street-wise attacker. You will be beaten up should you think you could defend yourself because you have won a trophy in a martial arts tournament! Street fights just do not happen as they do in ANY ring fight no matter how realistic they TRY to make them or call them. You have to do either sports or self-defense; the two can never be mixed.

Offensive/defensive sparring is the only way my schools teach fighting. The instructor wears all of the protective equipment he can find so he is not hurt. We even go so far as to wear large neck equipment so neck strikes are used without killing the instructor. I say that it is only the instructor with many years of training whom I would trust to act as the attacker because he or she will have the necessary control to be able to pull the attacks should they see the student is not going to defend against it! This is very important. However, we do not let the student whose time has come for this type of training, know that we will actually pull the attacks when necessary. The student believes he will be severely struck if he does not successfully defend himself. This is part of the mind training.

In any confrontation, it is the person with the strongest and most resolute mind who will win, depending upon relevant training, of course. I let the student believe he or she will be severely attacked with great force and if they do not successfully defend against me, they will be hurt. I then attack as they would be attacked in the street, totally illogically, without technique, like an animal. As soon as I register that the student has struck me in two or three deadly points, like the neck, I will stop the attack, as I have been defeated. This kind of training I have found to be the greatest training aid for any self-defense method. The students really feel intimidated as they would in the street, then they only have their mind to rely upon. They must summon up automatically more yang Qi (energy) than what I am landing upon them, only then will they be assured that their art will suffice in a real scrap.

In training sessions of form or Kata, we defend ourselves against an attacker often hundreds of times until the methods eventually become automatic.

However, here is the rub; you must be training in natural movement, that which is in tune with your natural flow of internal energy. Only these movements will ever become sub-conscious reflex actions. Movements which are jerky, stiff, not circular, will never become reflexive, as they are moves we must constantly think about. Many karate people around the world now are beginning to realize this and are changing their Katas slightly, so they represent that which was originally kept from them!

Your mind cannot tell the difference between what is imagined and what is real. As an example, my children only ever practice form thus far. But every now and again, I will simply attack them to test their sub-conscious learning. Without fail, each one of them is able to defend against my attacks, relative to their level of growth and strength, of course. But the sub-conscious is there, never having any kind of normal sparring. People who say they only practice what is 'real' and never practice form or Kata will never be good at self-defense. We only have to look at some of the great instructors, such as Bruce Lee. No one has ever come up to his level. Why? Because people tried to learn what he knew and not HOW he knew. Bruce Lee was well versed in a number of classical martial arts systems, including Taijiquan.

It would be exactly the same should I try to teach my own students and children what I know and not how I learned it. Sure, it is very hard not to teach what I now know, as it is exciting for me to teach the advanced stuff. But I must keep teaching basics to beginners, as they will never reach an advanced level if I only teach them how to fight using my system. In learning from the beginning, the classical internal systems, we eventually learn to invent our own system where the art becomes ours; we own the art, it does not own us. However, in order to own the art, it must own you for some time in the beginning.

Bruce Lee was correct in saying in as many words, that one must come across his own system, leaving all systems aside. But in order to do that, we must first of all learn a system! A baby cannot just get up and run, it must first of all learn a system of moving. When the baby grows up, it has its own system of moving like no other person. At this time, the baby has become the adult and has mastered the art of moving. It's exactly the same with the martial arts, we must first learn a system, preferably a classically proven system, and then we are able to convert that knowledge to our own system. This is the reason that I always tell my students, (at an advanced level), "don't do it like me, find your own way". However, in the beginning, they must do it exactly the same way I do it, as these are the basic building blocks of those forms that have been proven to lead you to greater things later. Many systems will not do this for you, however, as they are just not physically sound, they do not have natural movement.

When you are first of all imagining the opponent in front of you and performing the self-defense applications against his attacks, you must have it in your mind as vividly as possible. See what clothing the attacker is wearing; is he small or large, dark or light colored hair and complexion? Is he hairy or smooth? Really see that attacker as he attacks you but do not use tension in your forms as this will block the Qi and those movements will never become reflexive! This is what is wrong with the so-called dynamic tension Katas; they block the Qi. Be sure to put yourself into the picture as this really helps with long-term memory. This sounds like I am also talking about push hands? And I am. Push hands has all of the above elements and is the ideal exercise for gaining reflex self-defensive actions.

One last warning: Always be careful of what you think! Because what you think, your mind is actually doing! This applies to all facets of life and has a direct bearing upon your standard of life. Someone can be on the outside a saint and have a miserable life because what they are thinking on the inside is definitely not saintly! Your mind is the most powerful weapon you have in both self-defense and in your daily life. Use it wisely.

How Chang Yiu-Chun Taught
by Erle Montaigue

Most of what I now teach can in some way be attributed to Chang Yiu-chun, my primary teacher.

Although the physical parts of what I have learned are obviously very important, it is the 'Way', which Chang taught me that is the most crucial.

Chang believed that everything anyone needed to know was already in that person's mind somewhere and only had to be 'unlocked'. It is the job of a teacher to simply unlock that knowledge.

When I first began 'learning' with Chang, I was like most other Westerners in that I wanted a 'seeable' set of things I must learn in chronological order. I did not know why Chang would teach me one thing one day and then go to a completely different thing the next. In many ways, this is how I now teach and is the reason why many of my long distance students want a curriculum so they can learn in some 'human' set manner.

Chang knew we learn internally or sub-consciously and the brain is in ordered chaos at all times. So a set way of learning and teaching would not teach the student much in the way of internal material. In the same way, if we try to use a logical fighting art in real self-defense, we lose, because self-defense is not logical. So Chang believed the way we learn will also represent the way in which we will ultimately defend ourselves. If we learn in a logical chronological manner, then we will lose in a real self-defense situation. However, if we learn in an illogical manner, we have a much better chance of saving ourselves in a real fighting situation. The reverse applies to tournament fighting. If one wishes to win in tournaments and get trophies, then one must learn a Karate style which is logical in its teaching with logical movements. If, however, we try to use a tournament system in the street, we lose! It is impossible to mix the two, saying that you use the tournament stuff for the ring and the real stuff for self-defense! Self-defense is sub-conscious and if you learn and study so hard to win in the ring, then the sub-conscious brain will use that type of 'fighting' when attacked for real.

This is the reason that I do not teach a 'martial art', preferring to teach a self-defense system. I am not interested in teaching tournament fighting so men can show how good they are at 'fighting', to satisfy their own egos.

Chang would often simply push me all of a sudden to note how I reacted and how it moved my body. In the beginning, he would do this pushing, for instance, on my shoulder. He would not do it again for several days or even weeks and I often wondered why. He was waiting until I moved correctly and used my body in the correct manner before doing it again. His reasoning was that it takes some time for the sub-conscious brain to take in something new. To keep doing the incorrect movements in response to his pushes would cause the wrong information to go sub-conscious. It is the training itself that teaches us how to move and to react to any types of attack. So, if you have not done the training of form, and in particular push hands and power push hands, all of the 'fighting training' in the world will not help when you get into a realistic situation!

Much of my training with Chang was spent watching him do things. He believed this to be the most important part of my training and is now the reason I put out my videotapes, so students are able to see me doing it all. Of course, you have to be taught the physical movements in the beginning. However, once learned and more importantly, corrected, it is important to see your teacher doing it so your sub-conscious mind can take it all in and adjust your own body automatically. During these 'watching' times, it is important NOT to look for things you may be doing incorrectly, but rather just watch and take it in.

Doing is better than being taught! There would be times when Chang would not talk (his English was not good anyway). He would simply do his own movements having me watch or he would work with me, striking occasionally, pushing using short sharp attacks, showing me physically what I was doing wrong and obviously hoping that my sub-conscious mind was learning at a much higher level. In retrospect, I am glad I learned in this manner, as many years after Chang, all of the internal training began to come out in my own physical training. It was as if Chang was still teaching me, as the seeds had been sewn all those years ago.

Mr. Montaigue's Closing Comments:

Don't take your martial art too seriously. Do not live for your martial art, but rather 'martial art to live'. How could anyone take the martial arts seriously? Just look at the modern concept of the martial arts - the film stars, the fake fights, and totally unrealistic techniques. I know people who will stand up in a bus while passing their dojo and bow saying "books"! To take something that seriously, is to make it into something other than life, something apart from normal life, something we must do differently in order to practice - a special time of day, a special suit. Men in particular tend to take the martial arts seriously to the point of distraction because they are just so afraid of being beaten in a fight. So they practice their martial art purely for the physical self-defense area. How silly, training your whole life in a life art, waiting for the one time 'if you're unlucky', when you could be attacked! It's like the woman who spent her whole life learning how to move a matchbox using her mind! Why, when we can move it by pushing it and not have to waste an entire lifetime?

The martial arts are a way of life, something that is normal, a way of self-defense in both a physical sense and also in a microscopic sense - that of warding off disease and the healing of others. If we leave out any part of the martial arts training, then we lose the whole. If we practice the martial arts purely for the sake of the martial arts and self-defense, then we are not doing the martial arts! If we do the martial arts purely for the healing side, then we are not doing the martial arts. It would be like having yin without yang, black without white, hard without soft, and that is impossible.

Authors' note: Shinsei wishes to thank Mr. Montaigue for his support and recognition.

The Character of a Martial Artist
By Dr. Chris Dewey

Humility: Knowing one's place amongst mankind. For the martial artist, the first and most important quality is that of humility. Humility at its simplest is freedom from pride and arrogance. Without humility, growth, learning and development of a community spirit are hampered. At higher levels, humility is choosing to make one's needs and desires secondary to the training at hand and the needs of the larger community (your partner, your dojo, your martial art and your family). Humility enables us to grow as a community.

Awareness: Conscious cognizance of one's surroundings. For the martial artist, awareness comes in the form of the workings of the human body; technical excellence; the effects of one's actions upon others; the developmental needs of oneself and others; the intent of those around us; our environment and our situation, and consequently; opportunities for growth as an individual and a community. Without humility, there can be no true awareness of self or others. With awareness comes a zest for life, because one cannot help but see the opportunities in any situation.

Rectitude: A sense of, and strict adherence to, the correct. A martial artist should always seek the truth and discover what is correct and do what is right. There should be a sense of correctness in body, mind, spirit, action, training and technique. Rectitude encompasses the more specific qualities of honor, and diligence. An appropriate sense of rectitude is dependant upon both humility and awareness.

Mission: A goal or plan of action. As a martial artist, one should always have a view to the task at hand. A goal is a fulcrum without which neither movement nor growth can occur. There should be a specific goal for each action, each training session, each rank, and each stage of development in life.

Ministration: - A martial artist is, at all times an envoy. The martial artist has a calling much like that of a guide, and an errand to faithfully discharge. As martial artists we pass on all that we are and all that we learn. We must choose that behavior and knowledge which is appropriate to discharge to others and then give, give, give.

Order: Organization of mind, self, action; conformity to standards and rules. A martial artist is governed by order and discipline. We bow, we sit properly, and we show respect to each other and our instructors. This, in and of itself, is not enough. We must also learn to discipline our minds, our bodies and our spirits. Through the self-discipline and self-confidence that rectitude and order bring, we develop indomitable spirits. As we achieve self-discipline we re-organize ourselves according to our ideal self-image and open the door to true self-respect and self-esteem.

Rank in society: We bear the rank we wear, and if we give anything less at any time in our daily lives, then we are not fulfilling our mission. Others in our ranks look to us for direction and assistance, we owe them our respect, attention to duty and our very best performance. The external community looks to us as representatives of a brotherhood of martial artists and therefore our behavior as individuals reflects upon the whole community. A martial artist who does not live up to his rank both inside and outside the dojo is a detriment to all.

Brotherhood: We are a brotherhood of martial artists and have a responsibility to the order, each other and our martial community. A martial artist's individual code of conduct is a measure of how responsible he feels to the brotherhood. The greater the debt of honor, the greater the conduct should be.

Yielding: Bending, being pliant under external stresses, giving precedence. A true sense of humility, awareness, rectitude, mission and order give the martial artist a sense of perspective in life. There are times to fight and times in which giving way and going forward is the positive action to take. When we achieve a state of yielding to our own growth, we have conquered our greatest enemies: our own inadequacies and weaknesses.

Making an offering: A martial artist offers his inadequacies before the world and if he is humble and aware, he is given growth in return. A martial artist offers strength of character and is given respect and trust in return. A martial artist offers what he has learned, and in return learns from both his students and instructors.

Producing results: Apart from the obvious increase in skill, the final outcome of the diligent martial artist who perseveres is grace, efficiency, a community of well-being, an increased sense of self worth and a tendency toward a life of **Harmony**.

Passions to Live By
by Wayne Carman

(The following is used with permission from Wayne Carman's book, *Elvis's Karate Legacy*. It is condensed from the original.)

In this rich phase of my own life, I am blessed to serve as a martial arts teacher in the gorgeous scenery of Tennessee and Missouri. I continually add to and improve my philosophies and the way I instruct students. It is rewarding. I have developed a list of helpful guides, both for students and those outside the dojo. I call this my "Power of Passion List". Passion is a world that uses even its syllables and consonants to express its meaning, and it packs an even more insightful meaning, to me, that the word "Power" by itself. This list is a work in progress, one that you can use and add to as you journey life's roads and make it an adventure

Passion of Love - Ninety-nine percent of the people who hear this term automatically think of sex. Naturally, this is part of it, but think of the person who, through paralysis or illness, has lost this gift. I've known winners who overcame these setbacks by loving life with a passion! You don't have to be Mel Gibson and Julia Roberts in a romantic film to have passion of love. I love many things passionately – my wife, martial arts, new friendships. The Bible makes it plain that the greatest gift God gave humanity was love. When you love others more than yourself, true love comes back to you. It makes even simple things come alive. So develop a passion of love.

Passion of Giving – We know from experience that it is better to give than receive. Our first priority should be to help and love our fellow man. But I must admit, there is a benefit to the individual giving! We have all hade the experience of giving selflessly, and then watching in amazement as, seemingly from nowhere, that gift of giving comes back.

Passion of Believing – Whatever the mind can conceive, and really believe, you can achieve. This is my personal motto. I believe it because I live it. It is the essence of Elvis and his famous T.C.B theme: Do it now, or in a flash. The world's most successful people constantly beat this drum. They eat, sleep, drink, and work just like anyone, but they separate themselves from the average person by their belief in the goal.

Passion of Respect – Again, you must give in order to receive. I have followed this practice down through the years, and the result is that I have known and called friend some of the greats: Elvis Presley, Kang Rhee, Bill Wallace, Charlie Rich, and Ed Parker. I loved them all. Develop a passion of respect.

Passion of Commitment – I have learned this primarily from the great martial arts teachers I've studied under. Always try to finish what you start. Kang Rhee drummed into his students' heads that it's better to be the master of one thing than the master of no-thing. Develop a passion of commitment.

Passion of Positive Attitude – I believe in being honest, so I must tell you that I don't like false modesty. What you can conceive, you can achieve. Develop a passion of a positive attitude.

Passion of Discipline – I compare this to a great ocean liner without a rudder. The ship's opulence is worthless if it is only drifting or spinning in one spot. Attach the rudder (commitment) and you go places! This was one of Elvis' strengths. Sure, he was talented and good-looking and charismatic. But if you saw him in the studio, or the dojo, or onstage, you saw the discipline he learned from the martial arts. The man sweated. And he achieved. Develop a passion for discipline.

Passion of Power over Peer Pressure – This is killing our youth today. It is difficult to convince kids of the harm in going along with a bad crowd. Let the children know that it builds character to say NO! Invest in the children (Prov. 22:6). Develop a passion over peer pressure.

Passion of Consistency – Especially in the martial arts, it's better to do something a little each day than try to cram it all in one sitting. Read your Bible a little each day. Meditate a little each day. Smile a lot each day. Be consistent. Develop a passion of consistency.

Passion of Strategy – In the martial arts, a good fighter develops a strategy. A great example of the passion of strategy is the Gulf War of 1990-91. Remember when Gen. H. Norman Schwarzkopf explained his willing strategy to the Pentagon press corps after the 100-hour ground ward had ended. He highlighted with a pointer how he had controlled his troops' movements in battling the Iraqi's on their home turf. The general exhibited a fierce passion when he kept using the world "textbook" in explaining how his army "came in the back door" of Iraq when Saddam Hussein's elite Republican Guard expected the Americans to storm the beaches of the Persian Gulf. Instead, Schwarzkopf had his troops make an end-run around the enemy and destroy it completely in the desert. That is passion of strategy, and all highly successful people have it. Develop a passion of strategy.

Passion for Healthful Living – Our body is our temple. We have to take care of it. Don't abuse yourself. There is a need for healthy habits. Work out, eat right, get enough sleep – there must be a good balance. Remember, it's not how you start out but how you finish the race. Take care of yourself. Develop a passion for healthful living.

What is an Instructor?

By David Sheram

In order to be a good teacher, instructor, Sa Bum Nim, or Sensei, we must understand what it means to be this person of influence. Our goal in teaching is to be like the Master Teacher, Jesus Christ. A quality teacher must be Christ-like. The best training comes from modeling.
Matthew 5:16 Let your light so shine before men, that they may see your good works, and glorify your Father which is in heaven.

Our students will become like us. How we stand, they will stand. How we punch, they will punch. How we move, they will move. How we act, they will act. How we exemplify Jesus in our families, dojangs, business ethics, and personal lives, they will want to be the same way. We should deliver what we believe to be the truth.

Formal training is important but we need not only be martial arts instructors but we need to ensure that we are: born again, love people, prepared and apt to teach, always ready to learn, grounded in the Word, empowered with prayer and filled with the Spirit.
In Ephesians 6:14 the Apostle Paul told the church in Ephesus to, *"Stand therefore, having your loins girt about with truth"*. A quality instructor, the real Black Belt, must always wear the belt of truth. Honesty with students shows integrity. Our integrity is what we are even when no one else is around. We should be blameless, above reproach.

The mark of an excellent teacher is not so much the ability to impart information, but a contagious enthusiasm for learning.

In teaching, there are several points to remember:
Instructing is a challenge for you to use all of your knowledge, skill, and patience.

Beginners are very fragile. The first three classes are the most important.

Practice the K.I.S.S. Principle. (Keep It Short and Simple).

Every student is different. Have patience.

Use the student's name often. Be personable.

Discouragement and lack of confidence are the enemies of the student.

Change the class structure from time to time. Keep it interesting.

Maintain discipline at all times.

Know the rules and follow them, but remember, "Knowledge is knowing the rules. Experience is knowing the exceptions to the rules."

Any student who desires extra help should get it. It could come from you or an Assistant Instructor.

Remind the students of the rules often.

Take your time. Do it right. We are not in a race.

Don't look for perfection at first.

Pick out the main problem and correct it.

Always demonstrate in the same direction as the student.

At times, you must have a commanding voice.

Speak clearly and loud enough for the student to understand.

Keep the student active.

Use a variety of techniques.

Keep the class interesting and not boring.

Plan the workouts. Be organized.

Encourage and compliment them often.

Introduce them to other students.

Do not expect too much too soon.

Do not let them feel dumb, uncoordinated, or that they are not progressing.

A PA SA RYU PHILOSOPHY OF TEACHING AND MINISTRY
By Kang Rhee, an Interview with Daryl Covington

(Scriptures and words in parenthesis added by the authors of the book)

Perhaps one of the greatest philosophical influences in my life (Covington) has been Master Kang Rhee. A few days prior to writing this portion of the book, I sat anxiously in his office as he wrote in Han Gul on my newly obtained 4th Dan certificate.

"Dr. Eagle," he said. "It is a long journey from here to Washington D.C. There are many rest areas for one to stop and rest, all the while being on the journey. Your life is long to go, you need to slow down. Don't drive from here to Washington without stopping at each area." Those were words I needed to hear. For months I had been on the go - getting Shinsei recognized, pastoring a church, being Vice-President of Karate for Christ International, and the list goes on. I had indeed been "driving without stopping to rest."

Matthew 11:28 Come unto me, all ye that labour and are heavy laden, and I will give you rest.

The conversation continued, "The head of this system (Shinsei) can be the King of the Jungle, or the King of the Desert. The King of the Jungle has the respect of all the animals, and lives in harmony. The king of the Desert defeats and destroys, and then rules a wasteland alone".

Principles

Through Pa Sa Ryu, Master Rhee teaches the constant improvement and refinement of technique, and the development of the total person. "A Christian Martial Artist should be the best person in the world," he stated. "He is refining the body through Martial Arts, and having his soul refined by the Holy Spirit." Pa Sa Ryu is a contemporary martial art style combining both traditional and modern applications. It teaches logical techniques. "It is my belief that by teaching Martial Arts, I can guide my students to the realization of their fullest potential," Master Rhee states.

Ephesians 6:10 Finally, my brethren, be strong in the Lord, and in the power of his might.

Kang Rhee on Training for Adults

While the ability to defend against a physical attack is important, defense against stress and its effects is of greater importance. Kang Rhee desires the adult to become a master person. Training in the martial arts is an excellent means of stress management because it challenges and expands the mind while providing a heart-strengthening workout, all in an enthusiastic and exciting environment. Kang Rhee incorporates Karate, TaeKwonDo and Kung-Fu to create Pa Sa Ryu which is appropriate for all body styles and structures. The physical workout strengthens and elongates the musculature as coordination enhances manual dexterity. Opening the body to new experiences improves attitude, increases concentration skills, and develops leadership ability.

1Corinthians 9:25 And every man that striveth for the mastery is temperate in all things. Now they do it to obtain a corruptible crown; but we an incorruptible.

9:26 I therefore so run, not as uncertainly; so fight I, not as one that beateth the air:

9:27 But I keep under my body, and bring it into subjection: lest that by any means, when I have preached to others, I myself should be a castaway.

A JuJitsu Man Views Shinsei Hapkido
by Edward Graydon

(Edward Graydon - Shinsei Hapkido)

Jujitsu Defined

Jujitsu may also be spelled in any of the following ways: ju jutsu, jiu jitsu or ju jitsu. ju or jiu means gentleness, pliability, flexibility and yielding. Jitsu or Jutsu is simply, art or technique. Jujitsu is the gentle art of self-defense. It is that and so much more.

The techniques of Jujitsu were based on theoretical ideas. The range of techniques is infinite. One of the best known maxims of jujitsu is *"Ju yok go o sei suru"*, in English means" *flexibility masters hardness"* but was extended to *"softness can over come hardness and in yielding there is strength"*.

This is the principle of Jujitsu techniques; force is not directly opposed but yielded and redirected. You can never master an art; all you can do is learn to give a better expression of the art.

Jujitsu is a parent martial art, i.e. other martial arts or Way, to be accurate, were developed from it. Those thinking of participating in a martial art for the purpose of self-defense or exercise are weighted down with the differences between Judo and Jujitsu, and Aikido and some styles of Karate.

Anyone who sits in on a class cannot fail to see the close similarities between Jujitsu, Judo, Aikido and some forms of Karate. The greatest advantage of Jujitsu and the arts with their roots in Jujitsu, is they combine both types of techniques (striking and grappling), giving equal weight to both. It should never be forgotten that the roots of

Hapkido, and especially those of Shinsei, extend deeply into the fertile background of Jujitsu.

Philosophy

Because there are so many techniques, you may ask why they are all needed. No matter how good a medicine is, it is not effective for all illnesses. It is much wiser to have many different kinds of medicine for many types of illnesses. If people are taught to react reflexively in a certain way this can very easily be turned to their disadvantage. That is why many different medicines for many different illnesses.

The ethical attitudes embodied in the various martial arts differ from one another, but especially from Shinsei. Without a Godly based philosophy, the martial art becomes nothing more then physical training.

The role of the Shinsei instructor is one which carries great responsibility. It is not enough to merely instruct in techniques, but to direct them to a whole attitude of life, a Christian life. A life that is long-suffering, full of godly peace, joy, and love.

The nature of Jujitsu techniques toward self-defense rather than toward sport, makes physical power a less crucial factor then say, Judo, TaeKwonDo or Karate and consequently more alluring to those less well developed physically. The idea of competition is effectively impossible simply because of the destructive potential. With the absence of competitiveness undoubtedly results in a lower injury rate then in Judo, TaeKwonDo or Karate.

The real competition is with oneself. High quality Jujitsu flourishes in an atmosphere of co-operation between partners.

Mechanics of techniques

When an attacker attacks, he commits his energy to the direction of his attack. He is extending his center and energy in that direction.

You must use your energy to add to his energy and redirect it to a different direction.

If attacker is standing and is not resisting, it is easy to break his balance, to the left and right or backwards and forwards. This is achieved by a light push or pull, or by lifting or pressing. If attacker is resisting, you can break his balance just the same by pushing or pulling in the direction in which he is resisting.

You face an attacker who is 70% stronger and heavier than you. By meeting him straight on obviously you will lose.

Example: an attacker attempts to push you backwards, if, instead of resisting and trying to push back, you quickly step backwards faster then he is pushing, while you simultaneously grasp him. He is now at his weakest point Because of his position he has now lost 55% of his strength. If you now step sideways, again because of his position he has now lost 80% of his strength, while you still have 100% of your strength. If you try a technique and it doesn't work don't attempt to force the technique to work, quickly change to another technique. The use of power in Jujitsu is essential, but only when such power is not used in excess so it stands the test of Ju. The principle of Ju is the notion of pliability and yielding.

When new students first start participating in Jujitsu, they must strive to avoid reliance on physical strength, for this will be a stumbling block in their progress towards the skill of the technique. When they have obtained a level of skill in performing the technique, then the use of power is correct.

The principles of Jujitsu are balance, relaxation and power. When relaxed it is easy to maintain your balance and quickly change direction. You can also bring power to a point where it is needed without wasting your energy.

Covering up is not actually part of blocking. This is a bad habit many martial artists pick up from watching boxers. It certainly has a place in these sports but outside the ring there is no referee. You cannot and must not allow yourself to take punishment if attacked and you end up unable to defend yourself, it is better to cover particularly vulnerable parts as much as possible. Remember it is a dangerous assumption that a gang will stop kicking because they will feel you have had enough when you no longer represent a threat to them. Unfortunately, some people have been killed due to this thinking.

The blocks in other styles of Jujitsu differ from those of Shinsei, as they employ hard blocks. This is not to say Shinsei does not include hard blocks. Shinsei incorporates the four standard blocks, which are inwards, outwards, rising, downwards. These four blocks cover the whole body from the knees to the top of the head.

Hard Blocks

When using the hard block, it is most important to get the power of your whole body into the technique. Though the hands, wrist, forearms may be the focus for the techniques, it is necessary to twist the upper body forcefully in order to get the snap action into the block. One should think of the block as a punch, hence snap it into the block quickly and forcefully.

Soft Blocks

Incorporated in Shinsei are soft blocks utilizing the principle of Ju: pliable and yielding rather than opposing force with force. The block is a twisting action in order to redirect the force of attack, trapping it, possibly trying a striking counter throw, or straight into an immobilization technique. The principle of Shinsei is evade, invade, control.

Striking Techniques

Shinsei has many and varied techniques, but the really effective are divided into two groups, blows using the limbs of the upper body and those using the lower limbs.
This is very effective as self-defense in that these techniques can be combined with sweeps, locks, chokes, strangles, and throws with a suddenness impossible for the attacker to foresee.

Kicking Techniques

The legs are three to four times stronger than the arms; in addition, their reach is longer. Kicks can be used to target stomach, knees, shins, ankles, and feet.

Locking Techniques

Locking techniques can be applied to any part of the body, fingers, wrists, knees, elbows, shoulders, and spine. Joint locks can also be used for various functions, as in controlling an attacker through pain and possible damage. They can take the largest attacker off balance making it easier to throw him and can be used to disarm attackers with weapons, and to restrain or immobilize them.

In situations where self-defense is needed, locking-type techniques could be used with the intention of causing disabling injury, if applied with force. This type of injury could be dislocation, linear break or muscle tear.

Many attacks begin with the attacker's hand grasping some part of your body, before the strike follows. In this type of scenario, the best counter attack is to control the attacker's wrist. This is the key to controlling your attacker, this then can be moved up to his elbow then to the shoulder, thus controlling him completely.

Strangles and Chokes

The neck is a very vulnerable part of the human body and at risk to a number of forms of attack. Strangles and chokes may seem similar but in fact are considerably different both in use and purpose.

The strangle cuts the blood supply to the brain and is rarely painful yet has the effect of rendering the attacker unconscious without him being aware of it.

The choke is very nasty, causing both loss of breath and terror. The lock is very painful as a controlling technique. The choke is extremely dangerous.

If applied forcefully, a choke can easily cause severe damage to the windpipe and throat by crushing the trachea. When participating in the practice of chokes, the danger in the stranglehold must not be taken lightly at all, brain damage and death can both result from these techniques. If applying a choke, never jerk the neck.

Throwing Techniques

The throws used in Jujitsu are varied and some of the most extensive in the martial arts. Prohibited techniques in Judo are available and implemented in Shinsei. In Judo it is prohibited to precede a throw by a blow or to throw from a joint lock, both are included in the arsenal of Jujitsu, as Jujitsu is not a sport. When participating in throwing exercises, your partner's co-operation is required. However, if a technique is improperly used against you, and is not working, do not fall or tap out for the sake of appearances. Your partner will never learn. Never try to resist and do not try to move against the technique, you will find yourself in a more vulnerable position. Also, it could result in your partner using his strength to seriously injury you.

Self-Defense

Shinsei's techniques should only be used outside the training hall in a real situation, only if these two other options have failed.

1. The ability to run off and remove yourself from the area is not possible.
2. Your options of negotiation have failed.

Students practice to acquire all-around practical self-defense proficiency and must bear in mind the value of training on a regular basis. The difference between sport fighting and self-defense realistic effectiveness and practicality are the characteristics of genuine Shinsei techniques. It is an enjoyable pursuit imparting the student with benefits beyond a simple method of self-defense. It must, by no means, be forgotten that all styles of Jujitsu were originally fighting systems used on the battlefield. Shinsei is nothing less than weaponless warfare and depending on the circumstances should be remembered as just that.

Various techniques are taught in Shinsei for dealing with armed attacks. They include such items as iron bars, baseball bats, knives, guns, batons, and broken bottles. Some very visible but some are not so visible. Razor blades could be stitched into the lapels and cuffs of a man's jacket; a man who wears large heavy rings on all his fingers carries a weapon, these are legal knuckledusters; the man who wears steel toed boots-shoes turns a kick into a lethal blow.

Out in the real world of the streets, the surroundings are much more dangerous than in the training hall, in every sense of the word. Students train using equipment such as kicking bags and focus pads to develop power, speed and accuracy and effectiveness, in hitting to feel the effect or perhaps the lack of it. It is important that as students progress up the belt system they are introduced to randori (free fighting), not sport fighting but randori. Randori is very physical and shows the student any hollows in their foundation; it also takes them to a new level of confidence which "live practice" would have instilled in them.

Strategy

There are three main areas we need to look at when speaking about maneuvering methods: nullify the intention of the attacker; evasion by the defender; countering by the defender.

The "intention" of an attacker refers to that part of the body the attacker is attempting to target. A weapon may or may not be used to attack a target. The attacker may pick a target that is part of your body or something you hold dear. Whatever the case, he will concentrate on one target at a time.

In a genuine confrontation, your attacker will seek to put you in a position, which will only enhance his attack. In most situations an attacker's area of maximum effectiveness is normally the area directly in front of him, extending about 16 to 52 inches. This range is the effectiveness of his arms and legs striking you. If you are attacked from the rear the maximum effectiveness of his attack is from body contact to 52 inches. An item worth noting - if you do not hear the attack from the rear then it doesn't matter what you do or do not know.

Evasion by the defender means removing oneself to a point where the attacker does not reach his intended target, by moving straight in towards the attacker or straight back; moving sideways to the right or left; in a circular motion; moving either upward or downward; using a hard block or soft block.

Countering by the defender is the response to the attack by attacking the attacker. You now execute techniques on the attacker based on his method of attack, and your surroundings.

The maximum effectiveness for you to counter the same as above is straight in towards the attacker or straight back, sideways to the right or left, in a circular motion or either upward or downward. Maximum effectiveness of attack extends about 16 to 52 inches. Moving into your attacker can prevent strikes and kicks from being effective well as setting him up for a large variety of techniques. Moving away from him results in him overextending and making it easier for you to execute a technique.

The Art of War

Sun Tzu

(Public Domain Edition)

The Oldest Military Treatise in the World

Translated from the Chinese,
with an Introduction and Critical Notes

by

Lionel Giles, MA

Assistant

Department of Oriental Printed Books And Manuscripts

British Museum

1910

The Art of War by Sun Tzu

A work 2400 years old

may yet contain lessons

worth consideration

by the soldiers of today.

Joseph Lumpkin

The Art of War by Sun Tzu

Apologies for War

The Art of War

I Laying Plans

II Waging War

III Attack by Stratagem

IV Tactical Dispositions

V Energy

VI Weak Points and Strong

VII Maneuvering

VIII Variations in Tactics

IX The Army on the March

X Terrain

XI The Nine Situations

XII The Attack by Fire

XIII The Use of Spies

The Art of War by Sun Tzu
Apologies for War

Accustomed as we are to think of China as the greatest peace- loving
nation on earth, we are in some danger of forgetting that her
experience of war in all its phases has also been such as no modern
State can parallel. Her long military annals stretch back to a point at
which they are lost in the mists of time. She had built the Great Wall
and was maintaining a huge standing army along her frontier
centuries before the first Roman legionary was seen on the Danube.
What with the perpetual collisions of the ancient feudal States, the
grim conflicts with Huns, Turks and other invaders after the
centralization of government, the terrific upheavals which
accompanied the overthrow of so many dynasties, besides the
countless rebellions and minor disturbances that have flamed up and
flickered out again one by one, it is hardly too much to say that the
clash of arms has never ceased to resound in one portion or another
of the Empire.

No less remarkable is the succession of illustrious captains to whom
China can point with pride. As in all countries, the greatest are fond
of emerging at the most fateful crises of her history. Thus, Po Chfi
stands out conspicuous in the period when Chun was entering upon
her final struggle with the remaining independent states.

The stormy years which followed the break-up of the Chun dynasty
are illuminated by the transcendent genius of Han Hsin. When the
House of Han in turn is tottering to its fall, the great and baleful
figure of Ts'ao Ts'ao dominates the scene. And in the establishment
of the Tang dynasty, one of the mightiest tasks achieved by man, the
superhuman energy of Li Shih-min (afterwards the Emperor T'ai
Tsung) was seconded by the brilliant strategy of Li Ching. None of
these generals need fear comparison with the greatest names in the
military history of Europe. In spite of all this, the great body of
Chinese sentiment, from Lao Tzu downwards, and especially as
reflected in the standard literature of Confucianism, has been

consistently pacific and intensely opposed to militarism in any form. It is such an uncommon thing to find any of the literati defending warfare on principle, that I have thought it worthwhile to collect and translate a few passages in which the unorthodox view is upheld. The following, by Ssu-ma Chfien, shows that for all his ardent admiration of Confucius, he was yet no advocate of peace at any price: Military weapons are the means used by the Sage to punish violence and cruelty, to give peace to troublous times, to remove difficulties and dangers, and to succor those who are in peril.

Every animal with blood in its veins and horns on its head will fight when it is attacked. How much more so will man, who carries in his breast the faculties of love and hatred, joy and anger! When he is pleased, a feeling of affection springs up within him; when angry, his poisoned sting is brought into play. That is the natural law which governs his being. What then shall be said of those scholars of our time, blind to all great issues, and without any appreciation of relative values, who can only bark out their stale formulas about "virtue" and "civilization," condemning the use of military weapons? They will surely bring our country to impotence and dishonor and the loss of her rightful heritage; or, at the very least, they will bring about invasion and rebellion, sacrifice of territory and general enfeeblement. Yet they obstinately refuse to modify the position they have taken up.

The truth is that, just as in the family the teacher must not spare the rod, and punishments cannot be dispensed with in the State, so military chastisement can never be allowed to fall into abeyance in the Empire. All one can say is that this power will be exercised wisely by some, foolishly by others, and that among those who bear arms some will be loyal and others rebellious.

The next piece is taken from Tu Mu's preface to his commentary on Sun Tzu:

War may be defined as punishment, which is one of the functions of government. It was the profession of Chung Yu and Jan Chfiu, both disciples of Confucius. Nowadays, the holding of trials and hearing of litigation, the imprisonment of offenders and their execution by flogging in the market- place, are all done by officials. But the wielding of huge armies, the throwing down of fortified cities, the hauling of women and children into captivity, and the beheading of traitors — this is also work which is done by officials.

The objects of the rack and of military weapons are essentially the same. There is no intrinsic difference between the punishment of flogging and cutting off heads in war. For the lesser infractions of law, which are easily dealt with, only a small amount of force need be employed: hence the use of military weapons and wholesale decapitation. In both cases, however, the end in view is to get rid of wicked people, and to give comfort and relief to the good Chi-sun asked Jan Yu, saying: "Have you, Sir, acquired your military aptitude by study, or is it innate?" Jan Yu replied: "It has been acquired by study." "How can that be so," said Chi-sun, "seeing that you are a disciple of Confucius?" "It is a fact," replied Jan Yu; "I was taught by Confucius. It is fitting that the great Sage should exercise both civil and military functions, though to be sure my instruction in the art of fighting has not yet gone very far."

Now, who the author was of this rigid distinction between the "civil" and the "military," and the limitation of each to a separate sphere of action, or in what year of which dynasty it was first introduced, is more than I can say. But, at any rate, it has come about that the members of the governing class are quite afraid of enlarging on military topics, or do so only in a shamefaced manner. If any are bold enough to discuss the subject, they are at once set down as eccentric individuals of coarse and brutal propensities. This is an extraordinary instance in which, through sheer lack of reasoning, men unhappily lose sight of fundamental principles.

When the Duke of Chou was minister under Chfeng Wang, he regulated ceremonies and made music, and venerated the arts of scholarship and learning; yet when the barbarians of the River Huai revolted, he sallied forth and chastised them. When Confucius held office under the Duke of Lu, and a meeting was convened at Chia-ku, he said: "If pacific negotiations are in progress, warlike preparations should have been made beforehand." He rebuked and shamed the Marquis of Chfi, who cowered under him and dared not proceed to violence. How can it be said that these two great Sages had no knowledge of military matters? We have seen that the great Chu Hsi held Sun Tzu in high esteem. He also appeals to the authority of the Classics: Our Master Confucius, answering Duke Ling of Wei, said: "I have never studied matters connected with armies and battalions." Replying to K'ung Wen-tzu, he said: I have not been instructed about buff-coats and weapons."

But if we turn to the meeting at Chia-ku, we find that he used armed force against the men of Lai, so that the marquis of Chu was overawed. Again, when the inhabitants of Pi revolted, the ordered his officers to attack them, whereupon they were defeated and fled in confusion. He once uttered the words: "If I fight, I conquer." And Jan Yu also said: "The Sage exercises both civil and military functions." Can it be a fact that Confucius never studied or received instruction in the art of war? We can only say that he did not specially choose matters connected with armies and fighting to be the subject of his teaching.

Sun Hsing-yen, the editor of Sun Tzu, writes in similar strain:

Confucius said: "I am unversed in military matters."
He also said: "If I fight, I conquer." Confucius ordered ceremonies and regulated music.

Now war constitutes one of the five classes of State ceremonial, and must not be treated as an independent branch of study. Hence, the words "I am unversed in" must be taken to mean that there are things which even an inspired Teacher does not know. Those who have to lead an army and devise stratagems, must learn the art of war. But if one can command the services of a good general like Sun Tzu, who was employed by Wu Tzu-hsu, there is no need to learn it oneself. Hence the remark added by Confucius: "If I fight, I conquer." The men of the present day, however, willfully interpret these words of Confucius in their narrowest sense, as though he meant that books on the art of war were not worth reading. With blind persistency, they adduce the example of Chao Kua, who pored over his father's books to no purpose, as a proof that all military theory is useless. Again, seeing that books on war have to do with such things as opportunism in designing plans, and the conversion of spies, they hold that the art is immoral and unworthy of a sage. These people ignore the fact that the studies of our scholars and the civil administration of our officials also require steady application and practice before efficiency is reached. The ancients were particularly chary of allowing mere novices to botch their work.

Weapons are baneful and fighting perilous; and useless unless a general is in constant practice, he ought not to hazard other men's lives in battle. Hence it is essential that Sun Tzu's 13 chapters should be studied. Hsiang Liang used to instruct his nephew Chi in the art of war. Chi got a rough idea of the art in its general bearings, but would not pursue his studies to their proper outcome, the consequence being that he was finally defeated and overthrown. He did not realize that the tricks and artifices of war are beyond verbal computation. Duke Hsiang of Sung and King Yen of Hsu were brought to destruction by their misplaced humanity. The treacherous and underhand nature of war necessitates the use of guile and stratagem suited to the occasion. There is a case on record of Confucius himself having violated an extorted oath, and also of his having left the Sung State in disguise. Can we then recklessly arraign Sun Tzu for disregarding truth and honesty?

Laying Plans

[Ts'ao Kung, in defining the meaning of the Chinese for the title of this chapter, says it refers to the deliberations in the temple selected by the general for his temporary use, or as we should say, in his tent. See. ss. 26.]

1. Sun Tzu said: The art of war is of vital importance to the State.

2. It is a matter of life and death, a road either to safety or to ruin. Hence it is a subject of inquiry which can on no account be neglected.

3. The art of war, then, is governed by five constant factors, to be taken into account in one's deliberations, when seeking to determine the conditions obtaining in the field.

4. These are:

(1) The Moral Law;
(2) Heaven;
(3) Earth;
(4) The Commander;
(5) Method and discipline.

[It appears from what follows that Sun Tzu means by "Moral Law" a principle of harmony, not unlike the Tao (Tao Te Ching) of Lao Tzu in its moral aspect. One might be tempted to render it by "morale," were it not considered as an attribute of the ruler in ss. 13.]

5. The MORAL LAW causes the people to be in complete accord with their ruler,
6. so that they will follow him regardless of their lives, undismayed by any danger.

[Tu Yu quotes Wang Tzu as saying: "Without constant practice, the officers will be nervous and undecided when mustering for battle; without constant practice, the general will be wavering and irresolute when the crisis is at hand."]

7. HEAVEN signifies night and day, cold and heat, times and seasons.

[The commentators, I think, make an unnecessary mystery of two words here. Meng Shih refers to "the hard and the soft, waxing and waning" of Heaven. Wang Hsi, however, may be right in saying that what is meant is "the general economy of Heaven," including the five elements, the four seasons, wind and clouds, and other phenomena.]

8. EARTH comprises distances, great and small; danger and security; open ground and narrow passes; the chances of life and death.

9. The COMMANDER stands for the virtues of wisdom, sincerely, benevolence, courage and strictness.

[The five cardinal virtues of the Chinese are
(1) humanity or benevolence;
(2) uprightness of mind;
(3) self- respect, self-control, or "proper feeling;"
(4) wisdom;
(5) sincerity or good faith. Here "wisdom" and "sincerity" are put before "humanity or benevolence," and the two military virtues of "courage" and "strictness" substituted for "uprightness of mind" and "self-respect, self-control, or 'proper feeling.'"]

10. By METHOD AND DISCIPLINE are to be understood the marshaling of the army in its proper subdivisions, the graduations of rank among the officers, the maintenance of roads by which supplies may reach the army, and the control of military expenditure.

11. These five heads should be familiar to every general: he who knows them will be victorious; he who knows them not will fail.

12. Therefore, in your deliberations, when seeking to determine the military conditions, let them be made the basis of a comparison, in this wise:

13.
(1) Which of the two sovereigns is imbued with the Moral law?

[that is to say, "is in harmony with his subjects."]

(2) Which of the two generals has most ability?

(3) With whom lie the advantages derived from Heaven and Earth?

[See ss. 7,8]

(4) On which side is discipline most rigorously enforced?

[Tu Mu alludes to the remarkable story of Ts'ao Ts'ao
(A.D. 155-220), who was such a strict disciplinarian that once, in accordance with his own severe regulations against injury to standing crops, he condemned himself to death for having allowed

him horse to shy into a field of corn! However, in lieu of losing his head, he was persuaded to satisfy his sense of justice by cutting off his hair. Ts'ao Ts'ao's own comment on the present passage is characteristically curt: "when you lay down a law, see that it is not disobeyed; if it is disobeyed the offender must be put to death."]

(5) Which army is stronger?

[Morally as well as physically. As Mei Yao-chf en puts it, freely rendered, "ESPIRIT DE CORPS and 'big battalions.'"]

(6) On which side are officers and men more highly trained?

[Tu Yu quotes Wang Tzu as saying: "Without constant practice, the officers will be nervous and undecided when mustering for battle; without constant practice, the general will be wavering and irresolute when the crisis is at hand."]

(7) In which army is there the greater constancy both in reward and punishment?

[On which side is there the most absolute certainty that merit will be properly rewarded and misdeeds summarily punished?]

14. By means of these seven considerations I can forecast victory or defeat.

15. The general that hearkens to my counsel and acts upon it, will conquer: − let such a one be retained in command! The general that hearkens not to my counsel nor acts upon it, will suffer defeat: − let such a one be dismissed!

[The form of this paragraph reminds us that SunTzu's treatise was composed expressly for the benefit of his patron H o Lu, king of the W u State.]

16. While heading the profit of my counsel, avail yourself also of any helpful circumstances over and beyond the ordinary rules.

17. According as circumstances are favorable, one should modify one's plans.

[Sun Tzu, as a practical soldier, will have none of the "bookish theoric." He cautions us here not to pin our faith to abstract principles; "for," as Chang Yu puts it, "while the main laws of strategy can be stated clearly enough for the benefit of all and sundry, you must be guided by the actions of the enemy in attempting to secure a favorable position in actual warfare." On the eve of the battle of Waterloo, Lord Uxbridge, commanding the cavalry, went to the Duke of Wellington in order to learn what his plans and calculations were for the morrow, because, as he explained, he might suddenly find himself Commander-in-chief and would be unable to frame new plans in a critical moment. The Duke listened quietly and then said: "Who will attack the first tomorrow — I or Bonaparte?" "Bonaparte," replied Lord Uxbridge. "Well," continued the Duke, "Bonaparte has not given me any idea of his projects; and as my plans will depend upon his, how can you expect me to tell you what mine are?"

18. All warfare is based on deception.

[The truth of this pithy and profound saying will be admitted by every soldier. Col. Henderson tells us that Wellington, great in so many military qualities, was especially distinguished by "the extraordinary skill with which he concealed his movements and deceived both friend and foe."]

19. Hence, when able to attack, we must seem unable; when using our forces, we must seem inactive; when we are near, we must make the enemy believe we are far away; when far away, we must make him believe we are near.

20. Hold out baits to entice the enemy. Feign disorder, and crush him.

[All commentators, except Chang Yu, say, "When he is in disorder, crush him." It is more natural to suppose that Sun Tzu is still illustrating the uses of deception in war.]

21. If he is secure at all points, be prepared for him. If he is in superior strength, evade him.

22. If your opponent is of choleric temper, seek to irritate him. Pretend to be weak, that he may grow arrogant.

[Wang Tzu, quoted by Tu Yu, says that the good tactician plays with his adversary as a cat plays with a mouse, first feigning weakness and immobility, and then suddenly pouncing upon him.]

23. If he is taking his ease, give him no rest.

[This is probably the meaning though Mei Yao-chf en has the note: "while we are taking our ease, wait for the enemy to tire himself out." The YU LAN has "Lure him on and tire him out."]

If his forces are united, separate them.

[Less plausible is the interpretation favored by most of the commentators: "If sovereign and subject are in accord, put division between them."]

24. Attack him where he is unprepared, appear where you are not expected.

25. These military devices, leading to victory, must not be divulged beforehand.

26. Now the general who wins a battle makes many calculations in his temple ere the battle is fought.

[Chang Yu tells us that in ancient times it was customary for a temple to be set apart for the use of a general who was about to take the field, in order that he might there elaborate his plan of campaign. "Words on Wellington," by Sir. W. Fraser.]

The general who loses a battle makes but few calculations beforehand. Thus do many calculations lead to victory, and few calculations to defeat: how much more no calculation at all! It is by attention to this point that I can foresee who is likely to win or lose.

II Waging War

[Ts'ao Kung has the note: "He who wishes to fight must first count the cost," which prepares us for the discovery that the subject of the chapter is not what we might expect from the title, but is primarily a consideration of ways and means.]

1. Sun Tzu said: In the operations of war, where there are in the field a thousand swift chariots, as many heavy chariots, and a hundred thousand mail-clad soldiers,

[The "swift chariots" were lightly built and, according to Chang Yu, used for the attack; the "heavy chariots" were heavier, and designed for purposes of defense. Li Chfuan, it is true, says that the latter were light, but this seems hardly probable. It is interesting to note the analogies between early Chinese warfare and that of the Homeric Greeks. In each case, the war-chariot was the important factor, forming as it did the nucleus round which was grouped a certain number of foot-soldiers. With regard to the numbers given here, we are informed that each swift chariot was accompanied by 75 footmen, and each heavy chariot by 25 footmen, so that the whole army would be divided up into a thousand battalions, each consisting of two chariots and a hundred men.] with provisions enough to carry them a thousand LI (LI is about one mile, although length varied with time), the expenditure at home and at the front, including entertainment of guests, small items such as glue and paint, and sums spent on chariots and armor, will reach the total of a thousand ounces of silver per day. Such is the cost of raising an army of 100,000 men.

2. When you engage in actual fighting, if victory is long in coming, then men's weapons will grow dull and their ardor will be damped. If you lay siege to a town, you will exhaust your strength.

3. Again, if the campaign is protracted, the resources of the State will not be equal to the strain.

4. Now, when your weapons are dulled, your ardor damped, your strength exhausted and your treasure spent, other chieftains will spring up to take advantage of your extremity. Then no man, however wise, will be able to avert the consequences that must ensue.

5. Thus, though we have heard of stupid haste in war, cleverness has never been seen associated with long delays.

[This concise and difficult sentence is not well explained by any of the commentators. Ts'ao Kung, Li Chfuan, Meng Shih, Tu Yu, Tu Mu and Mei Yao-chfen have notes to the effect that a general, though naturally stupid, may nevertheless conquer through sheer force of rapidity.

Ho Shih says: "Haste may be stupid, but at any rate it saves expenditure of energy and treasure; protracted operations may be very clever, but they bring calamity in their train." Wang Hsi evades the difficulty by remarking: "Lengthy operations mean an army growing old, wealth being expended, an empty exchequer and distress among the people; true cleverness insures against the occurrence of such calamities."

Chang Yu says: "So long as victory can be attained, stupid haste is preferable to clever dilatoriness."

Now Sun Tzu says nothing whatever, except possibly by implication, about ill-considered haste being better than ingenious but lengthy operations. What he does say is something much more guarded, namely that, while speed may sometimes be injudicious, tardiness can never be anything but foolish — if only because it means impoverishment to the nation. In considering the point raised here by Sun Tzu, the classic example of Fabius Cunctator will inevitably occur to the mind. That general deliberately measured the endurance of Rome against that of Hannibals's isolated army, because it seemed to him that the latter was more likely to suffer from a long campaign in a strange country. But it is quite a moot question whether his tactics would have proved successful in the long run. Their reversal it is true, led to Cannae; but this only establishes a negative presumption in their favor.]

6. There is no instance of a country having benefited from prolonged warfare.

7. It is only one who is thoroughly acquainted with the evils of war that can thoroughly understand the profitable way of carrying it on.

[That is, with rapidity. Only one who knows the disastrous effects of a long war can realize the supreme importance of rapidity in bringing it to a close. Only two commentators seem to favor this interpretation, but it fits well into the logic of the context, whereas the rendering, "He who does not know the evils of war cannot appreciate its benefits," is distinctly pointless.]

8. The skillful soldier does not raise a second levy, neither are his supply-wagons loaded more than twice.

[Once war is declared, he will not waste precious time in waiting for reinforcements, nor will he return his army back for fresh supplies, but crosses the enemy's frontier without delay. This may seem an audacious policy to recommend, but with all great strategists, from Julius Caesar to Napoleon Bonaparte, the value of time — that is, being a little ahead of your opponent has counted for more than either numerical superiority or the nicest calculations with regard to commissariat.]

9. Bring war material with you from home, but forage on the enemy. Thus the army will have food enough for its needs.

[The Chinese word translated here as "war material" literally means "things to be used", and is meant in the widest sense. It includes all the impedimenta of an army, apart from provisions.]

10. Poverty of the State exchequer causes an army to be maintained by contributions from a distance. Contributing to maintain an army

at a distance causes the people to be impoverished.

[The beginning of this sentence does not balance properly with the next, though obviously intended to do so. The arrangement, moreover, is so awkward that I cannot help suspecting some corruption in the text.

It never seems to occur to Chinese commentators that an emendation may be necessary for the sense, and we get no help from them there. The Chinese words Sun Tzu used to indicate the cause of the people's impoverishment clearly have reference to some system by which the husbandmen sent their contributions of corn to the army direct. But why should it fall on them to maintain an army in this way, except because the State or Government is too poor to do so?]

11. On the other hand, the proximity of an army causes prices to go up; and high prices cause the people's substance to be drained away.

[Wang Hsi says high prices occur before the army has left its own territory. Ts'ao Kung understands it of an army that has already crossed the frontier.]

12. When their substance is drained away, the peasantry will be afflicted by heavy exactions.

13. With this loss of substance and exhaustion of strength, the homes of the people will be stripped bare, and three-tenths of their income will be dissipated.

14. While government expenses for broken chariots, worn-out horses, breast-plates and helmets, bows and arrows, spears and shields, protective mantles, draught-oxen and heavy wagons, will amount to four- tenths of its total revenue.

[Tu Mu and Wang Hsi agree that the people are not mulcted (taxed) not of 3/10, but of 7/10, of their income. But this is hardly to be extracted from our text. Ho Shih has a characteristic tag: "The PEOPLE being regarded as the essential part of the State, and FOOD as the people's heaven, is it not right that those in authority should value and be careful of both?"]

15. Hence a wise general makes a point of foraging on the enemy. One cartload of the enemy's provisions is equivalent to twenty of one's own, and likewise a single PICUL of his provender is equivalent to twenty from one's own store.

[Because twenty cartloads will be consumed in the process of transporting one cartload to the front. A PICUL is a unit of measure equal to 133.3 pounds (65.5 kilograms).]

16. Now in order to kill the enemy, our men must be roused to anger; that there may be advantage from defeating the enemy, they must have their rewards.

[Tu Mu says: "Rewards are necessary in order to make the soldiers see the advantage of beating the enemy; thus, when you capture spoils from the enemy, they must be used as rewards, so that all your men may have a keen desire to fight, each on his own account."]

17. Therefore in chariot fighting, when ten or more chariots have been taken, those should be rewarded who took the first. Our own flags should be substituted for those of the enemy, and the chariots mingled and used in conjunction with ours. The captured soldiers should be kindly treated and kept.

18. This is called, using the conquered foe to augment one's own strength.

19. In war, then, let your great object be victory, not lengthy campaigns.

[As Ho Shih remarks: "War is not a thing to be trifled with." Sun Tzu here reiterates the main lesson which this chapter is intended to enforce."]

20. Thus it may be known that the leader of armies is the arbiter of the people's fate, the man on whom it depends whether the nation shall be in peace or in peril.

Joseph Lumpkin

III Attack by Stratagem

Sun Tzu said: In the practical art of war, the best thing of all is to take the enemy's country whole and intact; to shatter and destroy it is not so good. So, too, it is better to recapture an army entire than to destroy it, to capture a regiment, a detachment or a company entire than to destroy them.

[The equivalent to an army corps, according to Ssu-ma Fa, consisted nominally of 12500 men; according to Ts'ao Kung, the equivalent of a regiment contained 500 men, the equivalent to a detachment consists from any number between 100 and 500, and the equivalent of a company contains from 5 to 100 men. For the last two, however, Chang Yu gives the exact figures of 100 and 5 respectively.]

Hence to fight and conquer in all your battles is not supreme excellence; supreme excellence consists in breaking the enemy's resistance without fighting.

[Here again, no modern strategist but will approve the words of the old Chinese general. Moltke's greatest triumph, the capitulation of the huge French army at Sedan, was won practically without bloodshed.]

Thus the highest form of generalship is to balk (nullify) the enemy's plans; the next best is to prevent the junction of the enemy's forces;

[Perhaps the word "balk" falls short of expressing the full force of the Chinese word, which implies not an attitude of defense, whereby one might be content to foil the enemy's stratagems one after another, but an active policy of counter-attack. Ho Shih puts this very clearly in his note: "When the enemy has made a plan of attack against us, we must anticipate him by delivering our own attack first." See the Book of Five Rings]

[Isolating him from his allies. We must not forget that Sun Tzu, in speaking of hostilities, always has in mind the numerous states or principalities into which the China of his day was split up.]

the next in order is to attack the enemy's army in the field; [When he is already at full strength.] and the worst policy of all is to besiege walled cities.

1. The rule is, not to besiege walled cities if it can possibly be avoided.

[Another sound piece of military theory. Had the Boers acted upon it in 1899, and refrained from dissipating their strength before Kimberley, Mafeking, or even Ladysmith, it is more than probable that they would have been masters of the situation before the British were ready seriously to oppose them.]

The preparation of manlets (rolling body shields), tents, movable shelters, weapons, and various implements of war, will take up three whole months;

[It is not quite clear what the Chinese word, here translated as "mantlets", described. Ts'ao Kung simply defines them as "large shields," but we get a better idea of them from Li Chfuan, who says they were to protect the heads of those who were assaulting the city walls at close quarters. This seems to suggest a sort of Roman TESTUDO, ready made. Tu Mu says they were wheeled vehicles used in repelling attacks, but this is denied by Chfen Hao.
The name is also applied to turrets on city walls. Of the "movable shelters" we get a fairly clear description from several commentators. They were wooden missile-proof structures on four wheels, propelled from within, covered over with raw hides, and used in sieges to convey parties of men to and from the walls, for the purpose of filling up the encircling moat with earth. Tu Mu adds that they are now called "wooden donkeys."] and the piling up of mounds over against the walls will take three months more.

[These were great mounds or ramparts of earth heaped up to the level of the enemy's walls in order to discover the weak points in the defense, and also to destroy the fortified turrets mentioned in the preceding note.]

5. The general, unable to control his irritation, will launch his men to the assault like swarming ants, with the result that one-third of his men are slain, while the town still remains untaken. Such are the disastrous effects of a siege.

[This vivid simile of Ts'ao Kung is taken from the spectacle of an army of ants climbing a wall. The meaning is that the general, losing patience at the long delay, may make a premature attempt to storm the place before his engines of war are ready.]

[We are reminded of the terrible losses of the Japanese before Port Arthur, in the most recent siege which history has to record.]

6. Therefore the skillful leader subdues the enemy's troops without any fighting; he captures their cities without laying siege to them; he overthrows their kingdom without lengthy operations in the field.

[Chia Lin notes that he only overthrows the Government, but does no harm to individuals. The classical instance is Wu Wang, who after having put an end to the Yin dynasty was acclaimed "Father and mother of the people."]

7. With his forces intact he will dispute the mastery of the Empire, and thus, without losing a man, his triumph will be complete.

[Owing to the double meanings in the Chinese text, the latter part of the sentence is susceptible of quite a different meaning: "And thus, the weapon not being blunted by use, its keenness remains perfect."]

This is the method of attacking by stratagem.

8. It is the rule in war, if our forces are ten to the enemy's one, to surround him; if five to one, to attack him; [Straightway, without waiting for any further advantage.] if twice as numerous, to divide our army into two.

[Tu Mu takes exception to the saying; and at first sight, indeed, it appears to violate a fundamental principle of war.
Ts'ao Kung, however, gives a clue to Sun Tzu's meaning: "Being two to the enemy's one, we may use one part of our army in the regular way, and the other for some special diversion." Chang Yu thus further elucidates the point: "If our force is twice as numerous as that of the enemy, it should be split up into two divisions, one to meet the enemy in front, and one to fall upon his rear; if he replies to the frontal attack, he may be crushed from behind; if to the rearward attack, he may be crushed in front." This is what is meant by saying that 'one part may be used in the regular way, and the other for some special diversion.' Tu Mu does not understand that dividing one's army is simply an irregular, just as concentrating it is the regular, strategical method, and he is too hasty in calling this a mistake."]

9. If equally matched, we can offer battle; [Li Chf uan, followed by Ho Shih, gives the following paraphrase: "If attackers and attacked are equally matched in strength, only the able general will fight."] if slightly inferior in numbers, we can avoid the enemy;

[The meaning, "we can WATCH the enemy," is certainly a great improvement on the above; but unfortunately there appears to be no very good authority for the variant. Chang Yu reminds us that the saying only applies if the other factors are equal; a small difference in numbers is often more than counterbalanced by superior energy and discipline.] if quite unequal in every way, we can flee from him.

10. Hence, though an obstinate fight may be made by a small force, in the end it must be captured by the larger force.

11. Now the general is the bulwark of the State; if the bulwark is complete at all points; the State will be strong; if the bulwark is defective, the State will be weak.

[As Li Chfuan tersely puts it: "Gap indicates deficiency; if the general's ability is not perfect (i.e. if he is not thoroughly versed in his profession), his army will lack strength."]

12. There are three ways in which a ruler can bring misfortune upon his army:

13. (1) By commanding the army to advance or to retreat, being ignorant of the fact that it cannot obey. This is called hobbling the army.

[Li Chf uan adds the comment: "It is like tying together the legs of a thoroughbred, so that it is unable to gallop." One would naturally think of "the ruler" in this passage as being at home, and trying to direct the movements of his army from a distance. But the commentators understand just the reverse, and quote the saying of "Tai Kung: "A kingdom should not be governed from without, and army should not be directed from within." Of course it is true that, during an engagement, or when in close touch with the enemy, the general should not be in the thick of his own troops, but a little distance apart. Otherwise, he will be liable to misjudge the position as a whole, and give wrong orders.]

14. (2) By attempting to govern an army in the same way as he administers a kingdom, being ignorant of the conditions which obtain in an army. This causes restlessness in the soldier's minds.

[Ts'ao Kung's note is, freely translated: "The military sphere and the civil sphere are wholly distinct; you can't handle an army in kid gloves." And Chang Yu says: "Humanity and justice are the principles on which to govern a state, but not an army; opportunism and flexibility, on the other hand, are military rather than civil virtues to assimilate the governing of an army" — to that of a State, understood.]

15. (3) By employing the officers of his army without discrimination, [That is, he is not careful to use the right man in the right place.] through ignorance of the military principle of adaptation to circumstances. This shakes the confidence of the soldiers.

[I follow Mei Yao-chf en here. The other commentators refer not to the ruler, as in SS. 13, 14, but to the officers he employs. Thus Tu Yu says: "If a general is ignorant of the principle of adaptability, he must not be entrusted with a position of authority." Tu Mu quotes: "The skillful employer of men will employ the wise man, the brave man, the covetous man, and the stupid man. For the wise man delights in establishing his merit, the brave man likes to show his courage in action, the covetous man is quick at seizing advantages, and the stupid man has no fear of death."]

16. But when the army is restless and distrustful, trouble is sure to come from the other feudal princes. This is simply bringing anarchy into the army, and flinging victory away.

17. Thus we may know that there are five essentials for victory: (1) He will win who knows when to fight and when not to fight.

[Chang Yu says: If he can fight, he advances and takes the offensive; if he cannot fight, he retreats and remains on the defensive. He will invariably conquer who knows whether it is right to take the offensive or the defensive.]

(2) He will win who knows how to handle both superior and inferior forces.

[This is not merely the general's ability to estimate numbers correctly, as Li Chfuan and others make out. Chang Yu expounds the saying more satisfactorily: "By applying the art of war, it is possible with a lesser force to defeat a greater, and vice versa. The secret lies in an eye for locality, and in not letting the right moment slip. Thus Wu Tzu says: 'With a superior force, make for easy ground; with an inferior one, make for difficult ground.'"]
(Superior – large) (Inferior – small)

(3) He will win whose army is animated by the same spirit throughout all its ranks.

(4) He will win who, prepared himself, waits to take the enemy unprepared.

(5) He will win who has military capacity and is not interfered with by the sovereign.

[Tu Yu quotes Wang Tzu as saying: "It is the sovereign's function to

give broad instructions, but to decide on battle it is the function of
the general." It is needless to dilate on the military disasters which
have been caused by undue interference with operations in the field
on the part of the home government. Napoleon undoubtedly owed
much of his extraordinary success to the fact that he was not
hampered by central authority.]

18. Hence the saying: If you know the enemy and know yourself, you
need not fear the result of a hundred battles. If you know yourself
but not the enemy, for every victory gained you will also suffer a
defeat.

[Li Chf uan cites the case of Fu Chien, prince of Chun, who in 383
A.D. marched with a vast army against the Chin Emperor. When
warned not to despise an enemy who could command the services of
such men as Hsieh An and Huan Chf ung, he boastfully replied: "I
have the population of eight provinces at my back, infantry and
horsemen to the number of one million; why, they could dam up the
Yangtsze River itself by merely throwing their whips into the stream.
What danger have I to fear?" Nevertheless, his forces were soon after
disastrously routed at the Fei River, and he was obliged to beat a
hasty retreat.]

If you know neither the enemy nor yourself, you will succumb in
every battle.

[Chang Yu said: "Knowing the enemy enables you to take the
offensive, knowing yourself enables you to stand on the defensive."
He adds: "Attack is the secret of defense; defense is the planning of
an attack." It would be hard to find a better epitome of the root-
principle of war.]

IV Tactical Dispositions

[Ts'ao Kung explains the Chinese meaning of the words for the title
of this chapter: "marching and countermarching on the part of the

two armies with a view to discovering each other's condition." Tu Mu says: "It is through the dispositions of an army that its condition may be discovered. Conceal your dispositions, and your condition will remain secret, which leads to victory,; show your dispositions, and your condition will become patent, which leads to defeat." Wang Hsi remarks that the good general can "secure success by modifying his tactics to meet those of the enemy."]

Sun Tzu said: The good fighters of old first put themselves beyond the possibility of defeat, and then waited for an opportunity of defeating the enemy.

To secure ourselves against defeat lies in our own hands, but the opportunity of defeating the enemy is provided by the enemy himself.

[That is, of course, by a mistake on the enemy's part.]

Thus the good fighter is able to secure himself against defeat, [Chang Yu says this is done, "By concealing the disposition of his troops, covering up his tracks, and taking unremitting precautions."] but cannot make certain of defeating the enemy.

Hence the saying: One may KNOW how to conquer without being able to DO it.

Security against defeat implies defensive tactics; ability to defeat the enemy means taking the offensive.

[I retain the sense found in a similar passage in ss. 1-3, in spite of the fact that the commentators are all against me. The meaning they give, "He who cannot conquer takes the defensive," is plausible enough.]

6. Standing on the defensive indicates insufficient strength; attacking, a superabundance of strength.

7. The general who is skilled in defense hides in the most secret recesses of the earth; [Literally, "hides under the ninth earth," which

is a metaphor indicating the utmost secrecy and concealment, so that the enemy may not know his whereabouts."] he who is skilled in attack flashes forth from the topmost heights of heaven.

[Another metaphor, implying that he falls on his adversary like a thunderbolt, against which there is no time to prepare. This is the opinion of most of the commentators.]

Thus on the one hand we have ability to protect ourselves; on the other, a victory that is complete.

8. To see victory only when it is within the ken of the common herd is not the acme of excellence.

[As Ts'ao Kung remarks, "the thing is to see the plant before it has germinated," to foresee the event before the action has begun. Li Chfuan alludes to the story of Han Hsin who, when about to attack the vastly superior army of Chao, which was strongly entrenched in the city of Chfeng-an, said to his officers: "Gentlemen, we are going to annihilate the enemy, and shall meet again at dinner." The officers hardly took his words seriously, and gave a very dubious assent. But Han Hsin had already worked out in his mind the details of a clever stratagem, whereby, as he foresaw, he was able to capture the city and inflict a crushing defeat on his adversary."]

9. Neither is it the acme of excellence if you fight and conquer and the whole Empire says, "Well done!"

[True excellence being, as Tu Mu says: "To plan secretly, to move surreptitiously, to foil the enemy's intentions and balk his schemes, so that at last the day may be won without shedding a drop of blood." Sun Tzu reserves his approbation for things that "the world's coarse thumb And finger fail to plumb."]

10. To lift an autumn hair is no sign of great strength; ["Autumn" hair" is explained as the fur of a hare, which is finest in autumn, when it begins to grow afresh. The phrase is a very common one in Chinese writers.] to see the sun and moon is no sign of sharp sight; to hear the noise of thunder is no sign of a quick ear.

[Ho Shih gives as real instances of strength, sharp sight and quick hearing: Wu Huo, who could lift a tripod weighing 250 stone; Li Chu, who at a distance of a hundred paces could see objects no bigger than a mustard seed; and Shih K'uang, a blind musician who could hear the footsteps of a mosquito.]

11. What the ancients called a clever fighter is one who not only wins, but excels in winning with ease.

[The last half is literally "one who, conquering, excels in easy conquering." Mei Yao-chf en says: "He who only sees the obvious, wins his battles with difficulty; he who looks below the surface of things, wins with ease."]

12. Hence his victories bring him neither reputation for wisdom nor credit for courage.

[Tu Mu explains this very well: "Inasmuch as his victories are gained over circumstances that have not come to light, the world as large knows nothing of them, and he wins no reputation for wisdom; inasmuch as the hostile state submits before there has been any bloodshed, he receives no credit for courage."]

13. He wins his battles by making no mistakes.

[Chf en Hao says: "He plans no superfluous marches, he devises no futile attacks." The connection of ideas is thus explained by Chang Yu: "One who seeks to conquer by sheer strength, clever though he may be at winning pitched battles, is also liable on occasion to be vanquished; whereas he who can look into the future and discern conditions that are not yet manifest, will never make a blunder and therefore invariably win."]

Making no mistakes is what establishes the certainty of victory, for it means conquering an enemy that is already defeated.

14. Hence the skillful fighter puts himself into a position which makes defeat impossible, and does not miss the moment for defeating the enemy.

[A "counsel of perfection" as Tu Mu truly observes. "Position" need not be confined to the actual ground occupied by the troops. It includes all the arrangements and preparations which a wise general will make to increase the safety of his army.]

15. Thus it is that in war the victorious strategist only seeks battle after the victory has been won, whereas he who is destined to defeat first fights and afterwards looks for victory.

[Ho Shih thus expounds the paradox: "In warfare, first lay plans which will ensure victory, and then lead your army to battle; if you will not begin with stratagem but rely on brute strength alone, victory will no longer be assured."]

16. The consummate leader cultivates the moral law, and strictly adheres to method and discipline; thus it is in his power to control success.

17. In respect of military method, we have, firstly, Measurement; secondly, Estimation of quantity; thirdly, Calculation; fourthly, Balancing of chances; fifthly, Victory.

18. Measurement owes its existence to Earth; Estimation of quantity to Measurement; Calculation to Estimation of quantity; Balancing of chances to Calculation; and Victory to Balancing of chances.

[It is not easy to distinguish the four terms very clearly in the Chinese. The first seems to be surveying and measurement of the ground, which enable us to form an estimate of the enemy's strength, and to make calculations based on the data thus obtained; we are thus led to a general weighing-up, or comparison of the enemy's chances with our own; if the latter turn the scale, then victory ensues. The chief difficulty lies in third term, which in the Chinese some commentators take as a calculation of NUMBERS, thereby making it nearly synonymous with the second term. Perhaps the second term should be thought of as a consideration of the enemy's general position or condition, while the third term is the estimate of his numerical strength. On the other hand, Tu Mu says: "The question of relative strength having been settled, we can bring the varied resources of cunning into play." Ho Shih seconds this interpretation, but weakens it. However, it points to the third term as being a calculation of numbers.]

19. A victorious army opposed to a routed one, is as a pound's weight placed in the scale against a single grain.

[Literally, "a victorious army is like an I (20 oz.) weighed against a SHU (1/24 oz.); a routed army is a SHU weighed against an I." The point is simply the enormous advantage which a disciplined force, flushed with victory, has over one demoralized by defeat." Legge, in his note on Mencius, I. 2. ix. 2, makes the I to be 24 Chinese ounces, and corrects Chu Hsi's statement that it equaled 20 oz. only. But Li Chfuan of the Tang dynasty here gives the same figure as Chu Hsi.]

20. The onrush of a conquering force is like the bursting of pent-up waters into a chasm a thousand fathoms deep.

V Energy

Sun Tzu said: The control of a large force is the same principle as the control of a few men: it is merely a question of dividing up their numbers.

[That is, cutting up the army into regiments, companies, etc., with subordinate officers in command of each. Tu Mu reminds us of Han Hsin's famous reply to the first Han Emperor, who once said to him: "How large an army do you think I could lead?" "Not more than 100,000 men, your Majesty." "And you?" asked the Emperor. "Oh!" he answered, "the more the better."]

Fighting with a large army under your command is nowise different from fighting with a small one: it is merely a question of instituting signs and signals. To ensure that your whole host may withstand the brunt of the enemy's attack and remain unshaken - this is effected by maneuvers direct and indirect.

[We now come to one of the most interesting parts of Sun Tzu's treatise, the discussion of the CHENG and the CH'I." As it is by no means easy to grasp the full significance of these two terms, or to render them consistently by good English equivalents; it may be as well to tabulate some of the commentators' remarks on the subject before proceeding further. Li Chf uan: "Facing the enemy is CHENG, making lateral diversion is CH X I. Chia Lin: "In presence of the enemy, your troops should be arrayed in normal fashion, but in order to secure victory abnormal maneuvers must be employed." Mei Yao-chf en: "CH X I is active, CHENG is passive; passivity means waiting for an opportunity, activity beings the victory itself." Ho Shih: "We must cause the enemy to regard our straightforward attack as one that is secretly designed, and vice versa; thus CHENG may also be Cl-H, and ChTI may also be CHENG." He instances the famous exploit of Han Hsin, who when marching ostensibly against Lin-chin (now Chao-i in Shensi), suddenly threw a large force across the Yellow River in wooden tubs, utterly disconcerting his opponent. [Chfien Han Shu, ch. 3.] Here, we are told, the march on Lin-chin was CHENG, and the surprise maneuver was CH'I." Chang Yu gives the following summary of opinions on the words: "Military writers do not agree with regard to the meaning of CH X I and CHENG. Wei Liao Tzu [4th cent. B.C.] says: 'Direct warfare favors frontal attacks, indirect warfare attacks from the rear.' Ts'ao Kung says: 'Going straight out to join battle is a direct operation; appearing on the enemy's rear is an indirect maneuver.' Li Wei-kung [6th and 7th cent. A.D.] says: 'In war, to march straight ahead is CHENG; turning movements, on the other hand, are CH X I.' These writers simply regard CHENG as CHENG, and ChTI as CITI; they do not note that the two are mutually interchangeable and run into each other like the two sides of a circle [see infra, ss. 11]. A comment on the "Tang Emperor "Tai Tsung goes to the root of the matter: 'A CH'I maneuver may be CHENG, if we make the enemy look upon it as CHENG; then our real attack will be CH X I, and vice versa. The whole secret lies in confusing the enemy, so that he cannot fathom our real intent.'" To put it perhaps a little more clearly: any attack or other operation is CHENG, on which the enemy has had his attention fixed; whereas that is CH X I," which takes him by surprise or comes from an unexpected quarter. If the enemy perceives a movement which is meant to be CH'I," it immediately becomes CHENG."]

That the impact of your army may be like a grindstone dashed against an egg - this is effected by the science of weak points and strong.

In all fighting, the direct method may be used for joining battle, but indirect methods will be needed in order to secure victory.

[Chang Yu says: "Steadily develop indirect tactics, either by pounding the enemy's flanks or falling on his rear." A brilliant example of "indirect tactics" which decided the fortunes of a campaign was Lord Roberts' night march round the Peiwar Kotal in the second Afghan war. [1]

Indirect tactics, efficiently applied, are inexhaustible as Heaven and Earth, unending as the flow of rivers and streams; like the sun and moon, they end but to begin anew; like the four seasons, they pass away to return once more.

[Tu Yu and Chang Yu understand this of the permutations of CI-TI and CHENG." But at present Sun Tzu is not speaking of CHENG at all, unless, indeed, we suppose with Cheng Yu-hsien that a clause relating to it has fallen out of the text. Of course, as has already been pointed out, the two are so inextricably interwoven in all military operations, that they cannot really be considered apart. Here we simply have an expression, in figurative language, of the almost infinite resource of a great leader.]

7. There are not more than five musical notes, yet the combinations of these five give rise to more melodies than can ever be heard.

8. There are not more than five primary colors (blue, yellow, red, white, and black), yet in combination
they produce more hues than can ever been seen.

9 There are not more than five cardinal tastes (sour, acrid, salt, sweet, bitter), yet combinations of them yield more flavors than can ever be tasted.

10. In battle, there are not more than two methods of attack - the direct and the indirect; yet these two in combination give rise to an

endless series of maneuvers.

11. The direct and the indirect lead on to each other in turn. It is like moving in a circle - you never come to an end. Who can exhaust the possibilities of their combination?

12. The onset of troops is like the rush of a torrent which will even roll stones along in its course.

13. The quality of decision is like the well-timed swoop of a falcon which enables it to strike and destroy its victim.

[The Chinese here is tricky and a certain key word in the context it is used defies the best efforts of the translator. Tu Mu defines this word as "the measurement or estimation of distance." But this meaning does not quite fit the illustrative simile in ss. 15. Applying this definition to the falcon, it seems to me to denote that instinct of SELF RESTRAINT which keeps the bird from swooping on its quarry until the right moment, together with the power of judging when the right moment has arrived. The analogous quality in soldiers is the highly important one of being able to reserve their fire until the very instant at which it will be most effective. When the "Victory" went into action at Trafalgar at hardly more than drifting pace, she was for several minutes exposed to a storm of shot and shell before replying with a single gun. Nelson coolly waited until he was within close range, when the broadside he brought to bear worked fearful havoc on the enemy's nearest ships.]

14. Therefore the good fighter will be terrible in his onset, and prompt in his decision.

[The word "decision" would have reference to the measurement of distance mentioned above, letting the enemy get near before striking. But I cannot help thinking that Sun Tzu meant to use the word in a figurative sense comparable to our own idiom "short and sharp." Cf. Wang Hsi's note, which after describing the falcon's mode of attack, proceeds: "This is just how the 'psychological moment' should be seized in war."]

15. Energy may be likened to the bending of a crossbow; decision, to

the releasing of a trigger.

[None of the commentators seem to grasp the real point of the simile of energy and the force stored up in the bent cross-bow until released by the finger on the trigger.]

16. Amid the turmoil and tumult of battle, there may be seeming disorder and yet no real disorder at all; amid confusion and chaos, your array may be without head or tail, yet it will be proof against defeat.

[Mei Yao-chf en says: "The subdivisions of the army having been previously fixed, and the various signals agreed upon, the separating and joining, the dispersing and collecting which will take place in the course of a battle, may give the appearance of disorder when no real disorder is possible. Your formation may be without head or tail, your dispositions all topsy-turvy, and yet a rout of your forces quite out of the question."]

17. Simulated disorder postulates perfect discipline, simulated fear postulates courage; simulated weakness postulates strength.

[In order to make the translation intelligible, it is necessary to tone down the sharply paradoxical form of the original. Ts'ao Kung throws out a hint of the meaning in his brief note: "These things all serve to destroy formation and conceal one's condition." But Tu Mu is the first to put it quite plainly: "If you wish to feign confusion in order to lure the enemy on, you must first have perfect discipline; if you wish to display timidity in order to entrap the enemy, you must have extreme courage; if you wish to parade your weakness in order to make the enemy over-confident, you must have exceeding strength."]

18. Hiding order beneath the cloak of disorder is simply a question of subdivision; [See supra, ss. 1.] concealing courage under a show of timidity presupposes a fund of latent energy; [The commentators strongly understand a certain Chinese word here differently than anywhere else in this chapter. Thus Tu Mu says: "seeing that we are favorably circumstanced and yet make no move, the enemy will believe that we are really afraid."] masking strength with weakness is

to be effected by tactical dispositions.

[Chang Yu relates the following anecdote of Kao Tsu, the first Han Emperor: "Wishing to crush the Hsiung-nu, he sent out spies to report on their condition. But the Hsiung- nu, forewarned, carefully concealed all their able-bodied men and well-fed horses, and only allowed infirm soldiers and emaciated cattle to be seen. The result was that spies one and all recommended the Emperor to deliver his attack. Lou Ching alone opposed them, saying: "When two countries go to war, they are naturally inclined to make an ostentatious display of their strength. Yet our spies have seen nothing but old age and infirmity. This is surely some ruse on the part of the enemy, and it would be unwise for us to attack." The Emperor, however, disregarding this advice, fell into the trap and found himself surrounded at Po-teng."]

19. Thus one who is skillful at keeping the enemy on the move maintains deceitful appearances, according to which the enemy will act.

[Ts'ao Kung's note is "Make a display of weakness and want." Tu Mu says: "If our force happens to be superior to the enemy's, weakness may be simulated in order to lure him on; but if inferior, he must be led to believe that we are strong, in order that he may keep off. In fact, all the enemy's movements should be determined by the signs that we choose to give him." Note the following anecdote of Sun Pin, a descendent of Sun Wu: In 341 B.C., the Chfi State being at war with Wei, sent T'ien Chi and Sun Pin against the general P'ang Chuan, who happened to be a deadly personal enemy of the later. Sun Pin said: "The Chfi State has a reputation for cowardice, and therefore our adversary despises us. Let us turn this circumstance to account." Accordingly, when the army had crossed the border into Wei territory, he gave orders to show 100,000 fires on the first night, 50,000 on the next, and the night after only 20,000. P'ang Chuan pursued them hotly, saying to himself: "I knew these men of Chfi were cowards: their numbers have already fallen away by more than half." In his retreat, Sun Pin came to a narrow defile, with he calculated that his pursuers would reach after dark. Here he had a tree stripped of its bark, and inscribed upon it the words: "Under this tree shall P'ang Chuan die." Then, as night began to fall, he placed a

strong body of archers in ambush near by, with orders to shoot
directly they saw a light. Later on, P'ang Chuan arrived at the spot,
and noticing the tree, struck a light in order to read what was written
on it. His body was immediately riddled by a volley of arrows, and
his whole army thrown into confusion. [The above is Tu Mu's
version of the story; the SHIH CHI, less dramatically but probably
with more historical truth, makes P'ang Chuan cut his own throat
with an exclamation of despair, after the rout of his army.]
He sacrifices something, that the enemy may snatch at it.

20. By holding out baits, he keeps him on the march; then with a
body of picked men he lies in wait for him.

[With an emendation suggested by Li Ching, this then reads, "He lies
in wait with the main body of his troops."]

21. The clever combatant looks to the effect of combined energy, and
does not require too much from individuals.

[Tu Mu says: "He first of all considers the power of his army in the
bulk; afterwards he takes individual talent into account, and uses
each men according to his capabilities. He does not demand
perfection from the untalented."]

Hence his ability to pick out the right men and utilize combined
energy.

22. When he utilizes combined energy, his fighting men become as it
were like unto rolling logs or stones. For it is the nature of a log or
stone to remain motionless on level ground, and to move when on a
slope; if four-cornered, to come to a standstill, but if round-shaped, to
go rolling down.

[Ts'au Kung calls this "the use of natural or inherent power."]

23. Thus the energy developed by good fighting men is as the
momentum of a round stone rolled down a mountain thousands of
feet in height. So much on the subject of energy.

[The chief lesson of this chapter, in Tu Mu's opinion, is the paramount importance in war of rapid evolutions and sudden rushes. "Great results," he adds, "can thus be achieved with small forces."]

[1] "Forty-one Years in India," chapter 46.

VI Weak Points and Strong

[Chang Yu attempts to explain the sequence of chapters as follows: "Chapter IV, on Tactical Dispositions, treated of the offensive and the defensive; chapter V, on Energy, dealt with direct and indirect methods. The good general acquaints himself first with the theory of attack and defense, and then turns his attention to direct and indirect methods. He studies the art of varying and combining these two methods before proceeding to the subject of weak and strong points. For the use of direct or indirect methods arises out of attack and defense, and the perception of weak and strong points depends again on the above methods. Hence the present chapter comes immediately after the chapter on Energy."]

1. Sun Tzu said: Whoever is first in the field and awaits the coming of the enemy, will be fresh for the fight; whoever is second in the field and has to hasten to battle will arrive exhausted.

2. Therefore the clever combatant imposes his will on the enemy, but does not allow the enemy's will to be imposed on him.

[One mark of a great soldier is that he fight on his own terms or fights not at all.]

3. By holding out advantages to him, he can cause the enemy to approach of his own accord; or, by inflicting damage, he can make it impossible for the enemy to draw near.

[In the first case, he will entice him with a bait; in the second, he will strike at some important point which the enemy will have to defend.]

4. If the enemy is taking his ease, he can harass him; [This passage may be cited as evidence against Mei Yao-Ch'en's interpretation of I. ss. 23.] if well supplied with food, he can starve him out; if quietly encamped, he can force him to move.

5. Appear at points which the enemy must hasten to defend; march swiftly to places where you are not expected.

6. An army may march great distances without distress, if it marches through country where the enemy is not.

[Ts'ao Kung sums up very well: "Emerge from the void
[q.d. like "a bolt from the blue"], strike at vulnerable points,
shunplaces that are defended, attack in unexpected quarters."]

7. You can be sure of succeeding in your attacks if you only attack places which are undefended.

[Wang Hsi explains "undefended places" as "weak points; that is to say, where the general is lacking in capacity, or the soldiers in spirit; where the walls are not strong enough, or the precautions not strict enough; where relief comes too late, or provisions are too scanty, or the defenders are variance amongst themselves."]

You can ensure the safety of your defense if you only hold positions that cannot be attacked.

[I.e., where there are none of the weak points mentioned above. There is rather a nice point involved in the interpretation of this later clause. Tu Mu, Chfen Hao, and Mei Yao-chf en assume the meaning to be: "In order to make your defense quite safe, you must defend EVEN those places that are not likely to be attacked;" and Tu Mu adds: "How much more, then, those that will be attacked." Taken thus, however, the clause balances less well with the preceding — always a consideration in the highly antithetical style which is natural to the Chinese. Chang Yu, therefore, seems to come nearer the mark in saying: "He who is skilled in attack flashes forth from the topmost heights of heaven [see IV. ss. 7], making it impossible for the enemy to guard against him. This being so, the places that I shall attack are precisely those that the enemy cannot defend He who is skilled in defense hides in the most secret recesses of the earth, making it impossible for the enemy to estimate his whereabouts. This being so, the places that I shall hold are precisely those that the enemy cannot attack."]

8. Hence that general is skillful in attack whose opponent does not know what to defend; and he is skillful in defense whose opponent does not know what to attack.

[An aphorism which puts the whole art of war in a nutshell.]

9. O divine art of subtlety and secrecy! Through you we learn to be invisible, through you inaudible; [Literally, "without form or sound," but it is said of course with reference to the enemy.] and hence we can hold the enemy's fate in our hands.

10. You may advance and be absolutely irresistible, if you make for the enemy's weak points; you may retire and be safe from pursuit if your movements are more rapid than those of the enemy.

11. If we wish to fight, the enemy can be forced to an engagement even though he be sheltered behind a high rampart and a deep ditch. All we need do is attack some other place that he will be obliged to relieve.

[Tu Mu says: "If the enemy is the invading party, we can cut his line of communications and occupy the roads by which he will have to return; if we are the invaders, we may direct our attack against the sovereign himself." It is clear that Sun Tzu, unlike certain generals in the late Boer war, was no believer in frontal attacks.]

12. If we do not wish to fight, we can prevent the enemy from engaging us even though the lines of our encampment be merely traced out on the ground. All we need do is to throw something odd and unaccountable in his way.

[This extremely concise expression is intelligibly paraphrased by Chia Lin: "even though we have constructed neither wall nor ditch." Li Chfuan says: "we puzzle him by strange and unusual dispositions;" and Tu Mu finally clinches the meaning by three illustrative anecdotes — one of Chu-ko Liang, who when occupying Yang-p'ing and about to be attacked by Ssu-ma I, suddenly struck his colors, stopped the beating of the drums, and flung open the city gates, showing only a few men engaged in sweeping and sprinkling the ground. This unexpected proceeding had the intended effect; for Ssu-ma I, suspecting an ambush, actually drew off his army and retreated. What Sun Tzu is advocating here, therefore, is nothing more nor less than the timely use of "bluff."]

13. By discovering the enemy's dispositions and remaining invisible ourselves, we can keep our forces concentrated, while the enemy's must be divided.

[The conclusion is perhaps not very obvious, but Chang Yu (after Mei Yao-chf en) rightly explains it thus: "If the enemy's dispositions are visible, we can make for him in one body; whereas, our own dispositions being kept secret, the enemy will be obliged to divide his forces in order to guard against attack from every quarter."]

14. We can form a single united body, while the enemy must split up into fractions. Hence there will be a whole pitted against separate parts of a whole, which means that we shall be many to the enemy's few.

15. And if we are able thus to attack an inferior force with a superior one, our opponents will be in dire straits.

16. The spot where we intend to fight must not be made known; for then the enemy will have to prepare against a possible attack at several different points; [Sheridan once explained the reason of General Grant's victories by saying that "while his opponents were kept fully employed wondering what he was going to do, HE was thinking most of what he was going to do himself."] and his forces being thus distributed in many directions, the numbers we shall have to face at any given point will be proportionately few.

17. For should the enemy strengthen his van, he will weaken his rear; should he strengthen his rear, he will weaken his van; should he strengthen his left, he will weaken his right; should he strengthen his

right, he will weaken his left. If he sends reinforcements everywhere, he will everywhere be weak.

[In Frederick the Great's INSTRUCTIONS TO HIS GENERALS we read: "A defensive war is apt to betray us into too frequent detachment. Those generals who have had but little experience attempt to protect every point, while those who are better acquainted with their profession, having only the capital object in view, guard against a decisive blow, and acquiesce in small misfortunes to avoid greater."]

18. Numerical weakness comes from having to prepare against possible attacks; numerical strength, from compelling our adversary to make these preparations against us.

[The highest generalship, in Col. Henderson's words, is "to compel the enemy to disperse his army, and then to concentrate superior force against each fraction in turn."]

19. Knowing the place and the time of the coming battle, we may concentrate from the greatest distances in order to fight.

[What Sun Tzu evidently has in mind is that nice calculation of distances and that masterly employment of strategy which enable a general to divide his army for the purpose of a long and rapid march, and afterwards to effect a junction at precisely the right spot and the right hour in order to confront the enemy in overwhelming strength. Among many such successful junctions which military history records, one of the most dramatic and decisive was the appearance of Blucher just at the critical moment on the field of Waterloo.]

20. But if neither time nor place be known, then the left wing will be impotent to succor the right, the right equally impotent to succor the left, the van unable to relieve the rear, or the rear to support the van. How much more so if the furthest portions of the army are anything under a hundred LI apart, and even the nearest are separated by several LI!

[The Chinese of this last sentence is a little lacking in precision, but the mental picture we are required to draw is probably that of an army advancing towards a given rendezvous in separate columns, each of which has orders to be there on a fixed date. If the general allows the various detachments to proceed at haphazard, without precise instructions as to the time and place of meeting, the enemy will be able to annihilate the army in detail. Chang Yu's note may be worth quoting here: "If we do not know the place where our opponents mean to concentrate or the day on which they will join battle, our unity will be forfeited through our preparations for defense, and the positions we hold will be insecure. Suddenly happening upon a powerful foe, we shall be brought to battle in a flurried condition, and no mutual support will be possible between wings, vanguard or rear, especially if there is any great distance between the foremost and hindmost divisions of the army."]

21. Though according to my estimate the soldiers of Yueh exceed our own in number, that shall advantage them nothing in the matter of victory. I say then that victory can be achieved.

[Alas for these brave words! The long feud between the two states ended in 473 B.C. with the total defeat of Wu by Kou Chien and its incorporation in Yueh. This was doubtless long after Sun Tzu's death. With his present assertion compare IV. ss. 4. Chang Yu is the only one to point out the seeming discrepancy, which he thus goes on to explain: "In the chapter on Tactical Dispositions it is said, 'One may KNOW how to conquer without being able to DO it,' whereas here we have the statement that 'victory' can be achieved.'
The explanation is, that in the former chapter, where the offensive and defensive are under discussion, it is said that if the enemy is fully prepared, one cannot make certain of beating him. But the present passage refers particularly to the soldiers of Yueh who, according to Sun Tzu's calculations, will be kept in ignorance of the time and place of the impending struggle. That is why he says here that victory can be achieved."]

22. Though the enemy be stronger in numbers, we may prevent him from fighting. Scheme so as to discover his plans and the likelihood of their success.

[An alternative reading offered by Chia Lin is: "Know beforehand all plans conducive to our success and to the enemy's failure."]

23. Rouse him, and learn the principle of his activity or inactivity.

[Chang Yu tells us that by noting the joy or anger shown by the enemy on being thus disturbed, we shall be able to conclude whether his policy is to lie low or the reverse. He instances the action of Cho-ku Liang, who sent the scornful present of a woman's head-dress to Ssu-ma I, in order to goad him out of his Fabian tactics.]

Force him to reveal himself, so as to find out his vulnerable spots.

24. Carefully compare the opposing army with your own, so that you may know where strength is superabundant and where it is deficient.

25. In making tactical dispositions, the highest pitch you can attain is to conceal them; [The piquancy of the paradox evaporates in translation. Concealment is perhaps not so much actual invisibility (see supra ss. 9) as "showing no sign" of what you mean to do, of the plans that are formed in your brain.] conceal your dispositions, and you will be safe from the prying of the subtlest spies, from the machinations of the wisest brains.

[Tu Mu explains: "Though the enemy may have clever and capable officers, they will not be able to lay any plans against us."]

26. How victory may be produced for them out of the enemy's own tactics — that is what the multitude cannot comprehend.

27. All men can see the tactics whereby I conquer, but what none can see is the strategy out of which victory is evolved.

[I.e., everybody can see superficially how a battle is won; what they cannot see is the long series of plans and combinations which has preceded the battle.]

28. Do not repeat the tactics which have gained you one victory, but let your methods be regulated by the infinite variety of circumstances.

[As Wang Hsi sagely remarks: "There is but one root- principle underlying victory, but the tactics which lead up to it are infinite in number." With this compare Col. Henderson: "The rules of strategy are few and simple. They may be learned in a week. They may be taught by familiar illustrations or a dozen diagrams. But such knowledge will no more teach a man to lead an army like Napoleon than a knowledge of grammar will teach him to write like Gibbon."]

29. Military tactics are like unto water; for water in its natural course runs away from high places and hastens downwards.

30. So in war, the way is to avoid what is strong and to strike at what is weak.

[Like water, taking the line of least resistance.]

31. Water shapes its course according to the nature of the ground over which it flows; the soldier works out his victory in relation to the foe whom he is facing.

32. Therefore, just as water retains no constant shape, so in warfare there are no constant conditions.

33. He who can modify his tactics in relation to his opponent and thereby succeed in winning, may be called a heaven-born captain.

34. The five elements (water, fire, wood, metal, earth) are not always equally predominant. They predominate alternately and have no invariable seat. The four seasons make way for each other in turn.

There are short days and long; the moon has its periods of waning and waxing.

[Cf. V. ss. 6. The purport of the passage is simply to illustrate the want of fixity in war by the changes constantly taking place in Nature. The comparison is not very happy, however, because the regularity of the phenomena which Sun Tzu mentions is by no means paralleled in war.]

See Col. Henderson's biography of Stonewall
Jackson, 1902 ed., vol. II, p. 490.

VII Maneuvering

Sun Tzu said: In war, the general receives his commands from the
sovereign. Having collected an army and concentrated his forces, he
must blend and harmonize the different elements thereof before
pitching his camp.

["Chang Yu says: "the establishment of harmony and confidence
between the higher and lower ranks before venturing into the field;"
and he quotes a saying of Wu Tzu (chap. 1 ad init.): "Without
harmony in the State, no military expedition can be undertaken;
without harmony in the army, no battle array can be formed." In an
historical romance Sun Tzu is represented as saying to Wu Yuan: "As
a general rule, those who are waging war should get rid of all the
domestic troubles before proceeding to attack the external foe."]

After that, comes tactical maneuvering, than which there is nothing
more difficult.

[I have departed slightly from the traditional interpretation of Ts'ao
Kung, who says: "From the time of receiving the sovereign's
instructions until our encampment over against the enemy, the
tactics to be pursued are most difficult." It seems to me that the
tactics or maneuvers can hardly be said to begin until the army has
sallied forth and encamped, and Chfien Hao's note gives color to this
view: "For levying, concentrating, harmonizing and entrenching an
army, there are plenty of old rules which will serve. The real
difficulty comes when we engage in tactical operations." Tu
Yu also observes that "the great difficulty is to be beforehand with
the enemy in seizing favorable position."]

The difficulty of tactical maneuvering consists in turning the devious
into the direct, and misfortune into gain.

[This sentence contains one of those highly condensed and somewhat enigmatical expressions of which Sun Tzu is so fond. This is how it is explained by Ts'ao Kung: "Make it appear that you are a long way off, then cover the distance rapidly and arrive on the scene before your opponent." Tu Mu says: "Hoodwink the enemy, so that he may be remiss and leisurely while you are dashing along with utmost speed." Ho Shih gives a slightly different turn: "Although you may have difficult ground to traverse and natural obstacles to encounter this is a drawback which can be turned into actual advantage by celerity of movement." Signal examples of this saying are afforded by the two famous passages across the Alps — that of Hannibal, which laid Italy at his mercy, and that of Napoleon two thousand years later, which resulted in the great victory of Marengo.]

Thus, to take a long and circuitous route, after enticing the enemy out of the way, and though starting after him, to contrive to reach the goal before him, shows knowledge of the artifice of DEVIATION.

[Tu Mu cites the famous march of Chao She in 270 B.C. to relieve the town of O-yu, which was closely invested by a Chun army. The King of Chao first consulted Lien P x o on the advisability of attempting a relief, but the latter thought the distance too great, and the intervening country too rugged and difficult. His Majesty then turned to Chao She, who fully admitted the hazardous nature of the march, but finally said: "We shall be like two rats fighting in a whole — and the pluckier one will win!" So he left the capital with his army, but had only gone a distance of 30 LI when he stopped and began throwing up entrenchments. For 28 days he continued strengthening his fortifications, and took care that spies should carry the intelligence to the enemy. The Chun general was overjoyed, and attributed his adversary's tardiness to the fact that the beleaguered city was in the Han State, and thus not actually part of Chao territory. But the spies had no sooner departed than Chao She began a forced march lasting for two days and one night, and arrive on the scene of action with such astonishing rapidity that he was able to occupy a commanding position on the "North hill" before the enemy had got wind of his movements. A crushing defeat followed for the Chun forces, who were obliged to raise the siege of O-yu in all haste and retreat across the border.]

5. Maneuvering with an army is advantageous; with an undisciplined multitude, most dangerous.

[I adopt the reading of the TUNG TIEN, Cheng Yu-hsien and the "TU SHU, since they appear to apply the exact nuance required in order to make sense. The commentators using the standard text take this line to mean that maneuvers may be profitable, or they may be dangerous: it all depends on the ability of the general.]

6. If you set a fully equipped army in march in order to snatch an advantage, the chances are that you will be too late. On the other hand, to detach a flying column for the purpose involves the sacrifice of its baggage and stores.

[Some of the Chinese text is unintelligible to the Chinese commentators, who paraphrase the sentence. I submit my own rendering without much enthusiasm, being convinced that there is some deep-seated corruption in the text. On the whole, it is clear that Sun Tzu does not approve of a lengthy march being undertaken without supplies. Cf. infra, ss. 11.]

7. Thus, if you order your men to roll up their buff-coats, and make forced marches without halting day or night, covering double the usual distance at a stretch, [The ordinary day's march, according to Tu Mu, was 30 LI; but on one occasion, when pursuing Liu Pei, Ts'ao Ts'ao is said to have covered the incredible distance of 300 _li_ within twenty-four hours.] doing a hundred LI in order to wrest an advantage, the leaders of all your three divisions will fall into the hands of the enemy.

8. The stronger men will be in front, the jaded ones will fall behind, and on this plan only one-tenth of your army will reach its destination.

[The moral is, as Ts'ao Kung and others point out: Don't march a hundred LI to gain a tactical advantage, either with or without impedimenta. Maneuvers of this description should be confined to short distances. Stonewall Jackson said: "The hardships of forced marches are often more painful than the dangers of battle." He did not often call upon his troops for extraordinary exertions. It was only when he intended a surprise, or when a rapid retreat was imperative, that he sacrificed everything for speed.]

9. If you march fifty LI in order to outmaneuver the enemy, you will lose the leader of your first division, and only half your force will reach the goal.

[Literally, "the leader of the first division will be TORN AWAY."]

10. If you march thirty LI with the same object, two- thirds of your army will arrive.

[In the T'UNG TIEN is added: "From this we may know the difficulty of maneuvering."]

11. We may take it then that an army without its baggage-train is lost; without provisions it is lost; without bases of supply it is lost.

[I think Sun Tzu meant "stores accumulated in depots." But Tu Yu says "fodder and the like," Chang Yu says "Goods in general," and Wang Hsi says "fuel, salt, foodstuffs, etc."]

12. We cannot enter into alliances until we are acquainted with the designs of our neighbors.

13. We are not fit to lead an army on the march unless we are familiar with the face of the country — its mountains and forests, its pitfalls and precipices, its marshes and swamps.

14. We shall be unable to turn natural advantage to account unless we make use of local guides.

15. In war, practice dissimulation, and you will succeed.

[In the tactics of Turenne, deception of the enemy, especially as to the numerical strength of his troops, took a very prominent position.]

16. Whether to concentrate or to divide your troops, must be decided

by circumstances.

17. Let your rapidity be that of the wind, [The simile is doubly appropriate, because the wind is not only swift but, as Mei Yao-chf en points out, "invisible and leaves no tracks."] your compactness that of the forest.

[Meng Shih comes nearer to the mark in his note: "When slowly marching, order and ranks must be preserved" — so as to guard against surprise attacks. But natural forest do not grow in rows, whereas they do generally possess the quality of density or compactness.]

18. In raiding and plundering be like fire, [Cf. SHIH CHING, IV. 3. iv. 6: "Fierce as a blazing fire which no man can check."] is immovability like a mountain.

[That is, when holding a position from which the enemy is trying to dislodge you, or perhaps, as Tu Yu says, when he is trying to entice you into a trap.]

19. Let your plans be dark and impenetrable as night, and when you move, fall like a thunderbolt.

[Tu Yu quotes a saying of T'ai Kung which has passed into a proverb: "You cannot shut your ears to the thunder or your eyes to the lighting — so rapid are they." Likewise, an attack should be made so quickly that it cannot be parried.]

20. When you plunder a countryside, let the spoil be divided amongst your men; [Sun Tzu wishes to lessen the abuses of indiscriminate plundering by insisting that all booty shall be thrown into a common stock, which may afterwards be fairly divided amongst all.] when you capture new territory, cut it up into allotments for the benefit of the soldiery.

[Chfen Hao says "quarter your soldiers on the land, and let them sow and plant it." It is by acting on this principle, and harvesting the lands they invaded, that the Chinese have succeeded in carrying out some of their most memorable and triumphant expeditions, such as that of Pan Chfao who penetrated to the Caspian, and in more recent years, those of Fu-k x ang-an and Tso Tsung-f ang.]

21. Ponder and deliberate before you make a move.

[Chang Yu quotes Wei Liao Tzu as saying that we must not break camp until we have gained the resisting power of the enemy and the cleverness of the opposing general. Cf. the "seven comparisons" in I. ss. 13.]

22. He will conquer who has learnt the artifice of deviation.

Such is the art of maneuvering.

[With these words, the chapter would naturally come to an end. But there now follows a long appendix in the shape of an extract from an earlier book on War, now lost, but apparently extant at the time when Sun Tzu wrote. The style of this fragment is not noticeable different from that of Sun Tzu himself, but no commentator raises a doubt as to its genuineness.]

23. The Book of Army Management says: [It is perhaps significant that none of the earlier commentators give us any information about this work. Mei Yao-Chfen calls it "an ancient military classic," and Wang Hsi, "an old book on war." Considering the enormous amount of fighting that had gone on for centuries before Sun Tzu's time between the various kingdoms and principalities of China, it is not in itself improbable that a collection of military maxims should have been made and written down at some earlier period.] On the field of battle, [Implied, though not actually in the Chinese.] the spoken word does not carry far enough: hence the institution of gongs and drums. Nor can ordinary objects be seen clearly enough: hence the institution of banners and flags.

24. Gongs and drums, banners and flags, are means whereby the ears and eyes of the host may be focused on one particular point.

[Chang Yu says: "If sight and hearing converge simultaneously on the same object, the evolutions of as many as a million soldiers will be like those of a single man."!]

25. The host thus forming a single united body, is it impossible either for the brave to advance alone, or for the cowardly to retreat alone.

[Chuang Yu quotes a saying: "Equally guilty are those who advance against orders and those who retreat against orders." Tu Mu tells a story in this connection of Wu Chfi, when he was fighting against the Chun State. Before the battle had begun, one of his soldiers, a man of matchless daring, sallied forth by himself, captured two heads from the enemy, and returned to camp. Wu Chfi had the man instantly executed, whereupon an officer ventured to remonstrate, saying: "This man was a good soldier, and ought not to have been beheaded." Wu Chfi replied: "I fully believe he was a good soldier, but I had him beheaded because he acted without orders."]

This is the art of handling large masses of men.

26. In night-fighting, then, make much use of signal- fires and drums, and in fighting by day, of flags and banners, as a means of influencing the ears and eyes of your army.

[Chf en Hao alludes to Li Kuang-pi's night ride to Ho- yang at the head of 500 mounted men; they made such an imposing display with torches, that though the rebel leader Shih Ssu-ming had a large army, he did not dare to dispute their passage.]

27. A whole army may be robbed of its spirit; ["In war," says Chang Yu, "if a spirit of anger can be made to pervade all ranks of an army at one and the same time, its onset will be irresistible. Now the spirit of the enemy's soldiers will be keenest when they have newly arrived on the scene, and it is therefore our cue not to fight at once, but to wait until their ardor and enthusiasm have worn off, and then strike. It is in this way that they may be robbed of their keen spirit." Li Chfuan and others tell an anecdote (to be found in the TSO CHUAN, year 10, ss. 1) of Ts'ao Kuei, a protege of Duke Chuang of Lu. The latter State was attacked by Chfi, and the duke was about to join battle at Chf ang-cho, after the first roll of the enemy's drums, when Ts'ao said: "Not just yet." Only after their drums had beaten for the third time, did he give the word for attack. Then they fought, and the men of Chfi were utterly defeated. Questioned afterwards by the

Duke as to the meaning of his delay, Ts'ao Kuei replied: "In battle, a courageous spirit is everything. Now the first roll of the drum tends to create this spirit, but with the second it is already on the wane, and after the third it is gone altogether. I attacked when their spirit was gone and ours was at its height. Hence our victory." Wu Tzu (chap. 4) puts "spirit" first among the "four important influences" in war, and continues: "The value of a whole army — a mighty host of a million men — is dependent on one man alone: such is the influence of spirit!"] a commander-in-chief may be robbed of his presence of mind.

[Chang Yu says: "Presence of mind is the general's most important asset. It is the quality which enables him to discipline disorder and to inspire courage into the panic- stricken." The great general Li Ching (A.D. 571-649) has a saying: "Attacking does not merely consist in assaulting walled cities or striking at an army in battle array; it must include the art of assailing the enemy's mental equilibrium."]

28. Now a solider's spirit is keenest in the morning; [Always provided, I suppose, that he has had breakfast. At the battle of the Trebia, the Romans were foolishly allowed to fight fasting, whereas Hannibal's men had breakfasted at their leisure. See Livy, XXI, liv. 8, lv. 1 and 8.] by noonday it has begun to flag; and in the evening, his mind is bent only on returning to camp.

29. A clever general, therefore, avoids an army when its spirit is keen, but attacks it when it is sluggish and inclined to return. This is the art of studying moods.

30. Disciplined and calm, to await the appearance of disorder and hubbub amongst the enemy: — this is the art of retaining self-possession.

31. To be near the goal while the enemy is still far from it, to wait at ease while the enemy is toiling and struggling, to be well-fed while the enemy is famished: — this is the art of husbanding one's strength.

32. To refrain from intercepting an enemy whose banners are in perfect order, to refrain from attacking an army drawn up in calm and confident array: this is the art of studying circumstances.

33. It is a military axiom not to advance uphill against the enemy, nor to oppose him when he comes downhill.

34. Do not pursue an enemy who simulates flight; do not attack soldiers whose temper is keen.

35. Do not swallow bait offered by the enemy.

[Li Crf uan and Tu Mu, with extraordinary inability to see a metaphor, take these words quite literally of food and drink that have been poisoned by the enemy. Chfen Hao and Chang Yu carefully point out that the saying has a wider application.]

Do not interfere with an army that is returning home.

[The commentators explain this rather singular piece of advice by saying that a man whose heart is set on returning home will fight to the death against any attempt to bar his way, and is therefore too dangerous an opponent to be tackled. Chang Yu quotes the words of Han Hsin: "Invincible is the soldier who hath his desire and returneth homewards." A marvelous tale is told of Ts'ao Ts'ao's courage and resource in ch. 1 of the SAN KUO CHI: In 198 A.D., he was besieging Chang Hsiu in Jang, when Liu Piao sent reinforcements with a view to cutting off Ts'ao's retreat. The latter was obligbed to draw off his troops, only to find himself hemmed in between two enemies, who were guarding each outlet of a narrow pass in which he had engaged himself. In this desperate plight Ts'ao waited until nightfall, when he bored a tunnel into the mountain side and laid an ambush in it. As soon as the whole army had passed by, the hidden troops fell on his rear, while Ts'ao himself turned and met his pursuers in front, so that they were thrown into confusion and annihilated. Ts'ao Ts'ao said afterwards: "The brigands tried to check my army in its retreat and brought me to battle in a desperate position: hence I knew how to overcome them."]

36. When you surround an army, leave an outlet free.

[This does not mean that the enemy is to be allowed to escape. The object, as Tu Mu puts it, is "to make him believe that there is a road to safety, and thus prevent his fighting with the courage of despair." Tu Mu adds pleasantly: "After that, you may crush him."]

Do not press a desperate foe too hard.

[Chfen Hao quotes the saying: "Birds and beasts when brought to bay will use their claws and teeth." Chang Yu says: "If your adversary has burned his boats and destroyed his cooking-pots, and is ready to stake all on the issue of a battle, he must not be pushed to extremities." Ho Shih illustrates the meaning by a story taken from the life of Yen- chfing. That general, together with his colleague Tu Chung- wei was surrounded by a vastly superior army of Khitans in the year 945 A.D. The country was bare and desert-like, and the little Chinese force was soon in dire straits for want of water. The wells they bored ran dry, and the men were reduced to squeezing lumps of mud and sucking out the moisture. Their ranks thinned rapidly, until at last Fu Yen- chfing exclaimed: "We are desperate men. Far better to die for our country than to go with fettered hands into captivity!" A strong gale happened to be blowing from the northeast and darkening the air with dense clouds of sandy dust. To Chung-wei was for waiting until this had abated before deciding on a final attack; but luckily another officer, Li Shou-cheng by name, was quicker to see an opportunity, and said: "They are many and we are few, but in the midst of this sandstorm our numbers will not be discernible; victory will go to the strenuous fighter, and the wind will be our best ally." Accordingly, Fu Yen-chung made a sudden and wholly unexpected onslaught with his cavalry, routed the barbarians and succeeded in breaking through to safety.]

37. Such is the art of warfare.

See Col. Henderson, op. cit. vol. I. p. 426.

For a number of maxims on this head, see "Marshal Turenne" (Longmans, 1907), p. 29.

VIII Variations in Tactics

[The heading means literally "The Nine Variations," but as Sun Tzu does not appear to enumerate these, and as, indeed, he has already told us (V SS. 6-11) that such deflections from the ordinary course are practically innumerable, we have little option but to follow Wang Hsi, who says that "Nine" stands for an indefinitely large number. "All it means is that in warfare we ought to very our tactics to the utmost degree do not know what Ts'ao Kung makes these Nine Variations out to be, but it has been suggested that they are connected with the Nine Situations" - of chapt. XI. This is the view adopted by Chang Yu. The only other alternative is to suppose that something has been lost — a supposition to which the unusual shortness of the chapter lends some weight.]

1. Sun Tzu said: In war, the general receives his commands from the sovereign, collects his army and concentrates his forces.

[Repeated from VII. ss. 1, where it is certainly more in place. It may have been interpolated here merely in order to supply a beginning to the chapter.]

2. When in difficult country, do not encamp. In country where high roads intersect, join hands with your allies. Do not linger in dangerously isolated positions.

[The last situation is not one of the Nine Situations as given in the beginning of chap. XI, but occurs later on (ibid, ss. 43. q.v.). Chang Yu defines this situation as being situated across the frontier, in hostile territory. Li Chfuan says it is "country in which there are no springs or wells, flocks or herds, vegetables or firewood;" Chia Lin, "one of gorges, chasms and precipices, without a road by which to advance."]

In hemmed-in situations, you must resort to stratagem. In desperate position, you must fight.

3. There are roads which must not be followed, ["Especially those

leading through narrow defiles," says Li Chfuan, "where an ambush is to be feared."] armies which must be not attacked, [More correctly, perhaps, "there are times when an army must not be attacked." Chf en Hao says: "When you see your way to obtain a rival advantage, but are powerless to inflict a real defeat, refrain from attacking, for fear of overtaxing your men's strength."] towns which must not be besieged, [Cf. III. ss. 4 Ts'ao Kung gives an interesting illustration from his own experience. When invading the territory of Hsu-chou, he ignored the city of Hua-pi, which lay directly in his path, and pressed on into the heart of the country. This excellent strategy was rewarded by the subsequent capture of no fewer than fourteen important district cities. Chang Yu says: "No town should be attacked which, if taken, cannot be held, or if left alone, will not cause any trouble." Hsun Ying, when urged to attack Pi-yang, replied: "The city is small and well-fortified; even if I succeed intaking it, it will be no great feat of arms; whereas if I fail, I shall make myself a laughing-stock." In the seventeenth century, sieges still formed a large proportion of war. It was Turenne who directed attention to the importance of marches, countermarches and maneuvers. He said: "It is a great mistake to waste men in taking a town when the same expenditure of soldiers will gain a province."] positions which must not be contested, commands of the sovereign which must not be obeyed.

[This is a hard saying for the Chinese, with their reverence for authority, and Wei Liao Tzu (quoted by Tu Mu) is moved to exclaim: "Weapons are baleful instruments, strife is antagonistic to virtue, a military commander is the negation of civil order!" The unpalatable fact remains, however, that even Imperial wishes must be subordinated to military necessity.]

The general who thoroughly understands the advantages that accompany variation of tactics knows how to handle his troops.

The general who does not understand these, may be well acquainted with the configuration of the country, yet he will not be able to turn his knowledge to practical account.

[Literally, "get the advantage of the ground," which means not only securing good positions, but availing oneself of natural advantages in every possible way. Chang Yu says: "Every kind of ground is characterized by certain natural features, and also gives scope for a certain variability of plan. How it is possible to turn these natural features to account unless topographical knowledge is supplemented by versatility of mind?"]

So, the student of war who is unversed in the art of war of varying his plans, even though he be acquainted with the Five Advantages, will fail to make the best use of his men.

[Chia Lin tells us that these imply five obvious and generally advantageous lines of action, namely: "if a certain road is short, it must be followed; if an army is isolated, it must be attacked; if a town is in a parlous condition, it must be besieged; if a position can be stormed, it must be attempted; and if consistent with military operations, the ruler's commands must be obeyed." But there are circumstances which sometimes forbid a general to use these advantages. For instance, "a certain road may be the shortest way for him, but if he knows that it abounds in natural obstacles, or that the enemy has laid an ambush on it, he will not follow that road. A hostile force may be open to attack, but if he knows that it is hard-pressed and likely to fight with desperation, he will refrain from striking," and so on.]

Hence in the wise leader's plans, considerations of advantage and of disadvantage will be blended together.

["Whether in an advantageous position or a disadvantageous one," says Ts'ao Kung, "the opposite state should be always present to your mind."]

8. If our expectation of advantage be tempered in this way, we may succeed in accomplishing the essential part of our schemes.

[Tu Mu says: "If we wish to wrest an advantage from the enemy, we must not fix our minds on that alone, but allow for the possibility of the enemy also doing some harm to us, and let this enter as a factor into our calculations."]

9. If, on the other hand, in the midst of difficulties we are always

ready to seize an advantage, we may extricate ourselves from misfortune.

[Tu Mu says: "If I wish to extricate myself from a dangerous position, I must consider not only the enemy's ability to injure me, but also my own ability to gain an advantage over the enemy. If in my counsels these two considerations are properly blended, I shall succeed in liberating myselfFor instance; if I am surrounded by the enemy and only think of effecting an escape, the nervelessness of my policy will incite my adversary to pursue and crush me; it would be far better to encourage my men to deliver a bold counter-attack, and use the advantage thus gained to free myself from the enemy's toils." See the story of Ts'ao Ts'ao, VII. ss. 35, note.]

10. Reduce the hostile chiefs by inflicting damage on them; [Chia Lin enumerates several ways of inflicting this injury, some of which would only occur to the Oriental mind: — "Entice away the enemy's best and wisest men, so that he may be left without counselors. Introduce traitors into his country, that the government policy may be rendered futile. Foment intrigue and deceit, and thus sow dissension between the ruler and his ministers. By means of every artful contrivance, cause deterioration amongst his men and waste of his treasure. Corrupt his morals by insidious gifts leading him into excess. Disturb and unsettle his mind by presenting him with lovely women." Chang Yu (after Wang Hsi) makes a different interpretation of Sun Tzu here: "Get the enemy into a position where he must suffer injury, and he will submit of his own accord."] and make trouble for them, [Tu Mu, in this phrase, in his interpretation indicates that trouble should be make for the enemy affecting their "possessions," or, as we might say, "assets," which he considers to be "a large army, a rich exchequer, harmony amongst the soldiers, punctual fulfillment of commands." These give us a whip-hand over the enemy.] and keep them constantly engaged; [Literally, "make servants of them." Tu Yu says "prevent the from having any rest."] hold out specious allurements, and make them rush to any given point.

[Meng Shih's note contains an excellent example of the idiomatic use of: "cause them to forget PIEN (the reasons for acting otherwise than on their first impulse), and hasten in our direction."]

11. The art of war teaches us to rely not on the likelihood of the enemy's not coming, but on our own readiness to receive him; not on the chance of his not attacking, but rather on the fact that we have made our position unassailable.

12. There are five dangerous faults which may affect a general:
(1) Recklessness, which leads to destruction; ["Bravery without forethought," as Ts'ao Kung analyzes it, which causes a man to fight blindly and desperately like a mad bull. Such an opponent, says Chang Yu, "must not be encountered with brute force, but may be lured into an ambush and slain." Cf. Wu Tzu, chap. IV. ad init.: "In estimating the character of a general, men are wont to pay exclusive attention to his courage, forgetting that courage is only one out of many qualities which a general should possess. The merely brave man is prone to fight recklessly; and he who fights recklessly, without any perception of what is expedient, must be condemned." Ssu-ma Fa, too, make the incisive remark: "Simply going to one's death does not bring about victory."]
(2) cowardice, which leads to capture; [Ts'ao Kung defines the Chinese word translated here as "cowardice" as being of the man "whom timidity prevents from advancing to seize an advantage," and Wang Hsi adds "who is quick to flee at the sight of danger." Meng Shih gives the closer paraphrase "he who is bent on returning alive," this is, the man who will never take a risk. But, as Sun Tzu knew, nothing is to be achieved in war unless you are willing to take risks. "Tai Kung said: "He who lets an advantage slip will subsequently bring upon himself real disaster." In 404 A.D., Liu Yu pursued the rebel Huan Hsuan up the Yangtsze and fought a naval battle with him at the island of Chfeng- hung. The loyal troops numbered only a few thousands, while their opponents were in great force. But Huan Hsuan, fearing the fate which was in store for him should be be overcome, had a light boat made fast to the side of his war- junk, so that he might escape, if necessary, at a moment's notice. The natural result was that the fighting spirit of his soldiers was utterly quenched, and when the loyalists made an attack from windward with fireships, all striving with the utmost ardor to be first in the fray, Huan Hsuan's forces were routed, had to burn all their baggage and fled for two days and nights without stopping. Chang Yu tells a somewhat similar story of Chao Ying-chfi, a general of the Chin State who during a battle with the army of Chfu in 597 B.C. had a boat

kept in readiness for him on the river, wishing in case of defeat to be the first to get across.]

(3) a hasty temper, which can be provoked by insults; [Tu Mu tells us that Yao Hsing, when opposed in 357 A.D. by Huang Mei, Teng Chfiang and others shut himself up behind his walls and refused to fight. Teng Chfiang said: "Our adversary is of a choleric temper and easily provoked; let us make constant sallies and break down his walls, then he will grow angry and come out. Once we can bring his force to battle, it is doomed to be our prey." This plan was acted upon, Yao Hsiang came out to fight, was lured as far as San-yuan by the enemy's pretended flight, and finally attacked and slain.]

(4) a delicacy of honor which is sensitive to shame; [This need not be taken to mean that a sense of honor is really a defect in a general. What Sun Tzu condemns is rather an exaggerated sensitiveness to slanderous reports, the thin-skinned man who is stung by opprobrium, however undeserved. Mei Yao-chfen truly observes, though somewhat paradoxically: "The seek after glory should be careless of public opinion."]

(5) over-solicitude for his men, which exposes him to worry and trouble. [Here again, Sun Tzu does not mean that the general is to be careless of the welfare of his troops. All he wishes to emphasize is the danger of sacrificing any important military advantage to the immediate comfort of his men. This is a shortsighted policy, because in the long run the troops will suffer more from the defeat, or, at best, the prolongation of the war, which will be the consequence. A mistaken feeling of pity will often induce a general to relieve a beleaguered city, or to reinforce a hard-pressed detachment, contrary to his military instincts. It is now generally admitted that our repeated efforts to relieve Ladysmith in the South African War were so many strategical blunders which defeated their own purpose. And in the end, relief came through the very man who started out with the distinct resolve no longer to subordinate the interests of the whole to sentiment in favor of a part. An old soldier of one of our generals who failed most conspicuously in this war, tried once, I remember, to defend him to me on the ground that he was always "so good to his men." By this plea, had he but known it, he was only condemning him out of Sun Tzu's mouth.]

13. These are the five besetting sins of a general, ruinous to the conduct of war.

14. When an army is overthrown and its leader slain, the cause will surely be found among these five dangerous faults. Let them be a subject of meditation.

"Marshal Turenne," p. 50.

IX. The Army on the March

[The contents of this interesting chapter are better indicated in ss. 1 than by this heading.]

1. Sun Tzu said: We come now to the question of encamping the army, and observing signs of the enemy. Pass quickly over mountains, and keep in the neighborhood of valleys.

[The idea is, not to linger among barren uplands, but to keep close to supplies of water and grass. Cf. Wu Tzu, ch. 3: "Abide not in natural ovens," i.e. "the openings of valleys." Chang Yu tells the following anecdote: Wu-tu Chfiang was a robber captain in the time of the Later Han, and Ma Yuan was sent to exterminate his gang. Chfiang having found a refuge in the hills, Ma Yuan made no attempt to force a battle, but seized all the favorable positions commanding supplies of water and forage. Chfiang was soon in such a desperate plight for want of provisions that he was forced to make a total surrender. He did not know the advantage of keeping in the neighborhood of valleys."]

2. Camp in high places, [Not on high hills, but on knolls or hillocks elevated above the surrounding country.] facing the sun.

[Tu Mu takes this to mean "facing south," and Chf en Hao "facing east."]

Do not climb heights in order to fight. So much for mountain warfare.

3. After crossing a river, you should get far away from it.

["In order to tempt the enemy to cross after you," according to Ts'ao Kung, and also, says Chang Yu, "in order not to be impeded in your evolutions." The T'UNG TIEN reads, "If THE ENEMY crosses a river," etc. But in view of the next sentence, this is almost certainly an interpolation.]

4. When an invading force crosses a river in its onward march, do not advance to meet it in mid-stream. It will be best to let half the army get across, and then deliver your attack.

[Li Chfuan alludes to the great victory won by Han Hsin over Lung Chu at the Wei River. Turning to the CI-TIEN HAN SHU, ch. 34, fol. 6 verso, we find the battle described as follows: "The two armies were drawn up on opposite sides of the river. In the night, Han Hsin ordered his men to take some ten thousand sacks filled with sand and construct a dam higher up. Then, leading half his army across, he attacked Lung Chu; but after a time, pretending to have failed in his attempt, he hastily withdrew to the other bank. Lung Chu wasmuch elated by this unlooked-for success, and exclaiming: "I felt sure that Han Hsin was really a coward!" he pursued him and began crossing the river in his turn. Han Hsin now sent a party to cut open the sandbags, thus releasing a great volume of water, which swept down and prevented the greater portion of Lung Chu's army from getting across. He then turned upon the force which had been cut off, and annihilated it, Lung Chu himself being amongst the slain. The rest of the army, on the further bank, also scattered and fled in all directions.]

5. If you are anxious to fight, you should not go to meet the invader near a river which he has to cross. [For fear of preventing his crossing.]

6. Moor your craft higher up than the enemy, and facing the sun.

[See supra, ss. 2. The repetition of these words in connection with water is very awkward. Chang Yu has the note: "Said either of troops marshaled on the river-bank, or of boats anchored in the stream itself; in either case it is essential to be higher than the enemy and facing the sun." The other commentators are not at all explicit.]

Do not move up-stream to meet the enemy. [Tu Mu says: "As water flows downwards, we must not pitch our camp on the lower reaches of a river, for fear the enemy should open the sluices and sweep us away in a flood. Chu-ko Wu-hou has remarked that 'in river warfare we must not advance against the stream,' which is as much as to say that our fleet must not be anchored below that of the enemy, for then they would be able to take advantage of the current and make short work of us." There is also the danger, noted by other commentators, that the enemy may throw poison on the water to be carried down to us.] So much for river warfare.

7. In crossing salt-marshes, your sole concern should be to get over them quickly, without any delay. [Because of the lack of fresh water, the poor quality of the herbage, and last but not least, because they are low, flat, and exposed to attack.]

8. If forced to fight in a salt-marsh, you should have water and grass near you, and get your back to a clump of trees. [Li Chfuan remarks that the ground is less likely to be treacherous where there are trees, while Tu Mu says that they will serve to protect the rear.] So much for operations in salt-marches.

9. In dry, level country, take up an easily accessible position with rising ground to your right and on your rear, [Tu Mu quotes T'ai Kung as saying: "An army should have a stream or a marsh on its left, and a hill or tumulus on its right."] so that the danger may be in front, and safety lie behind. So much for campaigning in flat country.

10. These are the four useful branches of military knowledge [Those, namely, concerned with (1) mountains, (2) rivers, (3) marshes, and (4) plains. Compare Napoleon's "Military Maxims," no. 1.] which enabled the Yellow Emperor to vanquish four several sovereigns.

[Regarding the "Yellow Emperor": Mei Yao-chf en asks, with some plausibility, whether there is an error in the text as nothing is known of Huang Ti having conquered four other Emperors. The SHIH CHI (ch. 1 ad init.) speaks only of his victories over Yen Ti and Chfih Yu. In the LIU T x AO it is mentioned that he "fought seventy battles and pacified the Empire." Ts'ao Kung's explanation is, that the Yellow Emperor was the first to institute the feudal system of vassals princes, each of whom (to the number of four) originally bore the title of Emperor. Li Chfuan tells us that the art of war originated under Huang Ti, who received it from his Minister Feng Hou.]

11. All armies prefer high ground to low. ["High Ground," says Mei Yao-chfen, "is not only more agreement and salubrious, but more convenient from a military point of view; low ground is not only damp and unhealthy, but also disadvantageous for fighting."] and sunny places to dark.

12. If you are careful of your men, [Ts'ao Kung says: "Make for fresh water and pasture, where you can turn out your animals to graze."] and camp on hard ground, the army will be free from disease of every kind, [Chang Yu says: "The dryness of the climate will prevent the outbreak of illness."] and this will spell victory.

13. When you come to a hill or a bank, occupy the sunny side, with the slope on your right rear. Thus you will at once act for the benefit of your soldiers and utilize the natural advantages of the ground.

14. When, in consequence of heavy rains up-country, a river which you wish to ford is swollen and flecked with foam, you must wait until it subsides.

15. Country in which there are precipitous cliffs with torrents running between, deep natural hollows, [The latter defined as "places enclosed on every side by steep banks, with pools of water at the bottom.] confined places, [Defined as "natural pens or prisons" or "places surrounded by precipices on three sides — easy to get into, but hard to get out of."] tangled thickets, [Defined as "places covered with such dense undergrowth that spears cannot be used."] quagmires

[Defined as "low-lying places, so heavy with mud as to be impassable for chariots and horsemen."] and crevasses, [Defined by Mei Yao-chfen as "a narrow difficult way between beetling cliffs." Tu Mu's note is "ground covered with trees and rocks, and intersected by numerous ravines and pitfalls." This is very vague, but Chia Lin explains it clearly enough as a defile or narrow pass, and Chang Yu takes much the same view. On the whole, the weight of the commentators certainly inclines to the rendering "defile." But the ordinary meaning of the Chinese in one place is "a crack or fissure" and the fact that the meaning of the Chinese elsewhere in the sentence indicates something in the nature of a defile, make me think that Sun Tzu is here speaking of crevasses.] should be left with all possible speed and not approached.

16. While we keep away from such places, we should get the enemy to approach them; while we face them, we should let the enemy have them on his rear.

17. If in the neighborhood of your camp there should be any hilly country, ponds surrounded by aquatic grass, hollow basins filled with reeds, or woods with thick undergrowth, they must be carefully routed out and searched; for these are places where men in ambush or insidious spies are likely to be lurking.

[Chang Yu has the note: "We must also be on our guard against traitors who may lie in close covert, secretly spying out our weaknesses and overhearing our instructions."]

18. When the enemy is close at hand and remains quiet, he is relying on the natural strength of his position.

[Here begin Sun Tzu's remarks on the reading of signs, much of which is so good that it could almost be included in a modern manual like Gen. Baden-Powell's "Aids to Scouting."]

19. When he keeps aloof and tries to provoke a battle, he is anxious for the other side to advance. [Probably because we are in a strong position from which he wishes to dislodge us. "If he came close up to us, says Tu Mu, "and tried to force a battle, he would seem to despise us, and there would be less probability of our responding to the challenge."]

20. If his place of encampment is easy of access, he is tendering a bait.

21. Movement amongst the trees of a forest shows that the enemy is advancing. [Ts'ao Kung explains this as "felling trees to clear a passage," and Chang Yu says: "Every man sends out scouts to climb high places and observe the enemy. If a scout sees that the trees of a forest are moving and shaking, he may know that they are being cut down to clear a passage for the enemy's march."]

The appearance of a number of screens in the midst of thick grass means that the enemy wants to make us suspicious. [Tu Yu's explanation, borrowed from Ts'ao Kung's, is as follows: "The presence of a number of screens or sheds in the midst of thick vegetation is a sure sign that the enemy has fled and, fearing pursuit, has constructed these hiding- places in order to make us suspect an ambush." It appears that these "screens" were hastily knotted together out of any long grass which the retreating enemy happened to come across.]

22. The rising of birds in theirflight is the sign of an ambuscade. [Chang Yu's explanation is doubtless right: "When birds that are flying along in a straight line suddenly shoot upwards, it means that soldiers are in ambush at the spot beneath."] Startled beasts indicate that a sudden attack is coming.

23. When there is dust rising in a high column, it is the sign of chariots advancing; when the dust is low, but spread over a wide area, it betokens the approach of infantry. ["High and sharp," or rising to a peak, is of course somewhat exaggerated as applied to dust. The commentators explain the phenomenon by saying that horses and chariots, being heavier than men, raise more dust, and also follow one another in the same wheel-track, whereas foot-soldiers would be marching in ranks, many abreast. According to Chang Yu, "every army on the march must have scouts some way in advance, who on sighting dust raised by the enemy, will gallop back and report it to the commander-in-chief." Cf. Gen. Baden-Powell: "As you move along, say, in a hostile country, your eyes should be looking afar for the enemy or any signs of him: figures, dust rising, birds getting up, glitter of arms, etc."]

When it branches out in different directions, it shows that parties have been sent to collect firewood. A few clouds of dust moving to and fro signify that the army is encamping. [Chang Yu says: "In apportioning the defenses for a cantonment, light horse will be sent out to survey the position and ascertain the weak and strong points all along its circumference. Hence the small quantity of dust and its motion."]

24. Humble words and increased preparations are signs that the enemy is about to advance.

["As though they stood in great fear of us," says Tu Mu. "Their object is to make us contemptuous and careless, after which they will attack us." Chang Yu alludes to the story of "Tien Tan of the Chf i-mo against the Yen forces, led by Chfi Chieh. In ch. 82 of the SHIH CHI we read: 'Tien Tan openly said: 'My only fear is that the Yen army may cut off the noses of their Chfi prisoners and place them in the front rank to fight against us; that would be the undoing of our city.' The other side being informed of this speech, at once acted on the suggestion; but those within the city were enraged at seeing their fellow-countrymen thus mutilated, and fearing only lest they should fall into the enemy's hands, were nerved to defend themselves more obstinately than ever. Once again Tien Tan sent back converted spies who reported these words to the enemy: "What I dread most is that the men of Yen may dig up the ancestral tombs outside the town, and by inflicting this indignity on our forefathers cause us to become faint-hearted.' Forthwith the besiegers dug up all the graves and burned the corpses lying in them.

And the inhabitants of Chi-mo, witnessing the outrage from the city-walls, wept passionately and were all impatient to go out and fight, their fury being increased tenfold. Tien Tan knew then that his soldiers were ready for any enterprise. But instead of a sword, he himself too a mattock in his hands, and ordered others to be distributed amongst his best warriors, while the ranks were filled up with their wives and concubines. He then served out all the remaining rations and bade his men eat their fill. The regular soldiers were told to keep out of sight, and the walls were manned with the old and weaker men and with women. This done, envoys were dispatched to the enemy's camp to arrange terms of surrender, whereupon the Yen army began shouting for joy. Fien Tan also collected 20,000 ounces of silver from the people, and got the wealthy citizens of Chi-mo to send it to the Yen general with the prayer that, when the town capitulated, he would allow their homes to be plundered or their women to be maltreated. Chfi Chieh, in high good humor, granted their prayer; but his army now became increasingly slack and careless. Meanwhile, Tien Tan got together a thousand oxen, decked them with pieces of red silk, painted their bodies, dragon-like, with colored stripes, and fastened sharp blades on their horns and well-greased rushes on their tails.

When night came on, he lighted the ends of the rushes, and drove the oxen through a number of holes which he had pierced in the walls, backing them up with a force of 5000 picked warriors. The animals, maddened with pain, dashed furiously into the enemy's camp where they caused the utmost confusion and dismay; for their tails acted as torches, showing up the hideous pattern on their bodies, and the weapons on their horns killed or wounded any with whom they came into contact. In the meantime, the band of 5000 had crept up with gags in their mouths, and now threw themselves on the enemy.

At the same moment a frightful din arose in the city itself, all those that remained behind making as much noise as possible by banging drums and hammering on bronze vessels, until heaven and earth were convulsed by the uproar. Terror-stricken, the Yen army fled in disorder, hotly pursued by the men of Chfi, who succeeded in slaying their general Chfi ChienThe result of the battle was the ultimate recovery of some seventy cities which had belonged to the Chfi State."]

Violent language and driving forward as if to the attack are signs that he will retreat.

25. When the light chariots come out first and take up a position on the wings, it is a sign that the enemy is forming for battle.

26. Peace proposals unaccompanied by a sworn covenant indicate a plot. [The reading here is uncertain. Li Chf uan indicates "a treaty confirmed by oaths and hostages." Wang Hsi and Chang Yu, on the other hand, simply say "without reason," "on a frivolous pretext."]

27. When there is much running about [Every man hastening to his proper place under his own regimental banner.] and the soldiers fall into rank, it means that the critical moment has come.

28. When some are seen advancing and some retreating, it is a lure.

29. When the soldiers stand leaning on their spears, they are faint from want of food.

30. If those who are sent to draw water begin by drinking themselves, the army is suffering from thirst. [As Tu Mu remarks: "One may know the condition of a whole army from the behavior of a single man."]

31. If the enemy sees an advantage to be gained and makes no effort to secure it, the soldiers are exhausted.

32. If birds gather on any spot, it is unoccupied.

[A useful fact to bear in mind when, for instance, as Chfen Hao says, the enemy has secretly abandoned his camp.]

Clamor by night betokens nervousness.

33. If there is disturbance in the camp, the general's authority is weak. If the banners and flags are shifted about, sedition is afoot. If the officers are angry, it means that the men are weary. [Tu Mu understands the sentence differently: "If all the officers of an army are angry with their general, it means that they are broken with fatigue" owing to the exertions which he has demanded from them.]

34. When an army feeds its horses with grain and kills its cattle for food, [In the ordinary course of things, the men would be fed on grain and the horses chiefly on grass.] and when the men do not hang their cooking-pots over the camp-fires, showing that they will not return to their tents, you may know that they are determined to fight to the death.

[I may quote here the illustrative passage from the HOU HAN SHU, ch. 71 , given in abbreviated form by the P X EI WEN YUN FU: "The rebel Wang Kuo of Liang was besieging the town of Chfen-ts'ang, and Huang-fu Sung, who was in supreme command, and Tung Cho were sent out against him. The latter pressed for hasty measures, but Sung turned a deaf ear to his counsel. At last the rebels were utterly worn out, and began to throw down their weapons of their own accord. Sung was not advancing to the attack, but Cho said: 'It is a principle of war not to pursue desperate men and not to press a retreating host.' Sung answered: 'That does not apply here. What I am about to attack is a jaded army, not a retreating host; with disciplined troops I am falling on a disorganized multitude, not a band of desperate men.' Thereupon he advances to the attack unsupported by his colleague, and routed the enemy, Wang Kuo being slain."]

35. The sight of men whispering together in small knots or speaking in subdued tones points to disaffection amongst the rank and file.

36. Too frequent rewards signify that the enemy is at the end of his resources; [Because, when an army is hard pressed, as Tu Mu says, there is always a fear of mutiny, and lavish rewards are given to keep the men in good temper.] too many punishments betray a condition of dire distress.

[Because in such case discipline becomes relaxed, and unwonted severity is necessary to keep the men to their duty.]

37. To begin by bluster, but afterwards to take fright at the enemy's numbers, shows a supreme lack of intelligence.

[I follow the interpretation of Ts'ao Kung, also adopted by Li Chf uan, Tu Mu, and Chang Yu. Another possible meaning set forth by Tu Yu, Chia Lin, Mei Tao-chfen and Wang Hsi, is: "The general who is first tyrannical towards his men, and then in terror lest they should mutiny, etc." This would connect the sentence with what went before about rewards and punishments.]

38. When envoys are sent with compliments in their mouths, it is a sign that the enemy wishes for a truce.

[Tu Mu says: "If the enemy open friendly relations be sending hostages, it is a sign that they are anxious for an armistice, either because their strength is exhausted or for some other reason." But it hardly needs a Sun Tzu to draw such an obvious inference.]

39. If the enemy's troops march up angrily and remain facing ours for a long time without either joining battle or taking themselves off again, the situation is one that demands great vigilance and circumspection.

[Ts'ao Kung says a maneuver of this sort may be only a ruse to gain time for an unexpected flank attack or the laying of an ambush.]

40. If our troops are no more in number than the enemy, that is amply sufficient; it only means that no direct attack can be made.

[Literally, "no martial advance." That is to say, CHENG tactics and frontal attacks must be eschewed, and stratagem resorted to instead.]

What we can do is simply to concentrate all our available strength, keep a close watch on the enemy, and obtain reinforcements.

[This is an obscure sentence, and none of the commentators succeed in squeezing very good sense out of it. I follow Li Chfuan, who appears to offer the simplest explanation: "Only the side that gets more men will win." Fortunately we have Chang Yu to expound its meaning to us in language which is lucidity itself: "When the numbers are even, and no favorable opening presents itself, although we may not be strong enough to deliver a sustained attack, we can find additional recruits amongst our sutlers and camp- followers, and then, concentrating our forces and keeping a close watch on the enemy, contrive to snatch the victory. But we must avoid borrowing foreign soldiers to help us." He then quotes from Wei Liao Tzu, ch. 3: "The nominal strength of mercenary troops may be 100,000, but their real value will be not more than half that figure."]

41. He who exercises no forethought but makes light of his opponents is sure to be captured by them.

[CIV en Hao, quoting from the TSO CHUAN, says: "If bees and scorpions carry poison, how much more will a hostile state! Even a puny opponent, then, should not be treated with contempt."]

42. If soldiers are punished before they have grown attached to you, they will not prove submissive; and, unless submissive, then will be practically useless. If, when the soldiers have become attached to you, punishments are not enforced, they will still be unless.

43. Therefore soldiers must be treated in the first instance with humanity, but kept under control by means of iron discipline.

[Yen Tzu [B.C. 493] said of Ssu-ma Jang-chu: "His civil virtues endeared him to the people; his martial prowess kept his enemies in awe." Cf. Wu Tzu, ch. 4 init.: "The ideal commander unites culture with a warlike temper; the profession of arms requires a combination of hardness and tenderness."]

This is a certain road to victory.

44. If in training soldiers commands are habitually enforced, the army will be well-disciplined; if not, its discipline will be bad.

45. If a general shows confidence in his men but always insists on his orders being obeyed, [Tu Mu says: "A general ought in time of peace to show kindly confidence in his men and also make his authority respected, so that when they come to face the enemy, orders may be executed and discipline maintained, because they all trust and look up to him." What Sun Tzu has said in ss. 44, however, would lead one rather to expect something like this: "If a general is always confident that his orders will be carried out," etc."] the gain will be mutual.

[Chang Yu says: "The general has confidence in the men under his command, and the men are docile, having confidence in him. Thus the gain is mutual" He quotes a pregnant sentence from Wei Liao Tzu, ch. 4: "The art of giving orders is not to try to rectify minor blunders and not to be swayed by petty doubts." Vacillation and fussiness are the surest means of sapping the confidence of an army.]

X Terrain

[Only about a third of the chapter, comprising ss. ss. 1- 13, deals with "terrain," the subject being more fully treated in ch. XI. The "six calamities" are discussed in SS. 14-20, and the rest of the chapter is again a mere string of desultory remarks, though not less interesting, perhaps, on that account.]

1. Sun Tzu said: We may distinguish six kinds of terrain, to wit: (1) Accessible ground; [Mei Yao-chf en says: "plentifully provided with roads and means of communications."] (2) entangling ground; [The same commentator says: "Net-like country, venturing into which you become entangled."] (3) temporizing ground; [Ground which allows you to "stave off" or "delay."] (4) narrow passes; (5) precipitous heights; (6) positions at a great distance from the enemy.

[It is hardly necessary to point out the faultiness of this classification. A strange lack of logical perception is shown in the Chinaman's unquestioning acceptance of glaring cross-divisions such as the above.]

2. Ground which can be freely traversed by both sides is called ACCESSIBLE.

3. With regard to ground of this nature, be before the enemy in occupying the raised and sunny spots, and carefully guard your line of supplies.

[The general meaning of the last phrase is doubtlessly, as Tu Yu says, "not to allow the enemy to cut your communications." In view of Napoleon's dictum, "the secret of war lies in the communications," [1] we could wish that Sun Tzu had done more than skirt the edge of this important subject here and in I. ss. 10, VII. ss. 11. Col. Henderson says: "The line of supply may be said to be as vital to the existence of an army as the heart to the life of a human being. Just as the duelist who finds his adversary's point menacing him with certain death, and his own guard astray, is compelled to conform to his adversary's movements, and to content himself with warding off his thrusts, so the commander whose communications are suddenly threatened finds himself in a false position, and he will be fortunate if he has not to change all his plans, to split up his force into more or less isolated detachments, and to fight with inferior numbers on ground which he has not had time to prepare, and where defeat will not be an ordinary failure, but will entail the ruin or surrender of his whole army."

Then you will be able to fight with advantage.

4. Ground which can be abandoned but is hard to re- occupy is called ENTANGLING.

5. From a position of this sort, if the enemy is unprepared, you may sally forth and defeat him. But if the enemy is prepared for your coming, and you fail to defeat him, then, return being impossible, disaster will ensue.

6. When the position is such that neither side will gain by making the first move, it is called TEMPORIZING ground.

[Tu Mu says: "Each side finds it inconvenient to move, and the situation remains at a deadlock."]

7. In a position of this sort, even though the enemy should offer us an attractive bait, [Tu Yu says, "turning their backs on us and pretending to flee." But this is only one of the lures which might induce us to quit our position.] it will be advisable not to stir forth, but rather to retreat, thus enticing the enemy in his turn; then, when part of his army has come out, we may deliver our attack with advantage.

8. With regard to NARROW PASSES, if you can occupy them first, let them be strongly garrisoned and await the advent of the enemy.

[Because then, as Tu Yu observes, "the initiative will lie with us, and by making sudden and unexpected attacks we shall have the enemy at our mercy."]

9. Should the army forestall you in occupying a pass, do not go after him if the pass is fully garrisoned, but only if it is weakly garrisoned.

10. With regard to PRECIPITOUS HEIGHTS, if you are beforehand with your adversary, you should occupy the raised and sunny spots, and there wait for him to come up.

[Ts'ao Kung says: "The particular advantage of securing heights and defiles is that your actions cannot then be dictated by the enemy." [For the enunciation of the grand principle alluded to, see VI. ss. 2]. Chang Yu tells the following anecdote of P'ei Hsing-chien (A.D. 619-682), who was sent on a punitive expedition against the Turkic tribes. "At night he pitched his camp as usual, and it had already been completely fortified by wall and ditch, when suddenly he gave orders that the army should shift its quarters to a hill near by. This was highly displeasing to his officers, who protested loudly against the extra fatigue which it would entail on the men. P x ei Hsing-chien, however, paid no heed to their remonstrances and had the camp moved as quickly as possible. The same night, a terrific storm came on, which flooded their former place of encampment to the depth of over twelve feet. The recalcitrant officers were amazed at the sight, and owned that they had been in the wrong. 'How did you know what was going to happen?' they asked. P'ei Hsing-chien replied: 'From this time forward be content to obey orders without asking unnecessary questions.' From this it may be seen," Chang Yu continues, "that high and sunny places are advantageous not only for fighting, but also because they are immune from disastrous floods."]

11. If the enemy has occupied them before you, do not follow him, but retreat and try to entice him away.

[The turning point of Li Shih-min's campaign in 621 A.D. against the two rebels, Tou Chien-te, King of Hsia, and Wang Shih-chfung, Prince of Cheng, was his seizure of the heights of Wu-lao, in spike of which Tou Chien-te persisted in his attempt to relieve his ally in Lo-yang, was defeated and taken prisoner. See CHIU "TANG, ch. 2, fol. 5 verso, and also ch. 54.]

12. If you are situated at a great distance from the enemy, and the strength of the two armies is equal, it is not easy to provoke a battle, [The point is that we must not think of undertaking a long and wearisome march, at the end of which, as Tu Yu says, "we should be exhausted and our adversary fresh and keen."] and fighting will be to your disadvantage.

13. These six are the principles connected with Earth.

[Or perhaps, "the principles relating to ground." See, however, I. ss. 8.]

The general who has attained a responsible post must be careful to study them.

14. Now an army is exposed to six several calamities, not arising from natural causes, but from faults for which the general is responsible. These are: (1) Flight; (2) insubordination; (3) collapse; (4) ruin; (5) disorganization; (6) rout.

15. Other conditions being equal, if one force is hurled against another ten times its size, the result will be the FLIGHT of the former.

16. When the common soldiers are too strong and their officers too weak, the result is INSUBORDINATION.

[Tu Mu cites the unhappy case of Tien Pu [HSIN TANG SHU, ch. 148], who was sent to Wei in 821 A.D. with orders to lead an army against Wang Ting-ts'ou. But the whole time he was in command, his soldiers treated him with the utmost contempt, and openly flouted his authority by riding about the camp on donkeys, several thousands at a time. Tien Pu was powerless to put a stop to this conduct, and when, after some months had passed, he made an attempt to engage the enemy, his troops turned tail and dispersed in every direction. After that, the unfortunate man committed suicide by cutting his throat.]

When the officers are too strong and the common soldiers too weak, the result is COLLAPSE. [Ts'ao Kung says: "The officers are energetic and want to press on, the common soldiers are feeble and suddenly collapse."]

17. When the higher officers are angry and insubordinate, and on meeting the enemy give battle on their own account from a feeling of resentment, before the commander-in-chief can tell whether or no he is in a position to fight, the result is RUIN.

[Wang Hsi's note is: "This means, the general is angry without cause, and at the same time does not appreciate the ability of his subordinate officers; thus he arouses fierce resentment and brings an avalanche of ruin upon his head."]

18. When the general is weak and without authority; when his orders are not clear and distinct; [Wei Liao Tzu (ch. 4) says: "If the commander gives his orders with decision, the soldiers will not wait to hear them twice; if his moves are made without vacillation, the soldiers will not be in two minds about doing their duty." General Baden-Powell says, italicizing the words: "The secret of getting successful work out of your trained men lies in one nutshell — in the clearness of the instructions they receive." [3] Cf. also Wu Tzu ch. 3: "the most fatal defect in a military leader is difference; the worst calamities that befall an army arise from hesitation."] when there are no fixed duties assigned to officers and men, [Tu Mu says: "Neither officers nor men have any regular routine."] and the ranks are formed in a slovenly haphazard manner, the result is utter DISORGANIZATION.

19. When a general, unable to estimate the enemy's strength, allows an inferior force to engage a larger one, or hurls a weak detachment against a powerful one, and neglects to place picked soldiers in the front rank, the result must be ROUT.

[Chang Yu paraphrases the latter part of the sentence and continues: "Whenever there is fighting to be done, the keenest spirits should be appointed to serve in the front ranks, both in order to strengthen the resolution of our own men and to demoralize the enemy." Cf. the primi ordines of Caesar ("De Bello Gallico," V. 28, 44, et al.).]

20. These are six ways of courting defeat, which must be carefully noted by the general who has attained a responsible post.

21. The natural formation of the country is the soldier's best ally; [Chf en Hao says: "The advantages of weather and season are not equal to those connected with ground."] but a power of estimating the adversary, of controlling the forces of victory, and of shrewdly calculating difficulties, dangers and distances, constitutes the test of a great general.

22. He who knows these things, and in fighting puts his knowledge into practice, will win his battles. He who knows them not, nor practices them, will surely be defeated.

23. If fighting is sure to result in victory, then you must fight, even though the ruler forbid it; if fighting will not result in victory, then you must not fight even at the ruler's bidding.

[Cf. VIII. ss. 3 fin. Huang Shih-kung of the Chun dynasty, who is said to have been the patron of Chang Liang and to have written the SAN LUEH, has these words attributed to him: "The responsibility of setting an army in motion must devolve on the general alone; if advance and retreat are controlled from the Palace, brilliant results will hardly be achieved. Hence the god-like ruler and the enlightened monarch are content to play a humble part in furthering their country's cause [lit., kneel down to push the chariot wheel]." This means that "in matters lying outside the zenana, the decision of the military commander must be absolute." Chang Yu also quote the saying: "Decrees from the Son of Heaven do not penetrate the walls of a camp."]

24. The general who advances without coveting fame and retreats without fearing disgrace, [It was Wellington, I think, who said that the hardest thing of all for a soldier is to retreat.] whose only thought is to protect his country and do good service for his sovereign, is the jewel of the kingdom.

[A noble presentiment, in few words, of the Chinese "happy warrior." Such a man, says Ho Shih, "even if he had to suffer punishment, would not regret his conduct."]

25. Regard your soldiers as your children, and they will follow you

into the deepest valleys; look upon them as your own beloved sons, and they will stand by you even unto death.

[Cf. I. ss. 6. In this connection, Tu Mu draws for us an engaging picture of the famous general Wu Chfi, from whose treatise on war I have frequently had occasion to quote: "He wore the same clothes and ate the same food as the meanest of his soldiers, refused to have either a horse to ride or a mat to sleep on, carried his own surplus rations wrapped in a parcel, and shared every hardship with his men. One of his soldiers was suffering from an abscess, and Wu Chfi himself sucked out the virus. The soldier's mother, hearing this, began wailing and lamenting. Somebody asked her, saying: 'Why do you cry? Your son is only a common soldier, and yet the commander-in-chief himself has sucked the poison from his sore.' The woman replied, 'Many years ago, Lord Wu performed a similar service for my husband, who never left him afterwards, and finally met his death at the hands of the enemy. And now that he has done the same for my son, he too will fall fighting I know not where.'" Li Chf uan mentions the Viscount of Chf u, who invaded the small state of Hsiao during the winter. The Duke of Shen said to him: "Many of the soldiers are suffering severely from the cold." So he made a round of the whole army, comforting and encouraging the men; and straightway they felt as if they were clothed in garments lined with floss silk.]

26. If, however, you are indulgent, but unable to make your authority felt; kind-hearted, but unable to enforce your commands; and incapable, moreover, of quelling disorder: then your soldiers must be likened to spoilt children; they are useless for any practical purpose.

[Li Ching once said that if you could make your soldiers afraid of you, they would not be afraid of the enemy. Tu Mu recalls an instance of stern military discipline which occurred in 219 A.D., when Lu Meng was occupying the town of Chiang-ling. He had given stringent orders to his army not to molest the inhabitants nor take anything from them by force.

Nevertheless, a certain officer serving under his banner, who happened to be a fellow-townsman, ventured to appropriate a bamboo hat belonging to one of the people, in order to wear it over his regulation helmet as a protection against the rain. Lu Meng considered that the fact of his being also a native of Ju-nan should not be allowed to palliate a clear breach of discipline, and accordingly he ordered his summary execution, the tears rolling down his face, however, as he did so. This act of severity filled the army with wholesome awe, and from that time forth, even articles dropped in the highway were not picked up.]

27. If we know that our own men are in a condition to attack, but are unaware that the enemy is not open to attack, we have gone only halfway towards victory.

[That is, Ts'ao Kung says, "the issue in this case is uncertain."]

28. If we know that the enemy is open to attack, but are unaware that our own men are not in a condition to attack, we have gone only halfway towards victory.

29. If we know that the enemy is open to attack, and also know that our men are in a condition to attack, but are unaware that the nature of the ground makes fighting impracticable, we have still gone only halfway towards victory.

30. Hence the experienced soldier, once in motion, is never bewildered; once he has broken camp, he is never at a loss.

[The reason being, according to Tu Mu, that he has taken his measures so thoroughly as to ensure victory beforehand. "He does not move recklessly," says Chang Yu, "so that when he does move, he makes no mistakes."]

31. Hence the saying: If you know the enemy and know yourself, your victory will not stand in doubt; if you know Heaven and know Earth, you may make your victory complete.

[Li Chfuan sums up as follows: "Given a knowledge of three things — the affairs of men, the seasons of heaven and the natural advantages of earth —, victory will invariably crown your battles."]

XI The Nine Situations

1. Sun Tzu said: The art of war recognizes nine varieties of ground:
(1) Dispersive ground;
(2) facile ground;
(3) contentious ground;
(4) open ground;
(5) ground of intersecting highways;
(6) serious ground;
(7) difficult ground;
(8) hemmed-in ground;
(9) desperate ground.

2. When a chieftain is fighting in his own territory, it is dispersive ground.

[So called because the soldiers, being near to their homes and anxious to see their wives and children, are likely to seize the opportunity afforded by a battle and scatter in every direction. "In their advance," observes Tu Mu, "they will lack the valor of desperation, and when they retreat, they will find harbors of refuge."]

3. When he has penetrated into hostile territory, but to no great distance, it is facile ground.

[Li Chfuan and Ho Shih say "because of the facility for retreating," and the other commentators give similar explanations. Tu Mu remarks: "When your army has crossed the border, you should burn your boats and bridges, in order to make it clear to everybody that you have no hankering after home."]

4. Ground the possession of which imports great advantage to either side, is contentious ground.

[Tu Mu defines the ground as ground "to be contended for." Ts'ao Kung says: "ground on which the few and the weak can defeat the

many and the strong," such as "the neck of a pass," instanced by Li
Chfuan. Thus, Thermopylae was of this classification because the
possession of it, even for a few days only, meant holding the entire
invading army in check and thus gaining invaluable time. Cf. Wu
Tzu, ch. V. ad in it .: "For those who have to fight in the ratio of one
to ten, there is nothing better than a narrow pass." When Lu Kuang
was returning from his triumphant expedition to Turkestan in 385
A.D., and had got as far as l-ho, laden with spoils, Liang Hsi,
administrator of Liang-chou, taking advantage of the death of Fu
Chien, King of Chun, plotted against him and was for barring his
way into the province. Yang Han, governor of Kao-chf ang,
counseled him, saying: "Lu Kuang is fresh from his victories in the
west, and his soldiers are vigorous and mettlesome. If we oppose
him in the shifting sands of the desert, we shall be no match for him,
and we must therefore try a different plan. Let us hasten to occupy
the defile at the mouth of the Kao-wu pass, thus cutting him off from
supplies of water, and when his troops are prostrated with thirst, we
can dictate our own terms without moving. Or if you think that the
pass I mention is too far off, we could make a stand against him at
the l-wu pass, which is nearer. The cunning and resource of Tzu-fang
himself would be expended in vain against the enormous strength of
these two positions." Liang Hsi, refusing to act on this advice, was
overwhelmed and swept away by the invader.]

Ground on which each side has liberty of movement is open ground.

[There are various interpretations of the Chinese adjective for this
type of ground. Ts'ao Kung says it means "ground covered with a
network of roads," like a chessboard. Ho Shih suggested: "ground on
which intercommunication is easy."]

Ground which forms the key to three contiguous states, [Ts'au Kung
defines this as: "Our country adjoining the enemy's and a third
country conterminous with both." Meng Shih instances the small
principality of Cheng, which was bounded on the north-east by Chfi,
on the west by Chin, and on the south by Chfu.] so that he who
occupies it first has most of the Empire at his command,
[The belligerent who holds this dominating position can constrain
most of them to become his allies.] is a ground of intersecting
highways, 7. When an army has penetrated into the heart of a hostile

country, leaving a number of fortified cities in its rear, it is serious ground.

[Wang Hsi explains the name by saying that "when an army has reached such a point, its situation is serious."]

8. Mountain forests, [Or simply "forests."] rugged steeps, marshes and fens — all country that is hard to traverse: this is difficult ground.

9. Ground which is reached through narrow gorges, and from which we can only retire by tortuous paths, so that a small number of the enemy would suffice to crush a large body of our men: this is hemmed in ground.

10. Ground on which we can only be saved from destruction by fighting without delay, is desperate ground.

[The situation, as pictured by Ts'ao Kung, is very similar to the "hemmed-in ground" except that here escape is no longer possible: "A lofty mountain in front, a large river behind, advance impossible, retreat blocked." Chfen Hao says: "to be on 'desperate ground' is like sitting in a leaking boat or crouching in a burning house." Tu Mu quotes from Li Ching a vivid description of the plight of an army thus entrapped: "Suppose an army invading hostile territory without the aid of local guides: — it falls into a fatal snare and is at the enemy's mercy. A ravine on the left, a mountain on the right, a pathway so perilous that the horses have to be roped together and the chariots carried in slings, no passage open in front, retreat cut off behind, no choice but to proceed in single file. Then, before there is time to range our soldiers in order of battle, the enemy is overwhelming strength suddenly appears on the scene. Advancing, we can nowhere take a breathing-space; retreating, we have no haven of refuge. We seek a pitched battle, but in vain; yet standing on the defensive, none of us has a moment's respite.

If we simply maintain our ground, whole days and months will crawl by; the moment we make a move, we have to sustain the enemy's attacks on front and rear. The country is wild, destitute of water and plants; the army is lacking in the necessaries of life, the horses are jaded and the men worn-out, all the resources of strength and skill unavailing, the pass so narrow that a single man defending it can check the onset of ten thousand; all means of offense in the hands of the enemy, all points of vantage already forfeited by ourselves: in this terrible plight, even though we had the most valiant soldiers and the keenest of weapons, how could they be employed with the slightest effect?" Students of Greek history may be reminded of the awful close to the Sicilian expedition, and the agony of the Athenians under Nicias and Demonsthenes. [See Thucydides, VII. 78 sqq.].]

11. On dispersive ground, therefore, fight not. On facile ground, halt not. On contentious ground, attack not.

[But rather let all your energies be bent on occupying the advantageous position first. So Ts'ao Kung. Li Chfuan and others, however, suppose the meaning to be that the enemy has already forestalled us, sot that it would be sheer madness to attack. In the SUN TZU HSU LU, when the King of Wu inquires what should be done in this case, Sun Tzu replies: "The rule with regard to contentious ground is that those in possession have the advantage over the other side. If a position of this kind is secured first by the enemy, beware of attacking him. Lure him away by pretending to flee — show your banners and sound your drums — make a dash for other places that he cannot afford to lose — trail brushwood and raise a dust — confound his ears and eyes — detach a body of your best troops, and place it secretly in ambuscade. Then your opponent will sally forth to the rescue."]

12. On open ground, do not try to block the enemy's way.

[Because the attempt would be futile, and would expose the blocking force itself to serious risks. There are two interpretations available here. I follow that of Chang Yu. The other is indicated in Ts'ao Kung's brief note: "Draw closer together" — i.e., see that a portion of your own army is not cut off.] On the ground of intersecting highways, join hands with your allies. [Or perhaps, "form alliances with neighboring states."]

13. On serious ground, gather in plunder.

[On this, Li Chfuan has the following delicious note: "When an army penetrates far into the enemy's country, care must be taken not to alienate the people by unjust treatment. Follow the example of the Han Emperor Kao Tsu, whose march into Chun territory was marked by no violation of women or looting of valuables. [Nota dene: this was in 207 BC, and may well cause us to blush for the Christian armies that entered Peking in 1900 AD] Thus he won the hearts of all. In the present passage, then, I think that the true reading must be, not 'plunder,' but 'do not plunder.'" Alas, I fear that in this instance the worthy commentator's feelings outran his judgment. Tu Mu, at least, has no such illusions. He says: "When encamped on 'serious ground,' there being no inducement as yet to advancefurther, and no possibility of retreat, one ought to take measures for a protracted resistance by bringing in provisions from all sides, and keep a close watch on the enemy."]

In difficult ground, keep steadily on the march.
[Or, in the words of VIII. ss. 2, "do not encamp.]

14. On hemmed-in ground, resort to stratagem.

[Ts'au Kung says: "Try the effect of some unusual artifice;" and Tu Yu amplifies this by saying: "In such a position, some scheme must be devised which will suit the circumstances, and if we can succeed in deluding the enemy, the peril may be escaped." This is exactly what happened on the famous occasion when Hannibal was hemmed in among the mountains on the road to Casilinum, and to all appearances entrapped by the dictator Fabius. The stratagem which Hannibal devised to baffle his foes was remarkably like that which T'ien Tan had also employed with success exactly 62 years before. [See IX. ss. 24, note.] When night came on, bundles of twigs were fastened to the horns of some 2000 oxen and set on fire, the terrified animals being then quickly driven along the mountain side towards the passes which were beset by the enemy. The strange spectacle of these rapidly moving lights so alarmed and discomfited the Romans that they withdrew from their position, and Hannibal's army passed safely through the defile. [See Polybius, III. 93, 94; Livy, XXII. 16 17.] On desperate ground, fight.

[For, as Chia Lin remarks: "if you fight with all your might, there is a chance of life; where as death is certain if you cling to your corner."]

15. Those who were called skillful leaders of old knew how to drive a wedge between the enemy's front and rear; [More literally, "cause the front and rear to lose touch with each other."] to prevent co-operation between his large and small divisions; to hinder the good troops from rescuing the bad, the officers from rallying their men.

16. When the enemy's men were united, they managed to keep them in disorder.

17. When it was to their advantage, they made a forward move; when otherwise, they stopped still.

[Mei Yao-chfen connects this with the foregoing: "Having succeeded in thus dislocating the enemy, they would push forward in order to secure any advantage to be gained; if there was no advantage to be gained, they would remain where they were."]

18. If asked how to cope with a great host of the enemy in orderly array and on the point of marching to the attack, I should say: "Begin by seizing something which your opponent holds dear; then he will be amenable to your will."

[Opinions differ as to what Sun Tzu had in mind. Ts'ao Kung thinks it is "some strategical advantage on which the enemy is depending." Tu Mu says: "The three things which an enemy is anxious to do, and on the accomplishment of which his success depends, are: (1) to capture our favorable positions; (2) to ravage our cultivated land; (3) to guard his own communications." Our object then must be to thwart his plans in these three directions and thus render him helpless. [Cf. III. ss. 3.] By boldly seizing the initiative in this way, you at once throw the other side on the defensive.]

19. Rapidity is the essence of war: [According to Tu Mu, "this is a summary of leading principles in warfare," and he adds: "These are the profoundest truths of military science, and the chief business of the general." The following anecdotes, told by Ho Shih, shows the importance attached to speed by two of China's greatest generals. In

227 A.D., Meng Ta, governor of Hsin- chfeng under the Wei Emperor Wen Ti, was meditating defection to the House of Shu, and had entered into correspondence with Chu-ko Liang, Prime Minister of that State.

The Wei general Ssu-ma I was then military governor of Wan, and getting wind of Meng Ta's treachery, he at once set off with an army to anticipate his revolt, having previously cajoled him by a specious message of friendly import. Ssu-ma's officers came to him and said: "If Meng Ta has leagued himself with Wu and Shu, the matter should be thoroughly investigated before we make a move." Ssu-ma I replied: "Meng Ta is an unprincipled man, and we ought to go and punish him at once, while he is still wavering and before he has thrown off the mask." Then, by a series of forced marches, be brought his army under the walls of Hsin-chfeng with in a space of eight days. Now Meng Ta had previously said in a letter to Chu-ko Liang: "Wan is 1200 LI from here. When the news of my revolt reaches Ssu-ma I, he will at once inform his imperial master, but it will be a whole month before any steps can be taken, and by that time my city will be well fortified. Besides, Ssu-ma I is sure not to come himself, and the generals that will be sent against us are not worth troubling about."

The next letter, however, was filled with consternation: "Though only eight days have passed since I threw off my allegiance, an army is already at the city-gates. What miraculous rapidity is this!" A fortnight later, Hsin-chfeng had fallen and Meng Ta had lost his head. [See CHIN SHU, ch. 1, f. 3.] In 621 A.D., Li Ching was sent from ICuei-chou in Ssu-chf uan to reduce the successful rebel Hsiao Hsien, who had set up as Emperor at the modern Ching-chou Fu in Hupeh. It was autumn, and the Yangtsze being then in flood, Hsiao Hsien never dreamt that his adversary would venture to come down through the gorges, and consequently made no preparations. But Li Ching embarked his army without loss of time, and was just about to start when the other generals implored him to postpone his departure until the river was in a less dangerous state for navigation. Li Ching replied: "To the soldier, overwhelming speed is of paramount importance, and he must never miss opportunities. Now is the time to strike, before Hsiao Hsien even knows that we have got an army together. If we seize the present moment when the river is in flood, we shall appear before his capital with startling suddenness, like the thunder which is heard before you have time to stop your ears against it. [See VII. ss. 19, note.]

This is the great principle in war. Even if he gets to know of our approach, he will have to levy his soldiers in such a hurry that they will not be fit to oppose us. Thus the full fruits of victory will be ours." All came about as he predicted, and Hsiao Hsien was obliged to surrender, nobly stipulating that his people should be spared and he alone suffer the penalty of death.] take advantage of the enemy's unreadiness, make your way by unexpected routes, and attack unguarded spots.

20. The following are the principles to be observed by an invading force: The further you penetrate into a country, the greater will be the solidarity of your troops, and thus the defenders will not prevail against you.

21. Make forays in fertile country in order to supply your army with food.

22. Carefully study the well-being of your men, [For "well-being", Wang Hsi means, "Pet them, humor them, give them plenty of food and drink, and look after them generally."] and do not overtax them. Concentrate your energy and hoard your strength.

[Chf en recalls the line of action adopted in 224 B.C. by the famous general Wang Chien, whose military genius largely contributed to the success of the First Emperor. He had invaded the Chf u State, where a universal levy was made to oppose him. But, being doubtful of the temper of his troops, he declined all invitations to fight and remained strictly on the defensive. In vain did the Chfu general try to force a battle: day after day Wang Chien kept inside his walls and would not come out, but devoted his whole time and energy to winning the affection and confidence of his men. He took care that they should be well fed, sharing his own meals with them, provided facilities for bathing, and employed every method of judicious indulgence to weld them into a loyal and homogenous body. After some time had elapsed, he told off certain persons to find out how the men were amusing themselves. The answer was, that they were contending with one another in putting the weight and long-jumping. When Wang Chien heard that they were engaged in these athletic pursuits, he knew that their spirits had been strung up to the required pitch and that they were now ready for fighting. By this time the Chfu army, after repeating their challenge again and again, had marched away eastwards in disgust. The Chf in general immediately broke up his camp and followed them, and in the battle that ensued they were routed with great slaughter. Shortly afterwards, the whole of Chfu was conquered by Chf in, and the king Fu-chf u led into captivity.]

Keep your army continually on the move,

[In order that the enemy may never know exactly where you are. It has struck me, however, that the true reading might be "link your army together."] and devise unfathomable plans.

23. Throw your soldiers into positions whence there is no escape, and they will prefer death to flight. If they will face death, there is nothing they may not achieve.

[Chang Yu quotes his favorite Wei Liao Tzu (ch. 3): "If one man were to run amok with a sword in the market-place, and everybody else tried to get our of his way, I should not allow that this man alone had courage and that all the rest were contemptible cowards. The truth is, that a desperado and a man who sets some value on his life do not meet on even terms."]

Officers and men alike will put forth their uttermost strength.

[Chang Yu says: "If they are in an awkward place together, they will surely exert their united strength to get out of it."]

24. Soldiers when in desperate straits lose the sense of fear. If there is no place of refuge, they will stand firm. If they are in hostile country, they will show a stubborn front. If there is no help for it, they will fight hard.

25. Thus, without waiting to be marshaled, the soldiers will be constantly on the qui vive; without waiting to be asked, they will do your will; [Literally, "without asking, you will get."] without

restrictions, they will be faithful; without giving orders, they can be trusted.

26. Prohibit the taking of omens, and do away with superstitious doubts. Then, until death itself comes, no calamity need be feared.

[The superstitious, "bound in to saucy doubts and fears," degenerate into cowards and "die many times before their deaths." Tu Mu quotes Huang Shih-kung: "'Spells and incantations should be strictly forbidden, and no officer allowed to inquire by divination into the fortunes of an army, for fear the soldiers' minds should be seriously perturbed.' The meaning is," he continues, "that if all doubts and scruples are discarded, your men will never falter in their resolution until they die."]

27. If our soldiers are not overburdened with money, it is not because they have a distaste for riches; if their lives are not unduly long, it is not because they are disinclined to longevity.

[Chang Yu has the best note on this passage: "Wealth and long life are things for which all men have a natural inclination. Hence, if they burn or fling away valuables, and sacrifice their own lives, it is not that they dislike them, but simply that they have no choice." Sun Tzu is slyly insinuating that, as soldiers are but human, it is for the general to see that temptations to shirk fighting and grow rich are not thrown in their way.]

28. On the day they are ordered out to battle, your soldiers may weep, [The word in the Chinese is "snivel." This is taken to indicate more genuine grief than tears alone.] those sitting up bedewing their garments, and those lying down letting the tears run down their cheeks.

[Not because they are afraid, but because, as Ts'ao Kung says, "all have embraced the firm resolution to do or die." We may remember that the heroes of the Iliad were equally childlike in showing their emotion. Chang Yu alludes to the mournful parting at the I River between Ching K x o and his friends, when the former was sent to attempt the life of the King of Chun (afterwards First Emperor) in 227 B.C. The tears of all flowed down like rain as he bade them farewell and uttered the following lines: "The shrill blast is blowing, Chilly the burn; Your champion is going — Not to return."]

But let them once be brought to bay, and they will display the courage of a Chu or a Kuei.

[Chu was the personal name of Chuan Chu, a native of the Wu State and contemporary with Sun Tzu himself, who was employed by Kung-tzu Kuang, better known as Ho Lu Wang, to assassinate his sovereign Wang Liao with a dagger which he secreted in the belly of a fish served up at a banquet. He succeeded in his attempt, but was immediately hacked to pieced by the king's bodyguard. This was in 515 B.C. The other hero referred to, Ts'ao Kuei (or Ts'ao Mo), performed the exploit which has made his name famous 166 years earlier, in 681 B.C. Lu had been thrice defeated by Chfi, and was just about to conclude a treaty surrendering a large slice of territory, when Ts'ao Kuei suddenly seized Huan Kung, the Duke of Chu, as he stood on the altar steps and held a dagger against his chest. None of the duke's retainers dared to move a muscle, and Ts'ao Kuei proceeded to demand full restitution, declaring the Lu was being unjustly treated because she was a smaller and a weaker state. Huan Kung, in peril of his life, was obliged to consent, whereupon Ts'ao Kuei flung away his dagger and quietly resumed his place amid the terrified assemblage without having so much as changed color. As was to be expected, the Duke wanted afterwards to repudiate the bargain, but his wise old counselor Kuan Chung pointed out to him the impolicy of breaking his word, and the upshot was that this bold stroke regained for Lu the whole of what she had lost in three pitched battles.]

29. The skillful tactician may be likened to the SHUAI- J AN. Now the SHUAI-J AN is a snake that is found in the Ch x ang mountains.

["Shuai-jan" means "suddenly" or "rapidly," and the snake in question was doubtless so called owing to the rapidity of its movements. Through this passage, the term in the Chinese has now come to beused in the sense of "military maneuvers."]

Strike at its head, and you will be attacked by its tail; strike at its tail, and you will be attacked by its head; strike at its middle, and you will be attacked by head and tail both.

30. Asked if an army can be made to imitate the SHUAI- J AN,

[That is, as Mei Yao-chfen says, "Is it possible to make the front and rear of an army each swiftly responsive to attack on the other, just as though they were part of a single living body?"]

I should answer, Yes. For the men of Wu and the men of Yueh are enemies; [Cf. VI. ss. 21.] yet if they are crossing a river in the same boat and are caught by a storm, they will come to each other's assistance just as the left hand helps the right.

[The meaning is: If two enemies will help each other in a time of common peril, how much more should two parts of the same army, bound together as they are by every tie of interest and fellow-feeling. Yet it is notorious that many a campaign has been ruined through lack of cooperation, especially in the case of allied armies.]

31. Hence it is not enough to put one's trust in the tethering of horses, and the burying of chariot wheels in the ground

[These quaint devices to prevent one's army from running away recall the Athenian hero Sophanes, who carried the anchor with him at the battle of Plataea, by means of which he fastened himself firmly to one spot. [See Herodotus, IX. 74.] It is not enough, says Sun Tzu, to render flight impossible by such mechanical means. You will not succeed unless your men have tenacity and unity of purpose, and, above all, a spirit of sympathetic cooperation. This is the lesson which can be learned from the SHUAI- JAN.]

32. The principle on which to manage an army is to set up one standard of courage which all must reach.

[Literally, "level the courage [of all] as though [it were that of] one." If the ideal army is to form a single organic whole, then it follows that the resolution and spirit of its component parts must be of the same quality, or at any rate must not fall below a certain standard. Wellington's seemingly ungrateful description of his army at Waterloo as "the worst he had ever commanded" meant no more than that it was deficient in this important particular — unity of spirit and courage. Had he not foreseen the Belgian defections and carefully kept those troops in the background, he would almost certainly have lost the day.]

33. How to make the best of both strong and weak — that is a question involving the proper use of ground.

[Mei Yao-chfen's paraphrase is: "The way to eliminate the differences of strong and weak and to make both serviceable is to utilize accidental features of the ground."
Less reliable troops, if posted in strong positions, will hold out as long as better troops on more exposed terrain. The advantage of position neutralizes the inferiority in stamina and courage. Col. Henderson says: "With all respect to the text books, and to the ordinary tactical teaching, I am inclined to think that the study of ground is often overlooked, and that by no means sufficient importance is attached to the selection of positions . and to the immense advantages that are to be derived, whether you are defending or attacking, from the proper utilization of natural features."]

34. Thus the skillful general conducts his army just as though he were leading a single man, willy-nilly, by the hand.

[Tu Mu says: "The simile has reference to the ease with which he does it."]

35. It is the business of a general to be quiet and thus ensure secrecy; upright and just, and thus maintain order.

36. He must be able to mystify his officers and men by false reports and appearances, [Literally, "to deceive their eyes and ears."] and thus keep them in total ignorance.

[Ts'ao Kung gives us one of his excellent apophthegms: "The troops must not be allowed to share your schemes in the beginning; they may only rejoice with you over their happy outcome." "To mystify, mislead, and surprise the enemy," is one of the first principles in war, as had been frequently pointed out. But how about the other process

— the mystification of one's own men? Those who may think that Sun Tzu is over-emphatic on this point would do well to read Col. Henderson's remarks on Stonewall Jackson's Valley campaign: "The infinite pains," he says, "with which Jackson sought to conceal, even from his most trusted staff officers, his movements, his intentions, and his thoughts, a commander less thorough would have pronounced useless" — etc. etc.

In the year 88 A.D., as we read in ch. 47 of the HOU HAN SHU, "Pan Chf ao took the field with 25,000 men from Khotan and other Central Asian states with the object of crushing Yarkand. The King of Kutcha replied by dispatching his chief commander to succor the place with an army drawn from the kingdoms of Wen-su, Ku-mo, and Wei-fou, totaling 50,000 men. Pan Chfao summoned his officers and also the King of Khotan to a council of war, and said: 'Our forces are now outnumbered and unable to make head against the enemy. The best plan, then, is for us to separate and disperse, each in a different direction.

The King of Khotan will march away by the easterly route, and I will then return myself towards the west. Let us wait until the evening drum has sounded and then start.' Pan Chfao now secretly released the prisoners whom he had taken alive, and the King of Kutcha was thus informed of his plans. Much elated by the news, the latter set off at once at the head of 10,000 horsemen to bar Pan Chfao's retreat in the west, while the King of Wen-su rode eastward with 8000 horse in order to intercept the King of Khotan. As soon as Pan Chfao knew that the two chieftains had gone, he called his divisions together, got them well in hand, and at cock-crow hurled them against the army of Yarkand, as it lay encamped. The barbarians, panic-stricken, fled in confusion, and were closely pursued by Pan Chfao. Over 5000 heads were brought back as trophies, besides immense spoils in the shape of horses and cattle and valuables of every description. Yarkand then capitulating, Kutcha and the other kingdoms drew off their respective forces. From that time forward, Pan Chfao's prestige completely overawed the countries of the west." In this case, we see that the Chinese general not only kept his own officers in ignorance of his real plans, but actually took the bold step of dividing his army in order to deceive the enemy.]

37. By altering his arrangements and changing his plans,
[Wang Hsi thinks that this means not using the same stratagem
twice.] he keeps the enemy without definite knowledge.

[Chang Yu, in a quotation from another work, says: "The axiom, that
war is based on deception, does not apply only to deception of the
enemy. You must deceive even your own soldiers. Make them follow
you, but without letting them know why."]

By shifting his camp and taking circuitous routes, he prevents the
enemy from anticipating his purpose.

38. At the critical moment, the leader of an army acts like one who
has climbed up a height and then kicks away the ladder behind him.
He carries his men deep into hostile territory before he shows his
hand.

[Literally, "releases the spring" (see V. ss. 15), that is, takes some
decisive step which makes it impossible for the army to return — like
Hsiang Yu, who sunk his ships after crossing a river. Chfen Hao,
followed by Chia Lin, understands the words less well as "puts forth
every artifice at his command."]

39. He burns his boats and breaks his cooking-pots; like a shepherd
driving a flock of sheep, he drives his men this way and that, and
nothing knows whither he is going.

[Tu Mu says: "The army is only cognizant of orders to advance or
retreat; it is ignorant of the ulterior ends of attacking and
conquering."]

40. To muster his host and bring it into danger: — this may be
termed the business of the general.

[Sun Tzu means that after mobilization there should be no delay in
aiming a blow at the enemy's heart. Note how he returns again and
again to this point. Among the warring states of ancient China,
desertion was no doubt a much more present fear and serious evil
than it is in the armies of today.]

41. The different measures suited to the nine varieties of ground; [Chang Yu says: "One must not be hide-bound in interpreting the rules for the nine varieties of ground.] the expediency of aggressive or defensive tactics; and the fundamental laws of human nature: these are things that must most certainly be studied.

42. When invading hostile territory, the general principle is, that penetrating deeply brings cohesion; penetrating but a short way means dispersion.

43. When you leave your own country behind, and take your army across neighborhood territory, you find yourself on critical ground.

[This "ground" is curiously mentioned in VIII. ss. 2, but it does not figure among the Nine Situations or the Six Calamities in chap. X. One's first impulse would be to translate it distant ground," but this, if we can trust the commentators, is precisely what is not meant here. Mei Yao- chfen says it is "a position not far enough advanced to be called 'facile,' and not near enough to home to be 'dispersive,' but something between the two." Wang Hsi says: "It is ground separated from home by an interjacent state, whose territory we have had to cross in order to reach it. Hence, it is incumbent on us to settle our business there quickly." He adds that this position is of rare occurrence, which is the reason why it is not included among the Nine Situations.]

When there are means of communication on all four sides, the ground is one of intersecting highways.

44. When you penetrate deeply into a country, it is serious ground. When you penetrate but a little way, it is facile ground.

45. When you have the enemy's strongholds on your rear, and narrow passes in front, it is hemmed-in ground. When there is no place of refuge at all, it is desperate ground.

46. Therefore, on dispersive ground, I would inspire my men with unity of purpose.

[This end, according to Tu Mu, is best attained by remaining on the defensive, and avoiding battle. Cf. supra, ss. 11.]

On facile ground, I would see that there is close connection between all parts of my army.

[As Tu Mu says, the object is to guard against two possible contingencies: "(1) the desertion of our own troops; (2) a sudden attack on the part of the enemy." Cf . VII. ss. 17. Mei Yao-chfen says: "On the march, the regiments should be in close touch; in an encampment, there should be continuity between the fortifications."]

47. On contentious ground, I would hurry up my rear.

[This is Ts'ao Kung's interpretation. Chang Yu adopts it, saying: "We must quickly bring up our rear, so that head and tail may both reach the goal." That is, they must not be allowed to straggle up a long way apart. Mei Yao-chfen offers another equally plausible explanation: "Supposing the enemy has not yet reached the coveted position, and we are behind him, we should advance with all speed in order to dispute its possession." Chf en Hao, on the other hand, assuming that the enemy has had time to select his own ground, quotes VI. ss. 1, where Sun Tzu warns us against coming exhausted to the attack. His own idea of the situation is rather vaguely expressed: "If there is a favorable position lying in front of you, detach a picked body of troops to occupy it, then if the enemy, relying on their numbers, come up to make a fight for it, you may fall quickly on their rear with your main body, and victory will be assured." It was thus, he adds, that Chao She beat the army of Chun. (See p. 57.)]

48. On open ground, I would keep a vigilant eye on my defenses. On ground of intersecting highways, I would consolidate my alliances.

49. On serious ground, I would try to ensure a continuous stream of supplies.

[The commentators take this as referring to forage and plunder, not, as one might expect, to an unbroken communication with a home base.]

On difficult ground, I would keep pushing on along the road.

50. On hemmed-in ground, I would block any way of retreat.

[Meng Shih says: "To make it seem that I meant to defend the position, whereas my real intention is to burst suddenly through the enemy's lines." Mei Yao-chfen says: "in order to make my soldiers fight with desperation." Wang Hsi says, "fearing lest my men be tempted to run away." Tu Mu points out that this is the converse of VII. ss. 36, where it is the enemy who is surrounded. In 532 A.D., Kao Huan, afterwards Emperor and canonized as Shen-wu, was surrounded by a great army under Erh-chu Chao and others. His own force was comparatively small, consisting only of 2000 horse and something under 30,000 foot. The lines of investment had not been drawn very closely together, gaps being left at certain points. But Kao Huan, instead of trying to escape, actually made a shift to block all the remaining outlets himself by driving into them a number of oxen and donkeys roped together. As soon as his officers and men saw that there was nothing for it but to conquer or die, their spirits rose to an extraordinary pitch of exaltation, and they charged with such desperate ferocity that the opposing ranks broke and crumbled under their onslaught.]

On desperate ground, I would proclaim to my soldiers the hopelessness of saving their lives.

[Tu Yu says: "Burn your baggage and impedimenta, throw away your stores and provisions, choke up the wells, destroy your cooking-stoves, and make it plain to your men that they cannot survive, but must fight to the death." Mei Yao-chfen says: "The only chance of life lies in giving up all hope of it." This concludes what Sun Tzu has to say about "grounds" and the "variations" corresponding to them.

Reviewing the passages which bear on this important subject, we cannot fail to be struck by the desultory and unmethodical fashion in which it is treated. Sun Tzu begins abruptly in VIII. ss. 2 to enumerate "variations" before touching on "grounds" at all, but only mentions five, namely nos. 7, 5, 8 and 9 of the subsequent list, and one that is not included in it. A few varieties of ground are dealt with in the earlier portion of chap. IX, and then chap. X sets forth six new grounds, with six variations of plan to match. None of these is mentioned again, though the first is hardly to be distinguished from ground no. 4 in the next chapter. At last, in chap. XI, we come to the Nine Grounds par excellence, immediately followed by the variations. This takes us down to ss. 14. In SS. 43-45, fresh definitions are provided for nos. 5, 6, 2, 8 and 9 (in the order given), as well as for the tenth ground noticed in chap. VIII; and finally, the nine variations are enumerated once more from beginning to end, all, with the exception of 5, 6 and 7, being different from those previously given.

Though it is impossible to account for the present state of Sun Tzu's text, a few suggestive facts maybe brought into prominence: (1) Chap. VIII, according to the title, should deal with nine variations, whereas only five appear. (2) It is an abnormally short of these are defined twice over, besides which there are two distinct lists of the corresponding variations. (4) The length of the chapter is disproportionate, being double that of any other except IX. I do not propose to draw any inferences from these facts, beyond the general conclusion that Sun Tzu's work cannot have come down to us in the shape in which it left his hands: chap. VIII is obviously defective and probably out of place, while XI seems to contain matter that has either been added by a later hand or ought to appear elsewhere.]

51. For it is the soldier's disposition to offer an obstinate resistance when surrounded, to fight hard when he cannot help himself, and to obey promptly when he has fallen into danger.

[Chang Yu alludes to the conduct of Pan Chfao's devoted followers in 73 A.D. The story runs thus in the HOU HAN SHU, ch. 47: "When Pan Chfao arrived at Shan-shan, Kuang, the King of the country, received him at first with great politeness and respect; but shortly afterwards his behavior underwent a sudden change, and he became remiss and negligent.

Pan Chfao spoke about this to the officers of his suite: 'Have you noticed,' he said, 'that Kuang's polite intentions are on the wane? This must signify that envoys have come from the Northern barbarians, and that consequently he is in a state of indecision, not knowing with which side to throw in his lot. That surely is the reason. The truly wise man, we are told, can perceive things before they have come to pass; how much more, then, those that are already manifest!' Thereupon he called one of the natives who had been assigned to his service, and set a trap for him, saying: 'Where are those envoys from the Hsiung-nu who arrived some day ago?' The man was so taken aback that between surprise and fear he presently blurted out the whole truth. Pan Chfao, keeping his informant carefully under lock and key, then summoned a general gathering of his officers, thirty-six in all, and began drinking with them.

When the wine had mounted into their heads a little, he tried to rouse their spirit still further by addressing them thus: 'Gentlemen, here we are in the heart of an isolated region, anxious to achieve riches and honor by some great exploit. Now it happens that an ambassador from the Hsiung-no arrived in this kingdom only a few days ago, and the result is that the respectful courtesy extended towards us by our royal host has disappeared. Should this envoy prevail upon him to seize our party and hand us over to the Hsiung-no, our bones will become food for the wolves of the desert. What are we to do?' With one accord, the officers replied: 'Standing as we do in peril of our lives, we will follow our commander through life and death.' For the sequel of this adventure, see chap. XII. ss. 1, note.]

52. We cannot enter into alliance with neighboring princes until we are acquainted with their designs. We are not fit to lead an army on the march unless we are familiar with the face of the country — its mountains and forests, its pitfalls and precipices, its marshes and swamps. We shall be unable to turn natural advantages to account unless we make use of local guides.

[These three sentences are repeated from VII. SS. 12- 14 — in order to emphasize their importance, the commentators seem to think. I prefer to regard them as interpolated here in order to form an antecedent to the following words. With regard to local guides, Sun Tzu might have added that there is always the risk of going wrong, either through their treachery or some misunderstanding such as Livy records (XXII. 13): Hannibal, we are told, ordered a guide to lead him into the neighborhood of Casinum, where there was an important pass to be occupied; but his Carthaginian accent, unsuited to the pronunciation of Latin names, caused the guide to understand Casilinum instead of Casinum, and turning from his proper route, he took the army in that direction, the mistake not being discovered until they had almost arrived.]

53. To be ignored of any one of the following four or five principles does not befit a warlike prince.

54. When a warlike prince attacks a powerful state, his generalship shows itself in preventing the concentration of the enemy's forces. He overawes his opponents, and their allies are prevented from joining against him.

[Mei Tao-chf en constructs one of the chains of reasoning that are so much affected by the Chinese: "In attacking a powerful state, if you can divide her forces, you will have a superiority in strength; if you have a superiority in strength, you will overawe the enemy; if you overawe the enemy, the neighboring states will be frightened; and if the neighboring states are frightened, the enemy's allies will be prevented from joining her." The following gives a stronger meaning: "If the great state has once been defeated (before she has had time to summon her allies), then the lesser states will hold aloof and refrain from massing their forces." Chfen Hao and Chang Yu take the sentence in quite another way. The former says: "Powerful though a prince may be, if he attacks a large state, he will be unable to raise enough troops, and must rely to some extent on external aid; if he dispenses with this, and with overweening confidence in his own strength, simply tries to intimidate the enemy, he will surely be defeated." Chang Yu puts his view thus: "If we recklessly attack a large state, our own people will be discontented and hang back. But if (as will then be the case) our display of military force is inferior by half to that of the enemy, the other chieftains will take fright and refuse to join us."]

55. Hence he does not strive to ally himself with all and sundry, nor does he foster the power of other states. He carries out his own secret designs, keeping his antagonists in awe.

[The train of thought, as said by Li Chfuan, appears to be this: Secure against a combination of his enemies, "he can afford to reject entangling alliances and simply pursue his own secret designs, his prestige enable him to dispense with external friendships."]

Thus he is able to capture their cities and overthrow their kingdoms.

[This paragraph, though written many years before the Chun State became a serious menace, is not a bad summary of the policy by which the famous Six Chancellors gradually paved the way for her final triumph under Shih Huang Ti. Chang Yu, following up his previous note, thinks that Sun Tzu is condemning this attitude of cold-blooded selfishness and haughty isolation.]

56. Bestow rewards without regard to rule, [Wu Tzu (ch. 3) less wisely says: "Let advance be richly rewarded and retreat be heavily punished."] issue orders [Literally, "hang" or post up."] without regard to previous arrangements; ["In order to prevent treachery," says Wang Hsi. The general meaning is made clear by Ts'ao Kung's quotation from the SSU-MA FA: "Give instructions only on sighting the enemy; give rewards when you see deserving deeds." Ts'ao Kung's paraphrase: "The final instructions you give to your army should not correspond with those that have been previously posted up." Chang Yu simplifies this into "your arrangements should not be divulged beforehand." And Chia Lin says: "there should be no fixity in your rules and arrangements." Not only is there danger in letting your plans be known, but war often necessitates the entire reversal of them at the last moment.] and you will be able to handle a whole army as though you had to do with but a single man.
[Cf. supra, ss. 34.]

57. Confront your soldiers with the deed itself; never let them know your design.

[Literally, "do not tell them words;" i.e. do not give your reasons for any order. Lord Mansfield once told a junior colleague to "give no reasons" for his decisions, and the maxim is even more applicable to a general than to a judge.]

When the outlook is bright, bring it before their eyes; but tell them nothing when the situation is gloomy.

58. Place your army in deadly peril, and it will survive; plunge it into desperate straits, and it will come off in safety.

[These words of Sun Tzu were once quoted by Han Hsin in explanation of the tactics he employed in one of his most brilliant battles, already alluded to on p. 28. In 204 B.C., he was sent against the army of Chao, and halted ten miles from the mouth of the Ching-hsing pass, where the enemy had mustered in full force. Here, at midnight, he detached a body of 2000 light cavalry, every man of which was furnished with a red flag. Their instructions were to make their way through narrow defiles and keep a secret watch on the enemy. "When the men of Chao see me in full flight," Han Hsin said, "they will abandon their fortifications and give chase. This must be the sign for you to rush in, pluck down the Chao standards and set up the red banners of Han in their stead." Turning then to his other officers, he remarked: "Our adversary holds a strong position, and is not likely to come out and attack us until he sees the standard and drums of the commander-in-chief, for fear I should turn back and escape through the mountains." So saying, he first of all sent out a division consisting of 10,000 men, and ordered them to form in line of battle with their backs to the River Ti. Seeing this maneuver, the whole army of Chao broke into loud laughter.

By this time it was broad daylight, and Han Hsin, displaying the generalissimo's flag, marched out of the pass with drums beating, and was immediately engaged by the enemy. A great battle followed, lasting for some time; until at length Han Hsin and his colleague Chang Ni, leaving drums and banner on the field, fled to the division on the river bank, where another fierce battle was raging. The enemy rushed out to pursue them and to secure the trophies, thus denuding their ramparts of men; but the two generals succeeded in joining the other army, which was fighting with the utmost desperation. The time had now come for the 2000 horsemen to play their part. As soon as they saw the men of Chao following up their advantage, they galloped behind the deserted walls, tore up the enemy's flags and replaced them by those of Han. When the Chao army looked back from the pursuit, the sight of these red flags struck them with terror.

Convinced that the Hans had got in and overpowered their king, they broke up in wild disorder, every effort of their leader to stay the panic being in vain. Then the Han army fell on them from both sides and completed the rout, killing a number and capturing the rest, amongst whom was King Ya himselfAfter the battle, some of Han Hsin's officers came to him and said: "In the ART OF WAR we are told to have a hill or tumulus on the right rear, and a river or marsh on the left front. [This appears to be a blend of Sun Tzu and T'ai Kung. See IX ss. 9, and note.]

You, on the contrary, ordered us to draw up our troops with the river at our back. Under these conditions, how did you manage to gain the victory?" The general replied: "I fear you gentlemen have not studied the Art of War with sufficient care. Is it not written there: 'Plunge your army into desperate straits and it will come off in safety; place it in deadly peril and it will survive'? Had I taken the usual course, I should never have been able to bring my colleague round. What says the Military Classic — 'Swoop down on the market-place and drive the men off to fight.' [This passage does not occur in the present text of Sun Tzu.] If I had not placed my troops in a position where they were obliged to fight for their lives, but had allowed each man to follow his own discretion, there would have been a general debandade, and it would have been impossible to do anything with them." The officers admitted the force of his argument, and said: "These are higher tactics than we should have been capable of." [See CH^IEN HAN SHU, ch. 34, ff. 4, 5.]]

59. For it is precisely when a force has fallen into harm's way that is capable of striking a blow for victory. [Danger has a bracing effect.]

60. Success in warfare is gained by carefully accommodating ourselves to the enemy's purpose. [Ts'ao Kung says: "Feign stupidity" — by an appearance of yielding and falling in with the enemy's wishes. Chang Yu's note makes the meaning clear: "If the enemy shows an inclination to advance, lure him on to do so; if he is anxious to retreat, delay on purpose that he may carry out his intention." The object is to make him remiss and contemptuous before we deliver our attack.]

61. By persistently hanging on the enemy's flank, [I understand the first four words to mean "accompanying the enemy in one direction." Ts'ao Kung says: "unite the soldiers and make for the enemy." But such a violent displacement of characters is quite indefensible.] we shall succeed in the long run [Literally, "after a thousand LI."] in killing the commander-in-chief. [Always a great point with the Chinese.]

62. This is called ability to accomplish a thing by sheer cunning.

63. On the day that you take up your command, block the frontier passes, destroy the official tallies, [These were tablets of bamboo or wood, one half of which was issued as a permit or passport by the official in charge of a gate. Cf. the "border-warden" of LUN YU III. 24, who may have had similar duties. When this half was returned to him, within a fixed period, he was authorized to open the gate and let the traveler through.] and stop the passage of all emissaries. [Either to or from the enemy's country.]

64. Be stern in the council-chamber, [Show no weakness, and insist on your plans being ratified by the sovereign.] so that you may control the situation.

[Mei Yao-chf en understands the whole sentence to mean: Take the strictest precautions to ensure secrecy in your deliberations.]

65. If the enemy leaves a door open, you must rush in.

66. Forestall your opponent by seizing what he holds dear, [Cf. supra, ss. 18.] and subtly contrive to time his arrival on the ground.

[Chf en Hao's explanation: "If I manage to seize a favorable position, but the enemy does not appear on the scene, the advantage thus obtained cannot be turned to any practical account. He who intends therefore, to occupy a position of importance to the enemy, must begin by making an artful appointment, so to speak, with his antagonist, and cajole him into going there as well." Mei Yao-chfen explains that this "artful appointment" is to be made through the medium of the enemy's own spies, who will carry back just the amount of information that we choose to give them.

Then, having cunningly disclosed our intentions, "we must manage, though starting after the enemy, to arrive before him (VII. ss. 4). We must start after him in order to ensure his marching thither; we must arrive before him in order to capture the place without trouble. Taken thus, the present passage lends some support to Mei Yao-chfen's interpretation of ss. 47.]

67. Walk in the path defined by rule, [Chia Lin says: "Victory is the only thing that matters, and this cannot be achieved by adhering to conventional canons." It is unfortunate that this variant rests on very slight authority, for the sense yielded is certainly much more satisfactory. Napoleon, as we know, according to the veterans of the old school whom he defeated, won his battles by violating every accepted canon of warfare.] and accommodate yourself to the enemy until you can fight a decisive battle.

[Tu Mu says: "Conform to the enemy's tactics until a favorable opportunity offers; then come forth and engage in a battle that shall prove decisive."]

68. At first, then, exhibit the coyness of a maiden, until the enemy gives you an opening; afterwards emulate the rapidity of a running hare, and it will be too late for the enemy to oppose you.

[As the hare is noted for its extreme timidity, the comparison hardly appears felicitous. But of course Sun Tzu was thinking only of its speed. The words have been taken to mean: You must flee from the enemy as quickly as an escaping hare; but this is rightly rejected by Tu Mu.]

XII The Attack by Fire

[Rather more than half the chapter (SS. 1-13) is devoted to the subject of fire, after which the author branches off into other topics.]

1. Sun Tzu said: There are five ways of attacking with fire. The first is

to burn soldiers in their camp; [So Tu Mu. Li Chf uan says: "Set fire to the camp, and kill the soldiers" (when they try to escape from the flames). Pan Chfao, sent on a diplomatic mission to the King of Shan- shan [see XI. ss. 51, note], found himself placed in extreme peril by the unexpected arrival of an envoy from the Hsiung- nu [the mortal enemies of the Chinese]. In consultation with his officers, he exclaimed: "Never venture, never win! The only course open to us now is to make an assault by fire on the barbarians under cover of night, when they will not be able to discern our numbers.

Profiting by their panic, we shall exterminate them completely; this will cool the King's courage and cover us with glory, besides ensuring the success of our mission.' the officers all replied that it would be necessary to discuss the matter first with the Intendant. Pan Chfao then fell into a passion: 'It is today,' he cried, 'that our fortunes must be decided! The Intendant is only a humdrum civilian, who on hearing of our project will certainly be afraid, and everything will be brought to light. An inglorious death is no worthy fate for valiant warriors.' All then agreed to do as he wished. Accordingly, as soon as night came on, he and his little band quickly made their way to the barbarian camp. A strong gale was blowing at the time. Pan Chfao ordered ten of the party to take drums and hide behind the enemy's barracks, it being arranged that when they saw flames shoot up, they should begin drumming and yelling with all their might. The rest of his men, armed with bows and crossbows, he posted in ambuscade at the gate of the camp. He then set fire to the place from the windward side, whereupon a deafening noise of drums and shouting arose on the front and rear of the Hsiung-nu, who rushed out pell-mell in frantic disorder.

Pan Chfao slew three of them with his own hand, while his companions cut off the heads of the envoy and thirty of his suite. The remainder, more than a hundred in all, perished in the flames. On the following day, Pan Chfao, divining his thoughts, said with uplifted hand: 'Although you did not go with us last night, I should not think, Sir, of taking sole credit for our exploit.' This satisfied Kuo Hsun, and Pan Chfao, having sent for Kuang, King of Shan-shan, showed him the head of the barbarian envoy.

The whole kingdom was seized with fear and trembling, which Pan

Chfao took steps to allay by issuing a public proclamation. Then, taking the king's sons as hostage, he returned to make his report to Tou Ku." HOU HAN SHU, ch. 47, ff. 1,2.] the second is to burn stores; [Tu Mu says: "Provisions, fuel and fodder."

In order to subdue the rebellious population of Kiangnan, Kao Keng recommended Wen Ti of the Sui dynasty to make periodical raids and burn their stores of grain, a policy which in the long run proved entirely successful.]
the third is to burn baggage trains; [An example given is the destruction of Yuan Shao's wagons and impedimenta by Ts'ao Ts'ao in 200 A.D.] the fourth is to burn arsenals and magazines;

[Tu Mu says that the things contained in "arsenals" and "magazines" are the same. He specifies weapons and other implements, bullion and clothing. Cf. VII. ss. 11.] the fifth is to hurl dropping fire amongst the enemy.

[Tu Yu says in the TUNG TIEN: "To drop fire into the enemy's camp. The method by which this may be done is to set the tips of arrows alight by dipping them into a brazier, and then shoot them from powerful crossbows into the enemy's lines."]

In order to carry out an attack, we must have means available.

[T'sao Kung thinks that "traitors in the enemy's camp" are referred to. But Chf en Hao is more likely to be right in saying: "We must have favorable circumstances in general, not merely traitors to help us." Chia Lin says: "We must avail ourselves of wind and dry weather."] the material for raising fire should always be kept in readiness.

[Tu Mu suggests as material for making fire: "dry vegetable matter, reeds, brushwood, straw, grease, oil, etc." Here we have the material cause. Chang Yu says: "vessels for hoarding fire, stuff for lighting fires."]

3. There is a proper season for making attacks with fire, and special days for starting a conflagration.

4. The proper season is when the weather is very dry; the special

days are those when the moon is in the constellations of the Sieve, the Wall, the Wing or the Cross-bar; [These are, respectively, the 7th, 14th, 27th, and 28th of the Twenty-eight Stellar Mansions, corresponding roughly to Sagittarius, Pegasus, Crater and Corvus.] for these four are all days of rising wind.

5. In attacking with fire, one should be prepared to meet five possible developments:

6. (1) When fire breaks out inside to enemy's camp, respond at once with an attack from without.

7. (2) If there is an outbreak of fire, but the enemy's soldiers remain quiet, bide your time and do not attack.

[The prime object of attacking with fire is to throw the enemy into confusion. If this effect is not produced, it means that the enemy is ready to receive us. Hence the necessity for caution.]

8. (3) When the force of the flames has reached its height, follow it up with an attack, if that is practicable; if not, stay where you are.

[Ts'ao Kung says: "If you see a possible way, advance; but if you find the difficulties too great, retire."]

9. (4) If it is possible to make an assault with fire from without, do not wait for it to break out within, but deliver your attack at a favorable moment.

[Tu Mu says that the previous paragraphs had reference to the fire breaking out (either accidentally, we may suppose, or by the agency of incendiaries) inside the enemy's camp. "But," he continues, "if the enemy is settled in a waste place littered with quantities of grass, or if he has pitched his camp in a position which can be burnt out, we must carry our fire against him at any seasonable opportunity, and not await on in hopes of an outbreak occurring within, for fear our opponents should themselves burn up the surrounding vegetation, and thus render our own attempts fruitless." The famous Li Ling once baffled the leader of the Hsiung-nu in this way.

The latter, taking advantage of a favorable wind, tried to set fire to the Chinese general's camp, but found that every scrap of combustible vegetation in the neighborhood had already been burnt down. On the other hand, Po-ts x ai, a general of the Yellow Turban rebels, was badly defeated in 184 A.D. through his neglect of this simple precaution. "At the head of a large army he was besieging Chf ang-she, which was held by Huang-fu Sung. The garrison was very small, and a general feeling of nervousness pervaded the ranks; so Huang-fu Sung called his officers together and said: "In war, there are various indirect methods of attack, and numbers do not count for everything. [The commentator here quotes Sun Tzu, V. SS. 5, 6 and 10.] Now the rebels have pitched their camp in the midst of thick grass which will easily burn when the wind blows. If we set fire to it at night, they will be thrown into a panic, and we can make a sortie and attack them on all sides at once, thus emulating the achievement of Tien Tan.' [See p. 90.]

That same evening, a strong breeze sprang up; so Huang-fu Sung instructed his soldiers to bind reeds together into torches and mount guard on the city walls, after which he sent out a band of daring men, who stealthily made their way through the lines and started the fire with loud shouts and yells. Simultaneously, a glare of light shot up from the city walls, and Huang-fu Sung, sounding his drums, led a rapid charge, which threw the rebels into confusion and put them to headlong flight." [HOU HAN SHU,
ch. 71.]

10. (5) When you start a fire, be to windward of it. Do not attack from the leeward.

[Chang Yu, following Tu Yu, says: "When you make a fire, the enemy will retreat away from it; if you oppose his retreat and attack him then, he will fight desperately, which will not conduce to your success." A rather more obvious explanation is given by Tu Mu: "If the wind is in the east, begin burning to the east of the enemy, and follow up the attack yourself from that side. If you start the fire on the east side, and then attack from the west, you will suffer in the same way as your enemy."]

11. A wind that rises in the daytime lasts long, but a night breeze soon falls.

[Cf. Lao Tzu's saying: "A violent wind does not last the space of a morning." (TAO TE CHING, chap. 23.) Mei Yao- chfen and Wang Hsi say: "A day breeze dies down at nightfall, and a night breeze at daybreak. This is what happens as a general rule." The phenomenon observed may be correct enough, but how this sense is to be obtained is not apparent.]

12. In every army, the five developments connected with fire must be known, the movements of the stars calculated, and a watch kept for the proper days.

[Tu Mu says: "We must make calculations as to the paths of the stars, and watch for the days on which wind will rise, before making our attack with fire." Chang Yu seems to interpret the text differently: "We must not only know how to assail our opponents with fire, but also be on our guard against similar attacks from them."]

13. Hence those who use fire as an aid to the attack show intelligence; those who use water as an aid to the attack gain an accession of strength.

14. By means of water, an enemy may be intercepted, but not robbed of all his belongings.

[Ts'ao Kung's note is: "We can merely obstruct the enemy's road or divide his army, but not sweep away all his accumulated stores." Water can do useful service, but it lacks the terrible destructive power of fire. This is the reason, Chang Yu concludes, why the former is dismissed in a couple of sentences, whereas the attack by fire is discussed in detail. Wu Tzu (ch. 4) speaks thus of the two elements: "If an army is encamped on low-lying marshy ground, from which the water cannot run off, and where the rainfall is heavy, it may be submerged by a flood. If an army is encamped in wild marsh lands thickly overgrown with weeds and brambles, and visited by frequent gales, it may be exterminated by fire."]

15. Unhappy is the fate of one who tries to win his battles and

succeed in his attacks without cultivating the spirit of enterprise; for the result is waste of time and general stagnation.

[This is one of the most perplexing passages in Sun Tzu. Ts'ao Kung says: "Rewards for good service should not be deferred a single day." And Tu Mu: "If you do not take opportunity to advance and reward the deserving, your subordinates will not carry out your commands, and disaster will ensue." For several reasons, however, and in spite of the formidable array of scholars on the other side, I prefer the interpretation suggested by Mei Yao-chfen alone, whose words I will quote: "Those who want to make sure of succeeding in their battles and assaults must seize the favorable moments when they come and not shrink on occasion from heroic measures: that is to say, they must resort to such means of attack of fire, water and the like. What they must not do, and what will prove fatal, is to sit still and simply hold to the advantages they have got."]

16. Hence the saying: The enlightened ruler lays his plans well ahead; the good general cultivates his resources.

[Tu Mu quotes the following from the SAN LUEH, ch. 2: "The warlike prince controls his soldiers by his authority, kits them together by good faith, and by rewards makes them serviceable. If faith decays, there will be disruption; if rewards are deficient, commands will not be respected."]

17. Move not unless you see an advantage; use not your troops unless there is something to be gained; fight not unless the position is critical.

[Sun Tzu may at times appear to be over-cautious, but he never goes so far in that direction as the remarkable passage in the TAO TE CHING, ch. 69. "I dare not take the initiative, but prefer to act on the defensive; I dare not advance an inch, but prefer to retreat a foot."]

18. No ruler should put troops into the field merely to gratify his own spleen; no general should fight a battle simply out of pique.

19. If it is to your advantage, make a forward move; if not, stay where you are.

[This is repeated from XI. ss. 17. Here I feel convinced that it is an interpolation, for it is evident that ss. 20 ought to follow immediately on ss. 18.]

20. Anger may in time change to gladness; vexation may be succeeded by content.

21. But a kingdom that has once been destroyed can never come again into being; [The Wu State was destined to be a melancholy example of this saying.] nor can the dead ever be brought back to life.

22. Hence the enlightened ruler is heedful, and the good general full of caution. This is the way to keep a country at peace and an army intact.

"Unless you enter the tiger's lair, you cannot get hold of the tiger's cubs."

XIII The Use of Spies

1. Sun Tzu said: Raising a host of a hundred thousand men and marching them great distances entails heavy loss on the people and a drain on the resources of the State. The daily expenditure will amount to a thousand ounces of silver. [Cf. II. ss. ss. 1, 13, 14.]

There will be commotion at home and abroad, and men will drop down exhausted on the highways.

[Cf. TAO TE CHING, ch. 30: "Where troops have been quartered, brambles and thorns spring up. Chang Yu has the note: "We may be reminded of the saying: 'On serious ground, gather in plunder.' Why then should carriage and transportation cause exhaustion on the highways? The answer is, that not victuals alone, but all sorts of munitions of war have to be conveyed to the army. Besides, the injunction to 'forage on the enemy' only means that when an army is deeply engaged in hostile territory, scarcity of food must be provided against. Hence, without being solely dependent on the enemy for corn, we must forage in order that there may be an uninterrupted flow of supplies. Then, again, there are places like salt deserts where provisions being unobtainable, supplies from home cannot be dispensed with."]

As many as seven hundred thousand families will be impeded in their labor.

[Mei Yao-chf en says: "Men will be lacking at the plough- tail." The allusion is to the system of dividing land into nine parts, each consisting of about 15 acres, the plot in the center being cultivated on behalf of the State by the tenants of the other eight. It was here also, so Tu Mu tells us, that their cottages were built and a well sunk, to be used by all in common. [See II. ss. 12, note.] In time of war, one of the families had to serve in the army, while the other seven contributed to its support. Thus, by a levy of 100,000 men (reckoning one able-bodied soldier to each family) the husbandry of 700,000 families would be affected.]

2. Hostile armies may face each other for years, striving for the victory which is decided in a single day. This being so, to remain in ignorance of the enemy's condition simply because one grudges the outlay of a hundred ounces of silver in honors and emoluments, ["For spies" is of course the meaning, though it would spoil the effect of this curiously elaborate exordium if spies were actually mentioned at this point.] is the height of inhumanity.

[Sun Tzu's agreement is certainly ingenious. He begins by adverting to the frightful misery and vast expenditure of blood and treasure which war always brings in its train. Now, unless you are kept informed of the enemy's condition, and are ready to strike at the right moment, a war may drag on for years. The only way to get this information is to employ spies, and it is impossible to obtain trustworthy spies unless they are properly paid for their services. But it is surely false economy to grudge a comparatively trifling amount for this purpose, when every day that the war lasts eats up an incalculably greater sum. This grievous burden falls on the shoulders of the poor, and hence Sun Tzu concludes that to neglect the use of spies is nothing less than a crime against humanity.]

3. One who acts thus is no leader of men, no present help to his sovereign, no master of victory.

[This idea, that the true object of war is peace, has its root in the national temperament of the Chinese. Even so far back as 597 B.C., these memorable words were uttered by Prince Chuang of the Chf u State: "The [Chinese] character for 'prowess' is made up of [the characters for] 'to stay' and 'a spear' (cessation of hostilities). Military prowess is seen in the repression of cruelty, the calling in of weapons, the preservation of the appointment of Heaven, the firm establishment of merit, the bestowal of happiness on the people, putting harmony between the princes, the diffusion of wealth."]

4. Thus, what enables the wise sovereign and the good general to strike and conquer, and achieve things beyond the reach of ordinary men, is FOREKNOWLEDGE.

[That is, knowledge of the enemy's dispositions, and what he means to do.]

5. Now this foreknowledge cannot be elicited from spirits; it cannot be obtained inductively from experience, [Tu Mu's note is: "[knowledge of the enemy] cannot be gained by reasoning from other analogous cases."] nor by any deductive calculation.

[Li Chfuan says: "Quantities like length, breadth, distance and magnitude, are susceptible of exact mathematical determination; human actions cannot be so calculated."]

6. Knowledge of the enemy's dispositions can only be obtained from other men.

[Mei Yao-chfen has rather an interesting note: "Knowledge of the spirit-world is to be obtained by divination; information in natural science may be sought by inductive reasoning; the laws of the universe can be verified by mathematical calculation: but the dispositions of an enemy are ascertainable through spies and spies alone."]

7. Hence the use of spies, of whom there are five classes: (1) Local spies; (2) inward spies; (3) converted spies; (4) doomed spies; (5) surviving spies.

8. When these five kinds of spy are all at work, none can discover the secret system. This is called "divine manipulation of the threads." It is the sovereign's most precious faculty.

[Cromwell, one of the greatest and most practical of all cavalry leaders, had officers styled 'scout masters,' whose business it was to collect all possible information regarding the enemy, through scouts and spies, etc., and much of his success in war was traceable to the previous knowledge of the enemy's moves thus gained."]

9. Having LOCAL SPIES means employing the services of the inhabitants of a district. [Tu Mu says: "In the enemy's country, win people over by kind treatment, and use them as spies."]

10. Having INWARD SPIES, making use of officials of the enemy.

[Tu Mu enumerates the following classes as likely to do good service in this respect: "Worthy men who have been degraded from office, criminals who have undergone punishment; also, favorite concubines who are greedy for gold, men who are aggrieved at being in subordinate positions, or who have been passed over in the distribution of posts, others who are anxious that their side should be defeated in order that they may have a chance of displaying their ability and talents, fickle turncoats who always want to have a foot in each boat. Officials of these several kinds," he continues, "should be secretly approached and bound to one's interests by means of rich presents. In this way you will be able to find out the state of affairs in the enemy's country, ascertain the plans that are being formed against you, and moreover disturb the harmony and create a breach between the sovereign and his ministers."

The necessity for extreme caution, however, in dealing with "inward spies," appears from an historical incident related by Ho Shih: "Lo Shang, Governor of I-Chou, sent his general Wei Po to attack the rebel Li Hsiung of Shu in his stronghold at P x i. After each side had experienced a number of victories and defeats, Li Hsiung had recourse to the services of a certain P'o-fai, a native of Wu-tu. He began to have him whipped until the blood came, and then sent him off to Lo Shang, whom he was to delude by offering to cooperate with him from inside the city, and to give a fire signal at the right moment for making a general assault.

Lo Shang, confiding in these promises, march out all his best troops, and placed Wei Po and others at their head with orders to attack at P x o- fai's bidding. Meanwhile, Li Hsiung's general, Li Hsiang, had prepared an ambuscade on their line of march; and Pxo- fai, having reared long scaling-ladders against the city walls, now lighted the beacon-fire. Wei Po's men raced up on seeing the signal and began climbing the ladders as fast as they could, while others were drawn up by ropes lowered from above. More than a hundred of Lo Shang's soldiers entered the city in this way, every one of whom was forthwith beheaded. Li Hsiung then charged with all his forces, both inside and outside the city, and routed the enemy completely." [This happened in 303 A.D. I do not know where Ho Shih got the story from. It is not given in the biography of Li Hsiung or that of his father Li Fe, CHIN SHU, ch. 120, 121.]

11. Having CONVERTED SPIES, getting hold of the enemy's spies and using them for our own purposes.

[By means of heavy bribes and liberal promises detaching them from the enemy's service, and inducing them to carry back false information as well as to spy in turn on their own countrymen. On the other hand, Hsiao Shih-hsien says that we pretend not to have detected him, but contrive to let him carry away a false impression of what is going on. Several of the commentators accept this as an alternative definition; but that it is not what Sun Tzu meant is conclusively proved by his subsequent remarks about treating the converted spy generously (ss. 21 sqq.). Ho Shih notes three occasions on which converted spies were used with conspicuous success: (1) by T'ien Tan in his defense of Chi-mo (see supra, p. 90); (2) by Chao She on his march to O-yu (see p. 57); and by the wily Fan Chu in 260 B.C., when Lien P x o was conducting a defensive campaign against Chun. The King of Chao strongly disapproved of Lien P'o's cautious and dilatory methods, which had been unable to avert a series of minor disasters, and therefore lent a ready ear to the reports of his spies, who had secretly gone over to the enemy and were already in Fan Chu's pay. They said: "The only thing which causes Chun anxiety is lest Chao Kua should be made general. Lien Pxo they consider an easy opponent, who is sure to be vanquished in the long run." Now this Chao Kua was a sun of the famous Chao She.

From his boyhood, he had been wholly engrossed in the study of war and military matters, until at last he came to believe that there was no commander in the whole Empire who could stand against him. His father was much disquieted by this overweening conceit, and the flippancy with which he spoke of such a serious thing as war, and solemnly declared that if ever Kua was appointed general, he would bring ruin on the armies of Chao. This was the man who, in spite of earnest protests from his own mother and the veteran statesman Lin Hsiang-ju, was now sent to succeed Lien P x o. Needless to say, he proved no match for the redoubtable Po Chfi and the great military power of Chun. He fell into a trap by which his army was divided into two and his communications cut; and after a desperate resistance lasting 46 days, during which the famished soldiers devoured one another, he was himself killed by an arrow, and his whole force, amounting, it is said, to 400,000 men, ruthlessly put to the sword.]

12. Having DOOMED SPIES, doing certain things openly for purposes of deception, and allowing our spies to know of them and report them to the enemy.

[Tu Yu gives the best exposition of the meaning: "We ostentatiously do thing calculated to deceive our own spies, who must be led to believe that they have been unwittingly disclosed. Then, when these spies are captured in the enemy's lines, they will make an entirely false report, and the enemy will take measures accordingly, only to find that we do something quite different.

The spies will thereupon be put to death." As an example of doomed spies, Ho Shih mentions the prisoners released by Pan Chf ao in his campaign against Yarkand. (See p. 132.) He also refers to T'ang Chien, who in 630 A.D. was sent by T x ai Tsung to lull the Turkish Kahn Chieh-li into fancied security, until Li Ching was able to deliver a crushing blow against him. Chang Yu says that the Turks revenged themselves by killing Tang Chien, but this is a mistake, for we read in both the old and the New Tang History (ch. 58, fol. 2 and ch. 89, fol. 8 respectively) that he escaped and lived on until 656. Li l-chi played a somewhat similar part in 203 B.C., when sent by the King of Han to open peaceful negotiations with Chfi. He has certainly more claim to be described a "doomed spy", for the king of Chfi, being subsequently attacked without warning by Han Hsin, and infuriated by what he considered the treachery of Li l-chi, ordered the unfortunate envoy to be boiled alive.]

13. SURVIVING SPIES, finally, are those who bring back news from the enemy's camp.

[This is the ordinary class of spies, properly so called, forming a regular part of the army. Tu Mu says: "Your surviving spy must be a man of keen intellect, though in outward appearance a fool; of shabby exterior, but with a will of iron. He must be active, robust, endowed with physical strength and courage; thoroughly accustomed to all sorts of dirty work, able to endure hunger and cold, and to put up with shame and ignominy." Ho Shih tells the following story of Ta'hsi Wu of the Sui dynasty: "When he was governor of Eastern Chun, Shen-wu of Chfi made a hostile movement upon Sha-yuan. The Emperor T'ai Tsu [? Kao Tsu] sent Ta- hsi Wu to spy upon the enemy. He was accompanied by two other men. All three were on horseback and wore the enemy's uniform. When it was dark, they dismounted a few hundred feet away from the enemy's camp and stealthily crept up to listen, until they succeeded in catching the passwords used in the army. Then they got on their horses again and boldly passed through the camp under the guise of night-watchmen; and more than once, happening to come across a soldier who was committing some breach of discipline, they actually stopped to give the culprit a sound cudgeling!

Thus they managed to return with the fullest possible information about the enemy's dispositions, and received warm commendation from the Emperor, who in consequence of their report was able to inflict a severe defeat on his adversary."]

14. Hence it is that which none in the whole army are more intimate

relations to be maintained than with spies.

[Tu Mu and Mei Yao-chf en point out that the spy is privileged to enter even the general's private sleeping-tent.]

None should be more liberally rewarded. In no other business should greater secrecy be preserved.

[Tu Mu gives a graphic touch: all communication with spies should be carried "mouth-to-ear." The following remarks on spies may be quoted from Turenne, who made perhaps larger use of them than any previous commander:

"Spies are attached to those who give them most, he who pays them ill is never served. They should never be known to anybody; nor should they know one another. When they propose anything very material, secure their persons, or have in your possession their wives and children as hostages for their fidelity. Never communicate anything to them but what is absolutely necessary that they should know.]

15. Spies cannot be usefully employed without a certain intuitive sagacity.

[Mei Yao-chfen says: "In order to use them, one must know fact from falsehood, and be able to discriminate between honesty and double-dealing." Wang Hsi in a different interpretation thinks more along the lines of "intuitive perception" and "practical intelligence." Tu Mu strangely refers these attributes to the spies themselves:
"Before using spies we must assure ourselves as to their integrity of character and the extent of their experience and skill." But he continues: "A brazen face and a crafty disposition are more dangerous than mountains or rivers; it takes a man of genius to penetrate such." So that we are left in some doubt as to his real opinion on the passage."]

16. They cannot be properly managed without benevolence and straightforwardness.

[Chang Yu says: "When you have attracted them by substantial offers, you must treat them with absolute sincerity; then they will work for you with all their might."]

17. Without subtle ingenuity of mind, one cannot make certain of the truth of their reports.

[Mei Yao-chfen says: "Be on your guard against the possibility of spies going over to the service of the enemy."]

18. Be subtle! be subtle! and use your spies for every kind of business.

19. If a secret piece of news is divulged by a spy before the time is ripe, he must be put to death together with the man to whom the secret was told. [Word for word, the translation here is: "If spy matters are heard before [our plans] are carried out," etc. Sun Tzu's main point in this passage is: Whereas you kill the spy himself "as a punishment for letting out the secret," the object of killing the other man is only, as Chfen Hao puts it, "to stop his mouth" and prevent news leaking any further. If it had already been repeated to others, this object would not be gained. Either way, Sun Tzu lays himself open to the charge of inhumanity, though Tu Mu tries to defend him by saying that the man deserves to be put to death, for the spy would certainly not have told the secret unless the other had been at pains to worm it out of him."]

20. Whether the object be to crush an army, to storm a city, or to assassinate an individual, it is always necessary to begin by finding out the names of the attendants, the aides-de-camp, [Literally "visitors", is equivalent, as Tu Yu says, to "those whose duty it is to keep the general supplied with information," which naturally necessitates frequent interviews with him.] and door-keepers and sentries of the general in command. Our spies must be commissioned to ascertain these.

[As the first step, no doubt towards finding out if any of these important functionaries can be won over by bribery.]

21. The enemy's spies who have come to spy on us must be sought out, tempted with bribes, led away and comfortably housed. Thus they will become converted spies and available for our service.

22. It is through the information brought by the converted spy that we are able to acquire and employ local and inward spies.

[Tu Yu says: "through conversion of the enemy's spies we learn the enemy's condition." And Chang Yu says: "We must tempt the converted spy into our service, because it is he that knows which of the local inhabitants are greedy of gain, and which of the officials are open to corruption."]

23. It is owing to his information, again, that we can cause the doomed spy to carry false tidings to the enemy.

[Chang Yu says, "because the converted spy knows how the enemy can best be deceived."]

24. Lastly, it is by his information that the surviving spy can be used on appointed occasions.

25. The end and aim of spying in all its five varieties is knowledge of the enemy; and this knowledge can only be derived, in the first instance, from the converted spy.

[As explained in ss. 22-24. He not only brings information himself, but makes it possible to use the other kinds of spy to advantage.]

Hence it is essential that the converted spy be treated with the utmost liberality.

26. Of old, the rise of the Yin dynasty [Sun Tzu means the Shang dynasty, founded in 1766 B.C. Its name was changed to Yin by P'an Keng in 1 401 . was due to I Chih [Better known as I Yin, the famous general and statesman who took part in Chfeng Tang's campaign against Chieh Kuei.] who had served under the Hsia. Likewise, the rise of the Chou dynasty was due to Lu Ya [Lu Shang rose to high office under the tyrant Chou Hsin, whom he afterwards helped to overthrow. Popularly known as Tai Kung, a title bestowed on him by Wen Wang, he is said to have composed a treatise on war, erroneously identified with the LIU T AO.] who had served under the Yin.

[There is less precision in the Chinese than I have thought it well to introduce into my translation, and the commentaries on the passage are by no means explicit. But, having regard to the context, we can hardly doubt that Sun Tzu is holding up I Chin and Lu Ya as illustrious examples of the converted spy, or something closely analogous. His suggestion is, that the Hsia and Yin dynasties were upset owing to the intimate knowledge of their weaknesses and shortcoming which these former ministers were able to impart to the other side. Mei Yao-chfen appears to resent any such aspersion on these historic names: "I Yin and Lu Ya," he says, "were not rebels against the Government. Hsia could not employ the former, hence Yin employed him. Yin could not employ the latter, hence Hou employed him. Their great achievements were all for the good of the people."

Ho Shih is also indignant: "How should two divinely inspired men such as I and Lu have acted as common spies? Sun Tzu's mention of them simply means that the proper use of the five classes of spies is a matter which requires men of the highest mental caliber like I and Lu, whose wisdom and capacity qualified them for the task. The above words only emphasize this point." Ho Shih believes then that the two heroes are mentioned on account of their supposed skill in the use of spies. But this is very weak.]

27. Hence it is only the enlightened ruler and the wise general who will use the highest intelligence of the army for purposes of spying and thereby they achieve great results.

[Tu Mu closes with a note of warning: "Just as water, which carries a boat from bank to bank, may also be the means of sinking it, so reliance on spies, while production of great results, is oft-times the cause of utter destruction."]

Spies are a most important element in water, because on them depends an army's ability to move.

[Chia Lin says that an army without spies is like a man with ears or eyes.]

THE SAYINGS OF WUTZU

INTRODUCTION

Now Wu, albeit clothed in the raiment of a scholar, was a man skilled in the art of war. And Wen, Lord of Wei, came unto him and said: " I am a man of peace, caring not for military affairs." And Wu said: " Your actions are witnesses of your mind; why do your words say not what is in your heart? "You do prepare and dress hides and leather through the four seasons, ornamenting them with red lacquer and the figures of panthers and elephants; which give not warmth in winter, neither in summer, coolness. Moreover, you make halberds, 24 feet long, and pikes 12 feet long, and leather (covered) chariots so large as to fill up the gateways, wheels with ornament, and naves capped with leather. Now, these are neither beautiful to the eye nor light in the chase; I know not for what use my lord makes these things. " But, although provided with these instruments of war, if the leader be not competent, a brooding hen might as well strike a badger, or a dog with young challenge the tiger: the spirit of encounter may be present, but there is no end but death. " In ancient times, the Prince Chengsang cultivated virtue, and put away military things, and his kingdom fell. " The Prince Yuhu put his trust in numbers, and delighted in war and was driven from the throne. " Therefore the enlightened ruler should ponder over these things; encourage learning and virtue in the kingdom, and be prepared against war from without.

All numbers connected with weapons were Yin, that is to say even or belonging to the negative principle of Chinese philosophy from their connection with death. " To hesitate before the enemy is not a cause for righteousness; remorse for the fallen is not true humanity." And when Lord Wen heard these words, he himself spread a seat, and his wife offered up a cup, and Wu was appointed general before the altar. Now, in the defense of Hsihe against different states there were fought seventy-six great fights, of which sixty-four were complete victories, and the remainder undecided. And the kingdom grew and stretched 1,000 leagues on every side, which was all due to the virtue of Wu.

I

Wu the Master said: The mighty rulers of old first trained their retainers, and then extended their regard to their outlying feudatories.

There are four discords:

Discord in the state: then never make war.

Discord in the army: then do not strike camp.

Discord in the camp: then do not advance to attack.

Discord in the battle array: then seek not to decide the issue.

Therefore, wise rulers who would employ their subjects in great endeavors, should first establish harmony among them. Lend not a ready ear to human counselors, but lay the matter before the altar; seek inside the turtle, (Divining the future by the cracks in the heated shell.) and consider well the time and season. Then, if all be well, commit ourselves to the undertaking. If the people know that their lord is careful of their lives, and laments their death beyond all else; then, in the time of danger, the soldiers advance, and, advancing, find glory in death; and in survival after retreat, dishonor. The Master said: The Way must follow the only true.

The back of a turtle was burnt, and the answer was ascertained by the manner in which the shell split. path: righteousness lies at the root of achievement and merit. The object of stratagem is to avoid loss and gain advantage. The object of government is to guard enterprise and to preserve the state. If conduct depart from the Way, and the undertaking accord not with righteousness, then disaster befalls the mighty. Therefore, wise men maintain order by keeping in the Way, and governing with righteousness; they move with discretion, and with benevolence they make the people amenable. If these four virtues be practiced, there is prosperity; if they be neglected, there is decay. For, when Lord Tang of Cheng defeated Lord Chieh, the people of Hsia rejoiced, and when Wu of Chou defeated Lord Chou, the people of Yin were not discomfited. And this was because it was ordained by Providence and human desire.

The Master said: In the government of a country and command of an army, the inculcation of propriety, stimulation of righteousness, and the promotion of a sense of shame are required. When men possess a sense of shame, they will attack with resolution when hi strength, and when few in number defend to the last. But while victory is easy in attack, it is difficult in defense. Now, of the fighting races below heaven; those who gained five victories have been worn out; those who have won four victories have been impoverished; three victories have given dominion; two victories have founded a kingdom; and upon one victory an empire has been established. For those who have gained power on earth by many victories are few; and those who have lost it, many.

The Master said: The causes of war are five: First, ambition; second, profit; third, overburdened hate; fourth, internal disorder; fifth, famine. Again, the natures of war are five: First, a righteous war; second, a war of might; third, a war of revenge; fourth, a war of tyranny; fifth, an unrighteous war. The prevention of tyranny and the restoration of order is just; to strike in reliance on numbers is oppression; to raise the standard for reasons of anger is a war of revenge; to quit propriety, and seize advantage is tyranny; when the state is disordered and the people worn out, to harbor designs, and set a multitude in motion, is a war of unrighteousness. There is a way of overcoming each of these five.

Righteousness is overcome by propriety; might by humanity; revenge by words; tyranny by deception; unrighteousness by strategy. Lord Wen asked and said: " I would know the way to control an army, to measure men, and make the country strong." Wu answered and said: "The enlightened rulers of antiquity respected propriety between sovereign and people; established etiquette between high and low; settled officials and citizens in close accord; gave instruction in accordance with custom; selected men of ability, and thereby provided against what should come to pass. " In ancient times, Prince Huan of Chi assembled 50,000 men at arms, and became chief among the princes; Prince Wen of Chin put 40,000 mighty men in the war, and gained his ambition; Prince Mu of Chin gathered together 30,000 invincible, and subdued his neighboring foes. Wherefore, the princes of powerful states must consider their people, and assemble the valiant and spirited men by companies. "Those who delight to attack, and to display their valor and fealty should be formed in companies. "Those skilful in scaling heights, or covering long distances, and who are quick and light of foot must be collected in companies. " Retainers who have lost their rank, and who are desirous of displaying their prowess before their superiors should be gathered into companies. " Those who have abandoned a castle, or deserted their trust, and are desirous of atoning for their misconduct, should be collected and formed into companies. " These five bodies form the flower of the army. With 3,000 of such troops, if they issue from within, an encompassing enemy can be burst asunder; if they enter from without, a castle can be overthrown."

Lord Wen asked and said: " I desire to know how to fix the battle array, render defense secure, and attack with certainty of victory." Wu answered and said: " To see with the eye is better than ready words. Yet, I say, if the wise men be put in authority and the ignorant in low places, then the army is already arranged. 'If the people be free from anxiety about their estates, and love their officials, then defense is already secure. " If all the lieges be proud of their lord, and think ill of neighboring states, then is the battle already won."

The Lord Wen once assembled a number of his subjects to discuss affairs of state: and none could equal him in wisdom, and when he left the council chamber his face was pleased. Then Wu advanced and said: In ancient times, Lord Chuang of Chu once consulted with his lieges, and none were like unto him in wisdom; and when the Lord left the council chamber his countenance was troubled. Then the Duke Shen asked and said: " Why is my Lord troubled? " And he answered: " I have heard that the world is never without sages, and that in every country there are wise men; that good advisers are the foundation of an empire; and friends of dominion. Now, if I, lacking wisdom, have no equal among the multitude of my officers, dangerous indeed is the state of Chu. It grieves me that whereas Prince Chuang of Chu was troubled in a like case my Lord should be pleased." And hearing this Lord Wen was inwardly troubled.

ESTIMATION OF THE ENEMY

II

Lord Wen said to Wu: " Chin threatens us on the west; Chu surrounds us on the south; Chao presses us in the north; Chi watches us in the east; Yen stops our rear, and Han is posted in our front. Thus, the armies of six nations encompass us on every side, and our condition is very unpropitious. Canst thou relieve my anxiety? " Wu answered and said: " The path of safety of a state lies first of all in vigilance. Now my Lord has already taken warning, wherefore misfortunes are yet distant.

6 'Let me state the habits of these six countries. The forces of Chi are weighty but without solidity; the soldiers of Chin are scattered, and fight each of his own accord: the army of Chu is well ordered, but cannot endure: the soldiers of Yen defend well, but are without dash: the armies of the three Chins are well governed, but cannot be used. "The nature of Chi is stubborn and the country rich, but prince and officials are proud and luxurious, and neglectful of the common people; government is loose and rewards not impartial; in one camp there are two minds; the front is heavy, but the rear is light. Therefore it is ponderous without stability. To attack it, the force must be divided into three parts, and, by threatening it on three sides, its front can be broken. " The nature of Chin is strong, the country rugged, and the government firm; rewards and punishments just, the people indomitable, and all have the fighting spirit; wherefore, when separated, each fights of his own accord. "To defeat this people, they must first be tempted by gain to leave their cause, so that the soldiers, greedy of profit, desert their general: then, taking advantage of their disobedience, their scattered forces can be chased, ambushes laid, favorable opportunities taken, and their general captured. " The nature of Chu is weak, its territory wide, the government weak, and the people exhausted; the troops are well ordered but of short endurance. "The way to defeat them is to assault their camp, throw it into confusion and crush their spirit, advance softly, and retire quickly; tire them out, avoid a serious encounter, and they may be defeated. "

The nature of Yen is straightforward; its people are cautious, loving courage and righteousness, and without guile; wherefore they defend but are not daring. ' The way to defeat them is to draw close and press them; tease them and pass to a distance; move quickly, and appear in the rear, thus causing bewilderment to their officers and fear in their ranks. Our chariots and horsemen will act with circumspection and avoid encounter. Thus their general can be captured. "The three Chins are the middle kingdom: their nature is peaceful and their rule just. Their people are tired of war; their troops are trained, but their leaders are despised; pay is small, and the soldiers lack the spirit of sacrifice, thus they are well governed but cannot be used. " The way to defeat them is to threaten them from afar. If a multitude attack defend; if they retreat pursue, and tire them out. "In every army there are mighty warriors with strength to lift the Censer, swifter of foot than the war horse; who can take the enemy's standard, or slay his general. If such men be selected, and set apart, cared for and honored, they are the life of the army. " Those who use the five arms l with skill, who are clever, strong and quick, and careless of the enemy, should be given rank and decoration, and used to decide the victory.

Their parents and families should be cared for, encouraged by rewards, and kept in fear of punishment. Give these men Halberd, shield, javelin, pike, and short pike, and consolidate the battle array; their presence causes endurance. "If these men be well selected, double their number can be defeated."

And Lord Wen said: " It is good ! " Wu the Master said: "In the estimation of the enemy there are eight cases when, without consulting the oracles, he may be attacked. " First, an enemy who, in great wind and cold, has risen early, started forth across ice and rivers, and braved stress and hardships. " Second, an enemy who, in the height of summer, and in great heat, has risen early, has travelled incessantly, is hungry and without water, and is striving to reach a distance. "Third, an enemy who has been en- camped long in one place, who is without provisions, when the farmers are vexed and indignant, who has suffered frequent calamities, and whose officers are unable to establish confidence. "Fourth, when the enemy's funds are exhausted, fuel and fodder scarce; when the heavens have been overcast by long continued rain; when there is the desire to loot, but no place to loot withal. "Fifth, when their numbers are few; when water is scarce; when men and horses are scourged by pestilence, and from no quarter or support is at hand. " Sixth, when night falls, and the way is yet far; when officers and men are worn out and fearful, weary and without food, and have laid aside their armor and are resting. " Seventh, when the general's authority is weak, the officials false, and the soldiers unsettled; when their army has been alarmed, and no help is forthcoming. " Eighth, when the battle formation is not yet fixed, or camp pitched; when climbing a hill, or passing through a difficult place; when half is hidden and half exposed. "An enemy in these situations may be smitten without hesitation.

" There are six enemies, that, without consulting oracles, should be avoided. " First, wide and vast territories, and a large and rich population.

" Second, where the officials care for the people, and bestow bountiful favors and rewards. " Third, where rewards are well deserved, punishment accurately apportioned, and operations undertaken only when the time is fitting. "Fourth, where merit is recognized and given rank, wise men appointed, and ability employed. "Fifth, where the troops are many and their weapons excellent. "Sixth, when help is at hand on every side, or from a powerful ally. " For, if the enemy excel in the foregoing, he must be avoided without hesitation. As it is written, if it be judged good, advance; if it be known to be difficult, retreat." And Lord Wen asked and said: " I desire to know how the interior of the enemy can be known from his outer appearance; the form of his camp by observing his advance, and how victory may be determined?" And Wu answered and said: " If the coming of the enemy be reckless like roaring waters, his banners and pennons disordered, and horses and men frequently looking behind, then ten can be struck with one. Panic will certainly seize them. " Before the various princes have assembled, before harmony has been established between lord and lieges, before ditches have been dug, or regulations established, and the army is alarmed; wishing to advance, but unable; wishing to retreat, but unable: then the force can strike twice their numbers, and in a hundred rights there is no fear of retreat."

Lord Wen asked: "How can the enemy be certainly defeated?" Chi answered and said: "Make certain of the enemy's real condition and quickly strike his weak point; strike an enemy who has just arrived from afar, before his ranks are arranged; or one who has eaten and has not completed his dispositions; or an enemy who is hurrying about, or is busily occupied; or has not made favorable use of the ground, or has let pass the opportunity; or one who has come a long distance, and those in rear are late and have not rested. "Strike an enemy who is half across waters; or who is on a difficult or narrow road; or whose flags and banners are in confusion; or who is frequently changing position; or whose general is not in accord with the soldiers; or who is fearful. "All such should be assaulted by the picked men; and the remainder of the army should be divided, and follow after them. They may be attacked at once without hesitation."

Ill
CONTROL OF THE ARMY

LORD WEN said: " What is of first importance in operations of war?" Wu answered and said: "Lightness, of which there are four natures, Weight, of which there are two natures, and Confidence must be clearly comprehended." And Wen said: " What are these? " And Wu answered: " If the way be easy, the horses are light of foot; if the horses be light of foot, the chariots travel freely; if the chariots travel easily, men can ride in them without difficulty; if the men be free to move, the fight prospers. If the difficult and easy ways be known, the horses are lightened; if the horses be fed at proper intervals, the chariots are swift; if there be plenty of oil on the axles of the chariots, the riders are quickly conveyed; if the spears be sharp and the armour strong, the men make the fight easy. "Large rewards in advance, heavy punishment in retreat, and impartiality in their bestowal are required. "He who well understands these things is the master of victory." And Lord Wen asked and said: " By what means can the army gain the victory? And Wu answered: "The foundation of victory is good government." Again, Wen asked and said: " Is it not determined by numbers? " And Wu replied: " If laws and orders be not clear; if rewards and punishments be not just; if the bell be sounded and they halt not, or drum be beaten and men do not advance; even if there be a hundred thousand men at arms, they are of no avail.

" Where there is order, then there is propriety at rest, and dignity in motion; none can withstand the attack, and retreat forbids pursuit; motion is regulated, and movements to right and left are made in answer to the signal; if the ranks be cut asunder, formation is preserved; if scattered, they are maintained; in fortune or in danger, there is unity; if a number be collected, they cannot be separated; they may be used but not wearied; in whatever situation they are placed, nothing under heaven can withstand them. The army may be called a father and his children." And Wu said: " In marching, movements and halts must be properly adjusted, suitable occasions for rationing not missed; the strength of men and horses not exhausted. If these three things be observed, the commands of the superior can be carried out; if the commands of the superior be carried out, order is maintained. If advances and halts be without method, vacillating or unsuitable, horses and men tired and weary neither unsaddled or housed it is because the orders cannot be obeyed; if the orders be set aside, there is disorder in the camp, and in battle defeat." Wu the Master said: " On that depository of corpses, the battle- field, if there be certain expectation of death, there is life; if there be happy expectation of life, there is death. The good general is like unto one sitting in a leaking ship, or lying under a burning roof; the wisest man cannot contrive against him; the strongest man cannot destroy his composure; and the enemy's onslaught can be withstood.

For procrastination is the greatest enemy of the general; disasters to the army are born of indecision." Wu the Master said: " Men meet their death from lack of ability or unskilfulness. Wherefore training is the first requirement of war. One man with a knowledge of war can teach ten; ten men skilled in war can teach one hundred; one hundred can teach one thousand; one thousand can teach ten thousand; and ten thousand men can train an army. "An enemy from a distance should be awaited, and struck at short range; an enemy that is tired should be met in good order; hunger should be opposed by full bellies; the battle formation should be round or square, the men should kneel or stand; go or remain; move to the right or left; advance or retire; concentrate or disperse; close or extend when the signal is given.

"All these changes must be learnt, and the weapons distributed. This is the business of the general." Wu the Master said: " In the teaching of war, spears are given to the short; bows and catapults to the tall; banners and standards to the strong; the bell and drum to the bold; fodder and provisions to the feeble; the arrangement of the plan to the wise. Men of the same district should be united; and groups and squads should help each other. At one beat of the drum the ranks are put in order; at two beats of the drum, formation will be made; at three beats of the drum, food will be issued; at four beats of .the drum, the men will prepare to march; at five beats of the drum, ranks will be formed; when the drums beat together, then the standards will be raised." And Lord Wen asked and said: " What is the way of marching and halting an army? "

And Wu answered: " Natural ovens and dragons' heads should be avoided. Natural ovens are the mouths of large valleys. Dragons' heads are the extremities of large mountains. The green dragons (banners) should be placed on the left, and the white tigers on the right; the red sparrows in front; the snakes and tortoises behind; the pole star (standard) above; and the soldiers will look to the standard. " When going forth to battle, the direction of the wind must be studied; if blowing in the direction of the enemy, the soldiers will be assembled and follow the wind; if a head wind, the position will be strengthened, and a wait made for the wind to change."

And Lord Wen asked and said: " In what way should horses be treated? " And Wu answered and said: " The places where they are kept should be made comfortable; fodder should be suitable and timely. In winter their stables should be warmed, and in summer sheltered from the heat; their coats clipped, their feet carefully pared, their attention directed so that they be not alarmed, their paces regulated, and their going and halting trained; horses and men should be in accord, and then the horses can be used. The harness, the saddle, bit, bridle, and reins must be strong; if the horse be without vice at the beginning, he can be used to the end; if the horse be hungry it is good; if his belly be full, his value decreases; if the sun be falling and the way still long, dismount frequently. For it is proper that the men be worked, but the horses must be used with discretion, so that they may be prepared should the enemy suddenly attack us. " If these things be well known, then there is free passage under heaven."

IV

QUALITIES OF THE GENERAL

Wu the Master said: " The leader of the army is one who is master of both arms and letters. He who is both brave and tender can be entrusted with troops. "In the popular estimation of generals, courage alone is regarded; nevertheless, courage is but one of the qualifications of the leader. Courage is heedless in en- counter; and rash encounter, which is ignorant of the consequences, cannot be called good. " There are five matters which leaders must carefully consider. " First, reason; second, preparation; third, determination; fourth, vigilance; fifth, simplicity. " With reason, a multitude can be controlled like a small number. "Preparedness sees an enemy outside the gate. " Determination before the enemy has no thought of life. " Even after a victory, vigilance behaves as before the first encounter. " Simplicity ensures few regulations, and preserves order. " When the leader receives his orders, he forthwith departs. Not until the enemy has been vanquished does he speak of return. This is the duty of the general.

" Wherefore, from the day of departure of the army, the general seeks glory in death, and dreams not of return in dishonor." Wu the Master said: " In war there are four important influences. " First, spirit; second, ground; third, opportunity; fourth, force.

" The military value of the nation's forces of one hundred times ten thousand fighting men depends upon the personality of one man alone; this is called the influence of spirit "When the road is steep and narrow, when there are famous mountains and fastnesses where ten men can defend and one thousand cannot pass them by; such is the influence of ground. "When spies have been skillfully sown, and mounted men pass to and from the enemy's camp, so that his masses are divided, his sovereign and ministers vexed with each other, and superiors and inferiors mutually censorious; this is the moment of opportunity. " When the linchpins are secure, the oars and sweeps ready for use in the boats, the armed men trained for war, and the horses exercised, we have what is called the influence of force. " He who understands these four matters has the qualifications of a general. Further- more, dignity, virtue, benevolence, courage, are needed to lead the troops, to calm the multitude, to put fear in the enemy, to remove doubts. When orders are issued, the subordinates do not defy them. Wheresoever the army is, that place the enemy avoids. If these four virtues be present, the country is strong; if they be not present, the country is overthrown. " Of such is the good general."

Wu the Master said: " The use of drums and bells is to attract the ear; of flags, standards, and banners to strike the eye; of laws and penalties to put fear in the heart. " To attract the ear the sound must be clear; to strike the eye the colors must be bright. The heart is awed by punishment, therefore punishment must be strict. " If these three matters be not ordered, the state may, peradventure, be preserved, but defeat by the enemy is certain. Therefore, as it has been said (if these three things be present), there is no departing from the commands of the general; when he orders, there is no going back from death."

Wu the Master said: " The secret of war is, first, to know who is the enemy's general, and to judge his ability. If our plans depend on his dispositions, then success will be achieved without toil. " If their general be stupid, and heedlessly trustful, he may be enticed by fraud; if he be avaricious and careless of his fame, he may be bribed with gifts. If he make unconsidered movements without plan, he should be tired out and placed in difficulties.

If the superiors be wealthy and proud, and the inferiors avaricious and resentful, they should be set against each other. An enemy that is undetermined, now advancing and then retreating, whose soldiers have naught wherein to put their trust, should be alarmed, and put to flight. " When an enemy thinks lightly of the general, and desires to return home, the easy roads should be blocked, and the difficult and narrow roads opened; await their coming and capture them. " If their advance be easy and retreat difficult, await their coming and then advance against them. " If their advance be difficult and retreat easy, then press and strike them,

" An army that is camped in marshy ground, where there are no water- courses, and long and frequent rains, should be inundated.
"An army that is camped in wild marshes, covered with dark and overhanging grass and brambles, and swept by frequent high winds, should be overthrown by fire.
"An army that has halted long without moving; whose general and soldiers have grown careless, and neglect precautions, should be approached by stealth, and taken by surprise."

Lord Wen asked, saying: " If the two armies be facing each other, and the name of the enemy's general unknown, in what manner can we discover it?" And Wu answered and said: "A brave man of low degree, lightly but well equipped, should be employed. He should think only of flight and naught of advantage. Then, if he observe the enemy's pursuit, if there be first a halt and then an advance, order is established. If we retreat and the enemy pursue, but pretend not to be able to overtake us, see an advantage but pretend not to be aware of it, then their general may be called a wise general, and conflict with him must be avoided. If their army be full of uproar; their banners and standards disordered, their soldiers going about or remaining of their own accord, some in line, others in column; if such an enemy be eager to pursue, and see an advantage which they are desperate to seize, then their general is a fool: even if there be a host, they may be taken."

V

SUITING THE OCCASION

LORD WEN asked and said: "If strong chariots, good horses, strong and valiant soldiers suddenly meet the enemy, and are thrown into confusion, and ranks broken, what should be done? " And Wu answered and said: " In general, the method of fighting is to effect order in daylight by means of flags and banners, pennons and batons; at night by gongs and drums, whistles and flutes. If a signal be made to the left, the troops move to the left; if to the right, they move to the right. Advance is made at the sound of the drum; halt at the sound of the gong; one blast of the whistle is for advance, two for the rally. If those who disobey be cut down, the forces are subject to authority. If officers and soldiers carry out orders, a superior enemy cannot exist; no position is impregnable in the attack," Lord Wen asked and said: "What is to be done if the enemy be many and we be few? "

And Wu answered and said: "Avoid such an enemy on open ground, and meet him in the narrow way; for, as it is written, if 1 is to stand against 1,000, there is naught better than a pass; if 10 are to hold against 100, there is nothing better than a steep place; if 1,000 are to strike 10,000, there is nothing better than a difficult place. If a small force, with beat of gong and drum, suddenly arise in a narrow way, even a host will be upset. Wherefore it is written: 'He who has a multitude seeks the plain, and he who has few seeks the narrow way.' '

And Lord Wu asked and said: " A mighty host, strong and courageous, which is on the defense with a mountain behind, a precipice between, high ground on the right, and a river on the left, with deep moats, and high walls, and which has artillery; whose retreat is like the removal of a mountain, advance like the hurricane, and whose supplies are in abundance, is an enemy against whom long defense is difficult.
In effect, what should be done in such a case? "

And Wu answered and said: " This indeed is a great question, whose issue depends, not upon the might of chariot and horse, but upon the schemes of a wise man. " Let 1,000 chariots and 10,000 horse, well equipped and with foot-men added to them, be divided into five armies, and a road allotted to each army. " Then if there be five armies, and each army take a different road, the enemy will be puzzled, and know not in what quarter to be prepared. If the enemy's defense be strong and united, send envoys quickly to him to discover his intention. If he listen to our advices, he will strike camp and withdraw. But, if he listen not to our advice, but strikes down the messenger, and burns his papers, then divide and attack from five quarters. If victorious, do not pursue; if defeated, flee to a distance. If feigning retreat, proceed slowly, and, if the enemy approach, strike swiftly. " One army will hold the enemy in front, with another cut his rear, two more with gags in their mouths 1 will attack his weak point, whether on the right or on the left. If five armies thus make alternate onslaughts, success is certain. " This is the way to strike strength."

And Lord Wen asked and said: " If the enemy draw near and encompass us, and we would retreat, but there is no way, and in our multitude there is fear, what should be done? " And Wu answered and said: " In such a case, if we be many and they be few, divide and fall upon them; if the enemy be many and we be few, use stratagem and act according to opportunity; and if opportunities be untiringly seized, even if the enemy be many, he will be reduced to subjection." Silently.

Lord Wen asked and said: " If, in a narrow valley with steep places on either side, the enemy be met, and they are many and we are few, what should be done? "

And Wu answered and said: " If they be met among hills, woods, in deep mountains, or wide fens, advance quickly, retire swiftly, and hesitate not. If the enemy be suddenly met among high mountains or deep valleys, be the first to strike the drum and fall upon them. Let bow and cross bow advance; shoot and capture; observe the state of their ranks; and, if there be confusion, do not hesitate to strike." Lord Wen asked and said: "If the enemy be suddenly met in a narrow place with high mountains on either side, and advance and retreat are alike impossible, what should be done in such a case? " And Wu answered and said: " This is called War in valleys where numbers are of no avail. The ablest officers should be collected, and set against the enemy.

Men light of foot and well armed should be placed in front; the chariots divided; the horsemen drawn up, and placed in ambush on four sides, with many leagues between, and without showing their weapons. Then, the enemy will certainly make his defense firm, and neither advance or retreat. Whereupon, the standards will be raised, and the ranks of banners shown, the mountains left, and camp pitched in the plain.

" The enemy will then be fearful, and should be challenged by chariot and horse, and allowed no rest. " This is the method of fighting in valleys." And Lord Wen asked and said: "If the enemy be met in a marsh where the water is out, so that the wheels of the chariots sink in, and the shafts be covered, and the chariots and horsemen overcome by the waters, when there are no boats or oars, and it is impossible either to advance or retreat, what should be done in such a case? "

And Wu answered and said: " This is called water fighting. Chariots and horsemen cannot be used, and they must be put for a time on one side. Go up to the top of a high place, and look out to the four quarters. Then the state of the waters will certainly be seen; their extent, and the deep places and shallows fully ascertained. Then, by stratagem, the enemy may be defeated. " If the enemy should cross the waters he should be engaged when half over." And Lord Wen asked and said: " If there has been long continued rain so that the horses sink, and the chariots cannot move; if the enemy appear from four quarters, and the forces are frightened, what is the course in such a case? " And Wu answered and said: "When wet and overcast, the chariots should halt; when fine and dry, they should arise. Seek height, and avoid low places; drive the strong chariots, and choose well the road on which to advance or halt. If the enemy suddenly arise, immediately pursue them."

Lord Wen asked and said: " If our fields and pastures be suddenly pillaged, and our oxen and sheep taken, what should be done? " And Wu answered and said: "Lawless enemies that arise are to be feared; defend well and do not reply. When, at sunset, they seek to withdraw, they will certainly be overladen and fearful. Striving to return quickly to their homes, connection will be lost. Then if they be pur- sued and attacked, they can be overthrown." Wu the Master said: " The way of attacking the enemy and investing his castle is as follows: " When the outlying buildings have been taken, and the assaulting parties enter the innermost sanctuary, make use of the enemy's officials, and take charge of their weapons. Let the army on no account fell trees or enter dwellings, cut the crops, slay the six domestic animals, or burn the barns; arid show the people that there desire. Those who wish to surrender, should be received and freed from anxiety."

VI

ENCOURAGEMENT OF THE TROOPS

AND Lord Wen asked and said: " If punishment be just and reward impartial, is victory thereby gained?" And Wu answered and said: " I cannot speak of all the things that concern justice and impartiality, but on these alone dependence cannot be placed. " If the people hear the word of command, or listen to the order with rejoicing; if, when the army be raised, and a multitude assembled, they go forth gladly to the fight; if, in the tumult of the fight, when blade crosses blade, the soldiers gladly die; upon these three things can the lord of the people place his trust." And Lord Wen said: " How can this be brought about? " And Wu answered and said: " Seek out merit, advance and reward it, and encourage those without fame."

Accordingly Lord Wen set seats in the garden of the palace in three rows, and made a feast unto his chief retainers. In the first row were set those of chief merit, and on the table were placed the best meats and precious utensils. Those of medium merit were set in the middle row, and the utensils on the table were fewer in number. Those without merit were set in the last row, and utensils of no value w r ere put before them. And when the feast was over, and they had all departed, the parents, wives, and children of those with merit were given presents outside the gates of the palace according to their degree. Further, messengers were sent yearly with gifts to condole with the parents of those who had lost a son in the service of the state, and to show that they were had in remembrance. And after this was carried out for three years, the people of Chin gathered an army, and came as far as the Western River. And when the soldiers of Wei heard this, without waiting for orders, they armed themselves and fell upon them; and they that went forth were 10,000 in number.

And Lord Wen called Wu and said: " The words that you spoke unto me, have they not indeed been carried out? " And Wu answered and said: ' I have heard that there are men, great and small; souls, grand and feeble. " As a trial, let 50,000 men, without merit, be collected, and placed under my command against the country of Chin. If we fail, the state will be the laughing-stock among the princes, and its power under heaven will be lost. If a desperate robber be hidden in a wide plain, and 1,000 men be pursuing him, their glances will be furtive like the owl, looking backward like the wolf, for they are in fear of harm from a sudden onslaught. " One desperate man can put fear in the hearts of a thousand. Now, if this host of 50,000 men become as a desperate thief, and are led against Chin, there is naught to fear." On hearing these words Lord Wen agreed, and adding further 500 chariots and 3,000 horse, the hosts of Chin were overthrown, all being due to the encouragement of the troops. On the day before the battle Wu gave orders to the forces, saying: " The army will attack the enemy's chariots, horse and foot, in accordance with our commands. If the chariots do not capture the enemy's chariots, or the horse those of the enemy's, or the foot the enemy's footmen, even if their army be overthrown, no merit will be gained." Therefore on the day of the battle, the orders were simple, and fear of Wei shook the heavens.

GoRiN No Sho

A Book OF Five Rings

Written by Miyamoto Musashi
Translated by Victor Harris
(Public Domain Edition)

Translator's Introduction

JAPAN DURING MUSASHI'S LIFETIME

Miyamoto Musashi was born in 1584, in a Japan struggling to recover from more than four centuries of internal strife. The traditional rule of the emperors had been overthrown in the twelfth century, and although each successive emperor remained the figurehead of Japan, his powers were very much reduced. Since that time, Japan had seen almost continuous civil war between the provincial lords, warrior monks and brigands, all fighting each other for land and power. In the fifteenth and sixteenth centuries the lords, called daimyo, built huge stone castles to protect themselves and their lands and castle towns outside the walls began to grow up. These wars naturally restricted the growth of trade and impoverished the whole country.

In 1573, however, one man, Oda Nobunaga, came to the fore in Japan. He became Shogun, or military dictator, and for nine years succeeded in gaining control of almost the whole of the country. When Nobunaga was assassinated in 1582, a commoner took over the government. Toyotomi Hideyoshi continued the work of unifying Japan which Nobunaga had begun, ruthlessly putting down any traces of insurrection. He revived the old gulf between the warriors of Japan— the samurai— and the commoners by introducing restrictions on the wearing of swords. "Hideyoshi's sword-hunt", as it was known, meant that only samurai were allowed to wear two swords; the short one which everyone could wear and the long one which distinguished the samurai from the rest of the population. Although Hideyoshi did much to settle Japan and increase trade with the outside world, by the time of his death in 1598 internal disturbances still had not been completely eliminated. The real isolation and unification of Japan began with the inauguration of the great Togugawa rule.

In 1603 Tokugawa Ieyasu, a former associate of both Hideyoshi and Nobunaga, formally became Shogun of Japan, after defeating Hideyoshi's son Hideyori at the battle of Seki ga Hara. Ieyasu established his government at Edo, present-day Tokyo, where he had a huge castle. His was a stable, peaceful government beginning a period of Japanese history which was to last until the Imperial Restoration of 1868, for although Ieyasu himself died in 1616 members of his family succeeded each other and the title Shogun became virtually an hereditary one for the Tokugawas. Ieyasu was determined to ensure his and his family's dictatorship. To this end, he paid lip-service to the emperor in Kyoto, who remained the titular head of Japan, while curtailing his duties and involvement in the government. The real threat to Ieyasu's position could only come from the lords, and he effectively decreased their opportunities for revolt by devising schemes whereby all lords had to live in Edo for alternate years and by placing great restrictions on travelling. He allotted land in exchange for oaths of allegiance, and gave the provincial castles around Edo to members of his own family. He also employed a network of secret police and assassins.

The Tokugawa period marks a great change in the social history of Japan. The Bureaucracy of the Tokugawas was all-pervading. Not only were education, law, government and class controlled, but even the costume and behavior of each class. The traditional class consciousness of Japan hardened into a rigid class structure. There were basically four classes of person: samurai, farmers, artisans and merchants. The samurai were the highest — in esteem if not in wealth — and included the lords, senior government officials, warriors, and minor officials and foot soldiers. Next in the hierarchy came the farmers, not because they were well thought of but because they provided the essential rice crops. Their lot was a rather unhappy one, as they were forced to give most of their crops to the lords and were not allowed to leave their farms. Then came the artisans and craftsmen, and last of all the merchants, who, though looked down upon, eventually rose to prominence because of the vast wealth they accumulated. Few people were outside this rigid hierarchy. Musashi belonged to the samurai class. We find the origins of the samurai class in the Kondei ("Stalwart Youth") system established in 792 AD, whereby the Japanese army — which had until then consisted mainly of spear-wielding foot soldiers — was revived by stiffening the ranks with permanent training officers recruited from among the young sons of the high families. These officers were mounted, wore armour, and used the bow and sword. In 782 the emperor Kammu started building Kyoto, and in Kyoto he built a training hall which exists to this day called the Butokuden, meaning "Hall of the virtues of war".

Within a few years of this revival the fierce Ainu, the aboriginal inhabitants of Japan who had until then confounded the army's attempt to move them from their wild lodgings, were driven far off to the northern island, Hokkaido. When the great provincial armies were gradually disbanded under Hideyoshi and Ieyasu, many out-of-work samurai roamed the country redundant in an era of peace. Musashi was one such samurai, a "ronin" or "wave man". There were still samurai retainers to the Tokugawas and provincial lords, but their numbers were few. The hordes of redundant samurai found themselves living in a society which was completely based on the old chivalry, but at the same time they were apart from a society in which there was no place for men at arms. They became an inverted class, keeping the old chivalry alive by devotion to military arts with the fervor only Japanese possess.

This was the time of the flowering in Kendo. Kendo, the Way of the sword, had always been synonymous with nobility in Japan. Since the founding of the samurai class in the eighth century, the military arts had become the highest form of study, inspired by the teachings of Zen and the feeling of Shinto. Schools of Kendo born in the early Muromachi period — approximately 1390 to 1600— were continued through the upheavals of the formation of the peaceful Tokugawa Shogunate, and survive to this day. The education of the sons of the Tokugawa Shoguns was by means of schooling in the Chinese classics and fencing exercises. Where a Westener might say "The pen is mightier than the sword", the Japanese would say "Bunbu Itchi", or "Pen and sword in accord". Today, prominent businessmen and political figures in Japan still practice the old traditions of Kendo schools, preserving the forms of several hundred years ago. To sum up, Musashi was a ronin at a time when the samurai were formally considered to be the elite, but actually had no means of livelihood unless they owned lands and castles. Many ronin put up their swords and became artisans, but others, like Musashi, pursued the ideal of the warrior searching for enlightenment through the perilous paths of Kendo. Duels of revenge and tests of skill were commonplace, and fencing schools multiplied. Two schools especially, the Itto school and the Yagyu school, were sponsored by the Tokugawas. The Itto school provided an unbroken line of Kendo teachers, and the Yagyu school eventually became the secret police of the Tokugawa bureaucracy.

KENDO

Traditionally, the fencing halls of Japan, called Dojo, were associated with shrines and temples, but during Musashi's lifetime numerous schools sprang up in the new castle towns. Each daimyo or lord, sponsored a Kendo school, where his retainers could be trained and his sons educated. The hope of every ronin was that he would defeat the students and master of a Dojo in combat, thus increasing his fame and bringing his name to the ears of one who might employ him. The samurai wore two swords thrust through the belt with the cutting edge uppermost. The longer sword was carried out of doors only, the shorter sword was worn at all times. For training, wooden swords and bamboo swords were often used. Dueling and other tests of arms were common, with both real and practice swords. These took place in fencing halls and before shrines, in the streets and within castle walls. Duels were fought to the death or until one of the contestants was disabled, but a few generations after Musashi's time the "shinai", a pliable bamboo sword, and later padded fencing armor, came to be widely used, so the chances of injury were greatly reduced. The samurai studied with all kinds of weapons: halberds, sticks, swords, chain and sickle, and others. Many schools using such weapons survive in traditional form in Japan today.
To train in Kendo one must subjugate the self, bear the pain of

grueling practice, and cultivate a level mind in the face of peril. But the Way of the sword means not only fencing training but also living by the code of honor of the samurai elite. Warfare was the spirit of the samurai's everyday life, and he could face death as if it were a domestic routine. The meaning of life and death by the sword was mirrored in the everyday conduct of the feudal Japanese, and he who realized the resolute acceptance of death at any moment in his everyday life was a master of the sword. It is in order to attain such an understanding that later men have followed the ancient traditions of the sword-fencing styles, and even today give up their lives for Kendo practice.

KENDO AND ZEN

The Way of the sword is the moral teaching of the samurai, fostered by the Confucianist philosophy which shaped the Tokugawa system, together with the native Shinto religion of Japan. The warrior courts of Japan from the Kamakura period to the Muromachi period encouraged the austere Zen study among the samurai, and Zen went hand in hand with the arts of war. In Zen there are no elaborations, it aims directly at the true nature of things. There are no ceremonies, no teachings, the prize of Zen is essentially personal. Enlightenment in Zen does not mean a change in behavior, but realization of the nature of ordinary life. The end point is the beginning, and the great virtue is simplicity. The secret teaching of the Itto Ryu school of Kendo, Kiriotoshi, is the first technique of some hundred or so. The teaching is "Ai Uchi", meaning to cut the opponent just as he cuts you. This is the ultimate timing ... it is lack of anger. It means to treat your enemy as an honored guest. It also means to abandon your life or throw away fear. The first technique is the last, the beginner and the master behave in the same way. Knowledge is a full circle.

The first of Musashi's chapter headings is Ground, for the basis of Kendo and Zen, and the last book is Void, for that understanding which can only be expressed as nothingness. The teachings of Kendo are like the fierce verbal forays to which the Zen student is subjected. Assailed with doubts and misery, his mind and spirit in a whirl, the student is gradually guided to realization and understanding by his teacher. The Kendo student practices furiously, thousands of cuts morning and night, learning fierce techniques of horrible war, until eventually sword becomes "no sword", intention becomes "no intention", a spontaneous knowledge of every situation. The first elementary teaching becomes the highest knowledge, and the master still continues to practice this simple training, his every prayer.

CONCERNING THE LIFE OF MIYAMOTO MUSASHI

Shinmen Musashi No Kami Fujiwara No Genshin, or as he is commonly known Miyamoto Musashi, was born in the village called Miyamoto in the province Mimasaka in 1584. "Musashi" is the name of an area southwest of Tokyo, and the appellation "No Kami" means noble person of the area, while "Fujiwara" is the name of the noble family foremost in Japan over a thousand years ago. Musashi's ancestors were a branch of the powerful Harima clan in Kyushu, the southern island of Japan. Hirada Shokan, his grandfather, was a retainer of Shinmen Iga No Kami Sudeshige, the lord of Takeyama castle, Hirada Shokan was highly thought of by his lord and eventually married his lord's daughter. When Musashi was seven, his father, Munisai, either died or abondoned the child. As his mother had died, Ben No Suke, as Musashi was known during his childhood, was left in the care of an uncle on his mother's side, a priest. So we find Musashi an orphan during Hideyoshi's campaigns of unification, son of a samurai in a violent unhappy land.

He was a boisterous youth, strong-willed and physically large for his age. Whether he was urged to pursue Kendo by his uncle, or whether his aggressive nature led him to it, we do not know, but it is recorded that he slew a man in single combat when he was just thirteen. The opponent was Arima Kihei, a samurai of the Shinto Ryu school of military arts, skilled with sword and spear. The boy threw the man to the ground, and beat him about the head with a stick when he tried to rise. Kihei died vomiting blood. Musashi's next contest was when he was sixteen, when he defeated Tadashima Akiyama. About this time, he left home to embark on the "Warrior Pilgrimage" which saw him victor in scores of contests and which took him to war six times, until he finally settled down at the age of fifty, having reached the end of his search for reason. There must have been many ronin

traveling the country on similar expeditions, some alone like
Musashi and some enjoying sponsorship, though not on the scale of
the pilgrimage of the famous swordsman Tsukahara Bokuden who
had traveled with a retinue of over one hundred men in the previous
century. This part of Musashi's life was spent living apart from
society while he devoted himself with a ferocious single-mindedness
to the search for enlightenment by the Way of the sword. Concerned
only with perfecting his skill, he lived as men need not live,
wandering over Japan soaked by the cold winds of winter, not
dressing his hair, not taking a wife, nor following any profession
save his study. It is said he never entered a bathtub lest he was
caught unawares without a weapon, and that his appearance was
uncouth and wretched.

 In the battle which resulted in Ieyasu succeeding Hideyoshi as
Shogun of Japan, Seki ga Hara, Musashi joined the ranks of the
Ashikaga army to fight against Ieyasu. He survived the terrible three
days during which seventy thousand people died, and also survived
the hunting down and massacre of the vanquished army He went up
to Kyoto, the capital, when he was twenty-one. This was the scene of
his vendetta against the Yoshioka family. The Yoshiokas had been
fencing instructors to the Ashikaga house for generations. Later
forbidden to teach Kendo by lord Tokugawa, the family became
dyers, and are dyers today. Munisai, Musashi's father, had been
invited to Kyoto some years before by the Shogun Ashikaga
Yoshiaka.

Munisai was a competant swordsman, and an expert with the "jitte",
a kind of iron truncheon with a tongue for catching sword blades.
The story has it thatMunisai fought three of the Yoshiokas, winning
two of the duels, and perhaps this has some bearing on Musashi's
behavior towards the family.

Yoshioka Seijiro, the head of the family, was the first to fight
Musashi, on the moor outside the city. Seijiro was armed with a real
sword, and Musashi with a wooden sword. Musashi laid Seijiro out
with a fierce attack and beat him savagely as he lay on the ground.
The retainers carried their lord home on a rain-shutter, where for
shame he cut off his samurai topknot. Musashi lingered on in the
capital, and his continued presence further irked the Yoshiokas. The

second brother, Denshichiro, applied to Musashi for a duel. As a military ploy, Musashi arrived late on the appointed day, and seconds after the start of the fight he broke his opponent's skull with one blow of his wooden sword. Denshichiro was dead. The house issued yet another challenge with Hanshichiro, the young son of Seijiro, as champion. Hanshichiro was a mere boy, not yet in his teens. The contest was to be held by a pine tree adjacent to rice fields. Musashi arrived at the meeting place well before the appointed time and waited in hiding for his enemy to come.

The child arrived dressed formally in war gear, with a party of well-armed retainers, determined to do away with Musashi. Musashi waited concealed in the shadows, and just as they were thinking that he had thought better of it and had decided to leave Kyoto, he suddenly appeared in the midst of them, and cut the boy down. Then, drawing both swords, he cut a path through them and made his escape. After that frightful episode Musashi wandered over Japan, becoming a legend in his own time. We find mention of his name and stories of his prowess in registers, dairies, on monuments, and in folk memory from Tokyo to Kyushu. He had more than sixty contests before he was twenty-nine, and won them all.

The earliest account of his contests appears in Niten Ki, or "Two Heavens Chronicle", a record compiled by his pupils a generation after his death. In the year of the Yoshioka affair, 1605, he visited the temple Hozoin in the south of the capital. Here he had a contest with Oku Hozoin, the Nichiren sect pupil of the Zen priest Hoin Inei. The priest was a spearman, but no match for Musashi who defeated him twice with his short wooden sword. Musashi stayed at the temple for some time studying fighting techniques and enjoying talks with the priests. There is still today a traditional spear fighting form practiced by the monks of Hozoin. It is interesting that in ancient times the word "Osho", which now means priest, used to mean "spear teacher".

Hoin Inei was pupil to Izumi Musashi no Kami, a master of Shinto Kendo. The priest used spears with cross-shaped blades kept outside the temple under the eaves and used in fire fighting. When Musashi was in Iga province he met a skilled chain and sickle fighter named Shishido Baikin. As Shishido twirled his chain Musashi drew a dagger and pierced his breast, advancing to finish him off. The watching pupils attacked Musashi but he frightened them away in four directions.

In Edo, a fighter named Muso Gonosuke visited Musashi requesting a duel. Musashi was cutting wood to make a bow, and granting Gonosuke's request stood up intending to use the slender wand he was cutting as a sword. Gonosuke made a fierce attack, but Musashi stepped straight in and banged him on the head. Gonosuke went away. Passing through Izumo province, Musashi visited lord Matsudaira and asked permission to fight with his strongest expert.

There were many good strategists in Izumo. Permission was granted against a man who used an eight foot long hexagonal wooden pole. The contest was held in the lord's library garden. Musashi used two wooden swords. He chased the samurai up the two wooden steps of the library veranda, thrust at his face on the second step, and hit him on both his arms as he flinched away. To the surprise of the assembled retainers, lord Matsudaira asked Musashi to fight him. Musashi drove the lord up the library steps as before, and when he tried to make a resolute fencing attitude Musashi hit his sword with the "Fire and Stones Cut", breaking it in two.

The lord bowed in defeat, and Musashi stayed for some time as his teacher. Musashi's most well-known duel was in the seventeenth year of Keicho, 1612, when he was in Ogura in Bunzen province. His opponent was Sasaki Kojiro, a young man who had developed a strong fencing technique known as Tsubame-gaeshi or "swallow counter", inspired by the motion of a swallow's tail in flight. Kojiro was retained by the lord of the province, Hosokawa Tadaoki. Musashi applied to Tadaoki for permission to fight Kojiro through the offices of one of the Hosokawa retainers who had been a pupil of Musashi's father, one Nagaoka Sato Okinaga. Permission was granted for the contest to be held at eight o'clock the next morning, and the place was to be an island some few miles from Ogura.

That night Musashi left his lodging and moved to the house of Kobayashi Tare Zaemon. This inspired a rumour that awe of Kojiro's subtle technique had made Musashi run away afraid for his life. The next day at eight o'clock Musashi could not be woken until a prompter came from the officials assembled on the island. He got up, drank the water they brought to him to wash with, and went straight down to the shore. As Sato rowed across to the island Musashi fashioned a paper string to tie back the sleeves of his kimono, and cut a wooden sword from the spare oar. When he had done this he lay down to rest. The boat neared the place of combat and Kojiro and the waiting officials were astounded to see the strange figure of Musashi, with his unkempt hair tied up in a towel, leap from the boat brandishing the long wooden oar and rush through the waves up the beach towards his enemy. Kojiro drew his long sword, a fine blade by Nagamitsu of Bizen, and threw away his scabbard. "You have no more need of that" said Musashi as he rushed forward with his sword held to one side. Kojiro was provoked into making the first cut and Musashi dashed upward at his blade, bringing the oar down on Kojiro's head. As Kojiro fell, his sword, which had cut the towel from Musashi's head, cut across the hem of his divided skirt. Musashi noted Kojiro's condition and bowed to the astounded officials before running back to his boat.

Some sources have it that after he killed Kojiro, Musashi threw down the oar and, nimbly leaping back several paces, drew both his swords and flourished them with a shout at his fallen enemy. It was about this time that Musashi stopped ever using real swords in duels. He was invincible, and from now on he devoted himself to the search for perfect understanding by way of Kendo.

In 1614 and again in 1615 he took the opportunity of once more experiencing warfare and siege, Ieyasu laid siege to Osaka castle where the supporters of the Ashikaga family were gathered in insurrection. Musashi joined the Tokugawa forces in both winter and summer campaigns, now fighting against those he had fought for as a youth at Seki ga Hara.

According to his own writing, he came to understand strategy when he was fifty or fifty-one in 1634. He and his adopted son Iori, the waif whom he had met in Dewa province on his travels, settled in Ogura in this year. Musashi was never again to leave Kyushu island.

The Hosokawa house had been entrusted with the command of the hot seat of Higo province, Kumamoto castle, and the new lord of Bunzen was an Ogasawara. Iori found employment under

Ogasawara Tadazane, and as a captain in Tadazane's army fought against the Christians in the Shimawara uprising of 1 638, when IVIusashi was about fifty-five. The lords of the southern provinces had always been antagonistic to the Tokugawas and were the instigators of intrigue with foreign powers and the Japanese Christians.

Musashi was a member of the field staff at Shimawara where the Christians were massacred. After this, Ieyasu closed the ports of Japan to foreign intercourse, and they remained closed for over two hundred years. After six years in Ogura, Musashi was invited to stay with Churl, the Hosokawa lord of Kumamoto castle, as a guest. He stayed a few years with lord Churi and spent his time teaching and painting.

In 1643, he retired to a life of seclusion in a cave called "Reigendo". Here he wrote Go Rin No Sho, addressed to his pupil Teruo Nobuyuki, a few weeks before his death on the nineteenth of May, 1645. Musashi is known to the Japanese as "Kensei", that is "Sword Saint".

Go Rin No Sho heads every Kendo bibliography, being unique among books of martial art in that it deals with both the strategy of warfare and the methods of single combat in exactly the same way. The book is not a thesis on strategy, it is in Musashi's words "a guide for men who want to learn strategy" and, as a guide always leads, so the contents are always beyond the student's understanding. The more one reads the book the more one finds in its pages.

It is Musashi's last will, the key to the path he trod. When, at twenty-eight or twenty-nine, he had become such a strong fighter, he did not settle down and build a school, replete with success, but became doubly engrossed with his study. In his last days even, he scorned the life of comfort with lord Hosokawa and lived two years alone in a mountain cave deep in contemplation. The behavior of this cruel, headstrong man was evidently most humble and honest.

Musashi wrote "When you have attained the Way of strategy there will be not one thing that you cannot understand" and "You will see the Way in everything". He did, in fact, become a master of arts and crafts. He produced masterpieces of ink painting, probably more highly valued by the Japanese than the ink paintings of any other. His works include cormorants, herons. Hotel the Shinto God, dragons, birds with flowers, bird in a dead tree, Daruma (Bodhidharma), and others. He was a fine calligrapher, evidenced by his piece "Senki" (Warspirit). There is a small wood sculpture of the Buddhist deity Fudo Myoo in private hands. A sculpture of Kwannon was lost recently. He made works in metal, and founded the school of sword guard makers who signed "Niten", after him (see below). He is said to have written poems and songs, but none of these survive.

It is said also that he was commissioned by the Shogun Iemitsu to paint the sunrise over Edo castle. His paintings are sometimes impressed with his seal, "Musashi", or his nom de plume "Niten". Niten means "Two Heavens", said by some to allude to his fighting attitude with a sword in each hand held above his head. In some places he established schools known as "Niten ryu", and in other places called it "Enmei ryu" (clear circle). He wrote "Study the Ways of all professions". It is evident that he did just that. He sought out not only great swordsmen but also priests, strategists, artists and craftsmen, eager to broaden his knowledge. Musashi writes about the various aspects of Kendo in such a way that it is possible for the beginner to study at beginner's level, and for Kendo masters to study the same words on a higher level. This applies not just to military strategy, but to any situation where plans and tactics are used.

Japanese businessmen have used Go Rin No Slio as a guide for business practice, making sales campaigns like military operations, using the same energetic methods. In the same way that Musashi seems to have been a horribly cruel man, yet was following logically an honest ideal, so successful business seems to most people to be without conscience. Musashi's life study is thus as relevant in the twentieth century as it was on the medieval battleground, and applies not just to the Japanese race but to all nations. I suppose you could sum up his inspiration as "humility and hard work".

Joseph Lumpkin

Introduction

I have been many years training in the Way of strategy/ called Ni
Ten Ichi Ryu, and now I think I will explain it in writing for the first
time. It is now during the first ten days of the tenth month in the
twentieth year of Kanei (1645). I have climbed mountain Iwato of
Higo in Kyushu to pay homage to heaven," pray to Kwannon," and
kneel before Buddha. I am a warrior of Harima province, Shinmen
Musashi No Kami Fujiwara No Geshin, age sixty years. From youth
my heart has been inclined toward the Way of strategy.

My first duel was when I was thirteen, I struck down a strategist of
the Shinto school, one Arima Kihei. When I was sixteen I struck
down an able strategist, Tadashima Akiyama. When I was twenty-
one I went up to the capital and met all manner of strategists, never
once failing to win in many contests. After that I went from province
to province dueling with strategists of various schools, and not once
failed to win even though I had as many as sixty encounters. This
was between the ages of thirteen and twenty-eight or twenty-nine.
When I reached thirty I looked back on my past. The previous
victories were not due to my having mastered strategy. Perhaps it
was natural ability, or the order of heaven, or that other schools'
strategy was inferior. After that I studied morning and evening
searching for the principle, and came to realize the Way of strategy
when I was fifty. Since then I have lived without following any
particular Way. Thus with the virtue of strategy I practice many arts
and abilities — all things with no teacher.' To write this book I did not
use the law of Buddha or the teachings of Confucius, neither old war
chronicles nor books on martial tactics. I take up my brush to explain
the true spirit' of this Ichi school (One Way):

1 The Character for Way is read "Michi" in Japanese or "Do" in
Chinese-based reading. It is equivalent to the Chinese "Tao" and
means the whole life of the warrior, his devotion to the sword, his
place in the Confucius-colored bureaucracy of the Tokugawa system.
It is the road of the cosmos, not just a set of ethics for the artist or
priest to live by, but the divine footprints of God pointing the Way.

2 Strategy: "Heiho" is a word of Chinese derivation meaning military strategy. "Hei" means soldier and "Ho" means method or form.

3 Homage to heaven: "Ten" or heaven means the Shinto religion, Shinto — a word compounding the two characters "Kami" (God) and "Michi" (Way)-is the old religion of Japan. In Shinto there are many Holies, gods of steel and fermentation, place and industry, and so-on, and the first gods, ancestors to the Imperial line.

Kwannon: God(dess) of mercy in Buddhism.

Arima Kihei: of the Shinto school.

All things with no teacher: There had been traditions instituted for the arts in the Muromachi period, system of grades and licenses and seniority, and these were perpetuated perhaps more rigidly under the Tokugawa bureaucracy. Musashi studied various arts in various schools, but when after his enlightenment he pursued his studies he had become separate from traditional guidance. He writes his final words in the book of the Void: "Then you will come to think of things in a wide sense, and taking the Void as the Way, you will see the Way as Void."

' Spirit "Shin" or "Kokoro" has been translated "heart", "soul", or "spirit". It could be put as feeling, manner. It has always been said "The sword is the soul of the samurai." as it is mirrored in the Way of heaven and Kwannon.

The time is the night of the tenth day of the tenth month, at the hour of the tiger (3-5 a.m.) 'The hour of the fi'ger: Years, months and hours were named after the ancient Chinese Zodiacal time system.

The Ground Book

Stategy is the craft of the warrior. Commanders must enact the craft, and troopers should know this Way. There is no warrior in the world today who really understands the Way of strategy. There are various

Ways. There is the Way of Salvation by the law of Buddha, the Way of Confucius governing the Way of learning, the Way of healing as a doctor, as a poet teaching the Way of Waka, tea, archery," and many arts and skills. Each man practices as he feels inclined. It is said the warrior's is the twofold Way of pen and sword,' and he should have a taste for both Ways. Even if a man has no natural ability he can be a warrior by sticking assiduously to both divisions of the Way. Generally speaking, the Way of the warrior is resolute acceptance of death.'" Although not only warriors but priests, women, peasants

Waka: The thirty-one syllable poem. The word means "Song of Japan" or "Song in Harmony".

Tea: Tea drinking is studied in schools, just like sword-fencing. It is basically a ritual, based on simple refined rules, between a few persons in a small room.

Archery: The bow was the main weapon of the samurai of the Nara and Heian periods, and was later superseded by the sword. Archery is practiced as a ritual like tea and sword. Hachiman, the God of War, is often depicted as an archer, and the bow is frequently illustrated as part of the paraphernalia of the gods.

Pen and sword: "Bunbu Itchi" or "Pen and sword in accord" is often presented in brushed calligraphy. Young men during the Tokugawa period were educated solely in writing the Chinese classics and exercising in swordplay. Pen and sword, in fact, filled the life of the Japanese nobility.

Resolute acceptance of death: This idea can be summed up as the philosophy expounded in Ha Gakure or "Hidden Leaves", a book written in the seventeenth century by Yamamoto Tsunenori and a few other samurai of the province Nabeshima Han, present-day Saga. Live and fight like you are already dead.

Under the Tokugawas, the enforced logic of the Confucius-influenced system ensured stability among the samurai, but it also meant the passing of certain aspects of Bushido. Discipline for both samurai and commoners became lax. Yamamoto Tsunenori had been counsellor to Mitsushige, lord of Nabeshima Han, for many years, and upon his lord's death he wanted to commit suicide with his family in the traditional manner. This kind of suicide was strictly prohibited by the new legislation, and, full of remorse, Yamamoto retired in sadness to the boundary of Nabeshima Han. There he met others who had faced the same predicament, and together they wrote a lament of what they saw as the decadence of Bushido. Their criticism is a revealing comment on the changing face of Japan during Musashi's lifetime: "There is no way to describe what a warrior should do other than he should adhere to the Way of the warrior (Bushido). I find that all men are negligent of this. There are a few men who can quickly reply to the question "What is the Way of the Warrior?" This is because they do not know in their hearts. From this we can say they do not follow the Way of the warrior. By the Way of the warrior is meant death.

The Way of the warrior is death. This means choosing death whenever there is a choice between life and death. It means nothing more than this. It means to see things through, being resolved. Sayings like "To die with your intention unrealized is to die uselessly", and so on, are from the weak Kyoto, Osaka Bushido. They are unresolved as to whether to keep to their original plan when faced with the choice of life and death. Every man wants to live. They theorize with staying alive in mind. "The man who lives on when he had failed in his intention is a coward" is a heartless definition. That to die having failed is to die uselessly is a mad point of view. This is not a shameful thing. It is the most important thing in the Way of the warrior. If you keep your spirit correct from morning to night, accustomed to the idea of death and resolved on death, and consider yourself as a dead body, thus becoming one with the Way of the warrior, you can pass through life with no possibility of failure and lowlier folk have been known to die readily in the cause of duty or out of shame, this is a different thing. The warrior is different in that studying the Way of strategy is based on overcoming men. By victory gained in crossing swords with individuals, or enjoining battle with large numbers, we can attain power and fame for ourselves or for our lord.

This is the virtue of strategy.

Joseph Lumpkin

The Way of Strategy

In China and Japan practitioners of the Way have been known as
"masters of strategy". Warriors must learn this Way. Recently there
have been people getting on in the world as strategists, but they are
usually just sword-fencers. The attendants of the Kashima Kantori
shrines'" of the province Hitachi received instruction from the gods,
and made schools based on this teaching, travelling from country to
country instructing men. This is the recent meaning of strategy. In
olden times strategy was listed among the Ten Abilities and Seven
Arts as a beneficial practice. It was certainly an art but as beneficial
practice it was not limited to sword-fencing. The true value of sword-
fencing cannot be seen within the confines of sword-fencing
technique. If we look at the world we see arts for sale. Men use
equipment to sell their own selves. As if with the nut and the flower,
the nut has become less than the flower. In this kind of Way of
strategy, both those teaching and those learning the way are
concerned with coloring and showing off their technique, trying to
hasten the bloom of the flower. They speak of "This Dojo" and "That
Dojo".'" They are looking for profit. Someone once said "Immature
strategy is the cause of grief". That was a true saying. perform your
office properly. "The servant must think earnestly of the business of
his employer. Such a fellow is a splendid retainer. In this house there
have been generations of splendid gentlemen and we are deeply
impressed by their warm kindness ... all our ancestors. This was
simply abandoning body and soul for the sake of their lord.
"Moreover, our house excels in wisdom and technical skill. What a
joyful thing if this can be used to advantage. "Even an unadaptable
man who is completely useless is a most trusted retainer if he does
nothing more than think earnestly of his lord's welfare. To think only
of the practical benefit of wisdom and technology is vulgar. "Some
men are prone to having sudden inspirations. Some men do not
quickly have good ideas but arrive at the answer by slow
consideration. Well, if we investigate the heart of the matter, even
though people's natural abilities differ, bearing in mind the Four
Oaths, when your thinking rises above concern for your own welfare,

wisdom which is independent of thought appears. Whoever thinks deeply on things, even though he may carefully consider the future, will usually think around the basis of his own welfare. By the result of such evil thinking he will only perform evil acts. It is very difficult for most silly fellows to rise above thinking of their own welfare. "So when you embark upon something, before you start, fix your intention on the Four Oaths and put selfishness behind you. Then you cannot fail.

"The Four Oaths: Never be late with respect to the Way of the warrior. Be useful to the lord. Be respectful to your parents. Get beyond love and grief: exist for the good of man."

Our lord: This refers to the daimyo, who retained numbers of samurai to fight for them.

Kashima Kantori shrines: The original schools of Kendo can be found in the traditions preserved in Shinto shrines.

Many of the school ancestors are entombed in the Kanto area, not far from Tokyo, where the Kashima and Kantori shrines still stand. Arima Kihei, the samurai whom Musashi killed at the age of thirteen, was a fencer of the Shinto school associated with the shrines. The Yagyu school was derived from the Kashima style. Shinto was a religion of industry in everyday life, and the War Gods enshrined at Kashima and Kantori are still invoked today as part of the everyday practice of the Shinto school.

Dojo: "Dojo" means "Way place", the room where something is studied.

There are four Ways in which men pass through life: as gentlemen, farmers, artisans, and merchants.

The way of the farmer. Using agricultural instruments, he sees springs through to autumns with an eye on the changes of season.

Second is the Way of the merchant. The wine maker obtains his ingredients and puts them to use to make his living. The Way of the merchant is always to live by taking profit. This is the Way of the merchant.

Thirdly, the gentleman warrior, carrying the weaponry of his Way. The Way of the warrior is to master the virtue of his weapons. If a gentleman dislikes strategy he will not appreciate the benefit of weaponry, so must he not have a little taste for this?

Fourthly, the Way of the artisan. The Way of the carpenter is to become proficient in the use of his tools, first to lay his plans with a true measure and then perform his work according to plan. Thus he passes through life. These are the four Ways of the gentleman, the farmer, the artisan and the merchant.

Comparing the Way of the Carpenter to Strategy

The comparison with carpentry is through the connection with houses. Houses of the nobility, houses of warriors, the Four houses," ruin of houses, thriving of houses, the style of the house, the tradition of the house, and the name of the house. The carpenter uses a master plan of the building, and the Way of strategy is similar in that there is a plan of campaign. If you want to learn the craft of war, ponder over this book. The teacher is as a needle, the disciple is as thread. You must practice constantly. Like the foreman carpenter, the commander must know natural rules, and the rules of the country, and the rules of houses. This is the Way of the foreman.

The foreman carpenter must know the architectural theory of towers and temples, and the plans of palaces, and must employ men to raise up houses. The Way of the foreman carpenter is the same as the Way of the commander of a warrior house.

In the construction of houses, choice of woods is made. Straight un-knotted timber of good appearance is used for the revealed pillars, straight timber with small defects is used for the inner pillars. Timber of the finest appearance, even if a little weak, is used for the thresholds, lintels, doors, and sliding doors," and so on. Good strong timber, though it be gnarled and knotted, can always be used discreetly in construction. Timber which is weak or knotted throughout should be used as scaffolding, and later for firewood.

The foreman carpenter allots his men work according to their ability. Floor layers, makers of sliding doors, thresholds and lintels, ceilings and so on. Those of poor ability lay the floor joist, and those of lesser ability carve wedges and do such miscellaneous work. If the foreman knows and deploys his men well the finished work will be good.

The foreman should take into account the abilities and limitations of his men, circulating among them and asking nothing unreasonable. He should know their morale and spirit, and encourage them when necessary. This is the same as the principle of the Four Ways:

Carpenter. All buildings in Japan, except for the walls of the great castles which appeared a few generations before Musashi's birth, were wooden. "Carpenter" means architect and builder.

Four Houses: There were four branches of the Fujiwara family, who dominated Japan in the Heian period. There are also four different schools of tea.

Warrior house: The warrior families who had been in control of Japan for most of her history kept private armies, each with its own commander.

Sliding doors: Japanese buildings made liberal use of sliding doors, detachable walls, and shutters made of wood which were put over door openings at night and in bad weather. strategy.

The Way of Strategy

Like a trooper (foot soldier), the carpenter sharpens his own tools. He carries his equipment in his tool box, and works under the direction of his foreman. He makes columns and girders with an axe, shapes floorboards and shelves with a plane, cuts fine openwork and carvings accurately, giving as excellent a finish as his skill will allow. This is the craft of carpenters.

When the carpenter becomes skilled and understands measures he can become a foreman. The carpenter's attainment is, having tools which will cut well, to make small shrines, writing shelves, tables, paper lanterns, chopping boards and pot-lids. These are the specialties of the carpenter.

(Small shrines: Small shrines to the Shinto gods are found in every Japanese home.)

Things are similar for the trooper. You ought to think deeply about this. The attainment of the carpenter is that his work is not warped, that the joints are not misaligned, and that the work is truly planed so that it meets well and is not merely finished in sections. This is essential. If you want to learn this Way, deeply consider the things written in this book one at a time. You must do sufficient research. Outline of the Five Books of this Book of Strategy The Way is shown in five books concerning different aspects.

These are Ground, Water, Fire, Tradition (Wind)," and Void."

Five books: Go Rin No Sho means a book of five rings. The "Go Dai" (Five Greats) of Buddhism are the five elements which make up the cosmos: ground, water, fire, wind, void. The "Go Rin" (Five Rings) of Buddhism are the five parts of the human body: head, left and right elbows, and left and right knees.

The body of the Way of strategy from the viewpoint of my Ichi school is explained in the Ground book. It is difficult to realize the true Way just through sword-fencing. Know the smallest things and the biggest things, the shallowest things and the deepest things. As if it were a straight road mapped out on the ground, the first book is called the Ground book.

Second is the Water book. With water as the basis, the spirit becomes like water. Water adopts the shape of its receptacle, it is sometimes a trickle and sometimes a wild sea. Water has a clear blue colour. By the clarity, things of Ichi school are shown in this book. If you master the principles of sword-fencing, when you freely beat one man, you beat any man in the world. The spirit of defeating a man is the same for ten million men. The strategist makes small things into big things, like building a great Buddha from a one foot model. I cannot write in detail how this is done. The principle of strategy is having one thing, to know ten thousand things. Things of Ichi school are written in this the Water book.

Third is the Fire book. This book is about fighting. The spirit of fire is fierce.

Like a trooper, the carpenter sharpens his own tools: Sharpening and polishing the Japanese sword is today a work undertaken only by a specialist, but perhaps the art was more widespread in the age of war. If a sword is imperfectly polished and the surface of the blade incorrectly shaped, even if it is a very sharp, fine weapon it will not cut at all well.

(Wind: The Japanese character for "wind" also means "style".)

(Void: The void, or Nothingness, is a Buddhist term for the illusionary nature of worldly things.)

Whether the fire be small or big; and so it is with battles. The Way of battles is the same for man to man fights and for ten thousand a side battles. You must appreciate that spirit can become big or small. What is big is easy to perceive: what is small is difficult to perceive. In short, it is difficult for large numbers of men to change position, so their movements can be easily predicted. An individual can easily change his mind, so his movements are difficult to predict. You must appreciate this. The essence of this book is that you must train day and night in order to make quick decisions. In strategy it is necessary to treat training as a part of normal life with your spirit unchanging. Thus combat in battle is described in the Fire book.

Fourthly, the Wind book. This book is not concerned with my Ichi school but with other schools of strategy. By Wind I mean old traditions, present-day traditions, and family traditions of strategy. Thus I clearly explain the strategies of the world. This is tradition. It is difficult to know yourself if you do not know others. To all Ways there are side-tracks. If you study a Way daily, and your spirit diverges, you may think you are obeying a good way, but objectively it is not the true Way. If you are following the true Way and diverge a little, this will later become a large divergence. You must realize this. Other strategies have come to be thought of as mere sword-fencing, and it is not unreasonable that this should be so. The benefit of my strategy, although it includes sword-fencing, lies in a separate principle. I have explained what is commonly meant by strategy in other schools in the Tradition (Wind) book.

Fifthly, the book of the Void. By Void I mean that which has no beginning and no end. Attaining this principle means not attaining the principle. The Way of strategy is the Way of nature. When you appreciate the power of nature, knowing the rhythm of any situation, you will be able to hit the enemy naturally and strike naturally. All this is the Way of the Void. I intend to show how to follow the true Way according to nature in the book of the Void.

The Name Ichi Ryu Ni To (One school-Two swords)

Warriors, both commanders and troopers, carry two swords at their belt. In olden times these were called the long sword and the sword; nowadays they are known as the sword and the companion sword. Let it suffice to say that in our land, whatever the reason, a warrior carries two swords at his belt. It is the Way of the warrior.

"Nito Ichi Ryu" shows the advantage of using both swords. The spear and halberd" are weapons that are carried out of doors. Students of the Ichi school Way of strategy should train from the start with the sword and long sword in either hand. This is the truth: when you sacrifice your life, you must make fullest use of your weaponry. It is false not to do so, and to die with a weapon yet undrawn.

Two swords: The samurai wore two swords thrust through the belt with the cutting edges upward on the left side. The shorter, or companion, sword was carried at all times, and the longer sword worn only out of doors. From time to time there were rules governing the style and length of swords. Samurai carried two swords but other classes were allowed only one sword for protection against brigands on the roads between towns. The samurai kept their swords at their bedsides, and there were racks for long swords inside the vestibule of every samurai home.

Spear and halberd: The techniques for spear and halberd fighting are the same as those of sword fighting. Spears were first popular in the Muromachi period, primarily as arms for the vast armies of common infantry, and later became objects of decoration for the processions of the daimyo to and from the capital in the Tokugawa period. The spear is used to cut and thrust, and is not thrown. The halberd and similar weapons with long curved blades were especially effective against cavalry, and came to be used by women who might have to defend their homes in the absence of menfolk. The art is widely studied by women today.

If you hold a sword with both hands, it is difficult to wield it freely to left and right, so my method is to carry the sword in one hand. This does not apply to large weapons such as the spear or halberd, but swords and companion swords can be carried in one hand. It is encumbering to hold a sword in both hands when you are on horseback, when running on uneven roads, on swampy ground, muddy rice fields, stony ground, or in a crowd of people. To hold the long sword in both hands is not the true Way, for if you carry a bow or spear or other arms in your left hand you have only one hand free for the long sword. However, when it is difficult to cut an enemy down with one hand, you must use both hands. It is not difficult to wield a sword in one hand; the Way to learn this is to train with two long swords, one in each hand. It will seem difficult at first, but everything is difficult at first. Bows are difficult to draw, halberds are difficult to wield; as you become accustomed to the bow so your pull will become stronger. When you become used to wielding the long sword, you will gain the power of the Way and wield the sword well.

As I will explain in the second book, the Water Book, there is no fast way of wielding the long sword. The long sword should be wielded broadly, and the companion sword closely. This is the first thing to realise. According to this Ichi school, you can win with a long weapon, and yet you can also win with a short weapon. In short, the Way of the Ichi school is the spirit of winning, whatever the weapon and whatever its size. It is better to use two swords rather than one when you are fighting a crowd and especially if you want to take a prisoner. These things cannot be explained in detail. From one thing, know ten thousand things. When you attain the Way of strategy there will not be one thing you cannot see. You must study hard.

The Benefit of the Two Characters Reading "Strategy"

Masters of the long sword are called strategists. As for the other military arts, those who master the bow are called archers, those who master the spear are called spearmen, those who master the gun are called marksmen, those who master the halberd are called halberdiers. But we do not call masters of the Way of the long sword "longswordsmen", nor do we speak of "companion swordsmen". Because bows, guns, spears and halberds are all warriors' equipment they are certainly part of strategy.

To master the virtue of the long sword is to govern the world and oneself, thus the long sword is the basis of strategy. The principle is "strategy by means of the long sword". If he attains the virtue of the long sword, one man can beat ten men. Just as one man can beat ten, so a hundred men can beat a thousand, and a thousand men can beat ten thousand. In my strategy, one man is the same as ten thousand, so this strategy is the complete warrior's craft.

The Way of the warrior does not include other Ways, such as Confucianism, Buddhism, certain traditions, artistic accomplishments and dancing.'" But even though these are not part of the Way, if you know the Way broadly you will see it in everything. Men must polish their particular Way.

The gun: The Japanese gun was matchlock, the form in which it was first introduced into the country by missionaries. The matchlock remained until the nineteenth century.

Dancing: There are various kinds of dancing. There are festival dances, such as the harvest dance, which incorporate local characteristics and are very colorful, sometimes involving many people. There is Noh theatre, which is enacted by a few performers using stylized dance-movements. There are also dances of fan and dances of sword.

The Benefit of Weapons in Strategy

There is a time and a place for use of weapons. The best use of the companion sword is in a confined space, or when you are engaged closely with an opponent.

The long sword can be used effectively in all situations. The halberd is inferior to the spear on the battlefield. With the spear you can take the initiative; the halberd is defensive. In the hands of one of two men of equal ability, the spear gives a little extra strength. Spear and halberd both have their uses, but neither is very beneficial in confined spaces. They cannot be used for taking a prisoner. They are essentially weapons for the field. Anyway, if you learn "indoor" techniques, you will think narrowly and forget the true Way. Thus you will have difficulty in actual encounters.

The bow is tactically strong at the commencement of battle, especially battles on a moor, as it is possible to shoot quickly from among the spearmen. However, it is unsatisfactory in sieges, or when the enemy is more than forty yards away. For this reason there are nowadays few traditional schools of archery. There is little use nowadays for this kind of skill. From inside fortifications, the gun has no equal among weapons. It is the supreme weapon on the field before the ranks clash, but once swords are crossed the gun becomes useless.

One of the virtues of the bow is that you can see the arrows in flight and correct your aim accordingly, whereas gunshot cannot be seen. You must appreciate the importance of this. Just as a horse must have endurance and no defects, so it is with weapons. Horses should walk strongly, and swords and companion swords should cut strongly. Spears and halberds must stand up to heavy use: bows and guns must be sturdy. Weapons should be hardy rather than decorative.

You should not have a favorite weapon. To become over-familiar with one weapon is as much a fault as not knowing it sufficiently well. You should not copy others, but use weapons, which you can handle properly. It is bad for commanders and troops to have likes and dislikes. These are things you must learn thoroughly.

Joseph Lumpkin

Timing in Strategy

There is timing in everything. Timing in strategy cannot be mastered without a great deal of practice. Timing is important in dancing and pipe or string music, for they are in rhythm only if timing is good. Timing and rhythm are also involved in the military arts, shooting bows and guns, and riding horses. In all skills and abilities there is timing. There is also timing in the Void. There is timing in the whole life of the warrior, in his thriving and declining, in his harmony and discord. Similarly, there is timing in the Way of the merchant, in the rise and fall of capital. All things entail rising and falling timing. You must be able to discern this. In strategy there are various timing considerations. From the outset you must know the applicable timing and the inapplicable timing, and from among the large and small things and the fast and slow timings find the relevant timing, first seeing the distance timing and the background timing. This is the main thing in strategy. It is especially important to know the background timing, otherwise your strategy will become uncertain.

Indoor techniques: Dojos were mostly where a great deal of formality and ritual was observed, safe from the prying eyes of rival schools.

You win in battles with the timing in the Void born of the timing of cunning by knowing the enemies' timing, and this using a timing, which the enemy does not expect. All the five books are chiefly concerned with timing. You must train sufficiently to appreciate all this. If you practice day and night in the above Ichi school strategy, your spirit will naturally broaden. Thus is large scale strategy and the strategy of hand to hand combat propagated in the world.

This is recorded for the first time in the five books of Ground, Water, Fire, Tradition (Wind), and Void. This is the Way for men who want to learn my strategy:

Do not think dishonestly.

The Way is in training.

Become acquainted with every art.

Know the Ways of all professions.

Distinguish between gain and loss in worldly matters.

Develop intuitive judgment and understanding for everything.

Perceive those things which cannot be seen.

Pay attention even to trifles.

Do nothing which is of no use.

It is important to start by setting these broad principles in your heart, and train in the Way of strategy. If you do not look at things on a large scale it will be difficult for you to master strategy.

If you learn and attain this strategy you will never lose even to twenty or thirty enemies. More than anything to start with you must set your heart on strategy and earnestly stick to the Way. You will come to be able to actually beat men in fights, and to be able to win with your eye. Also by training you will be able to freely control your own body, conquer men with your body, and with sufficient training you will be able to beat ten men with your spirit. When you have reached this point, will it not mean that you are invincible?

Moreover, in large scale strategy the superior man will manage many subordinates dexterously, bear himself correctly, govern the country and foster the people, thus preserving the ruler's discipline. If there is a Way involving the spirit of not being defeated, to help oneself and gain honor, it is the Way of strategy.

The second year of Shoho (1645), the fifth month, the twelfth day.

Teruo Magonojo" SHINMEN MUSASHI

(Teruo Magonojo: The pupil, sometimes called Teruo Nobuyuki, to whom Musashi addressed Go Rin No Sho.)

The Water Book

The spirit of the Ni Ten Ichi school of strategy is based on water, and this Water Book explains methods of victory as the long-sword form of the Ichi school. Language does not extend to explaining the Way in detail, but it can be grasped intuitively. Study this book; read a word then ponder on it. If you interpret the meaning loosely you will mistake the Way. The principles of strategy are written down here in terms of single combat, but you must think broadly so that you attain an understanding for ten-thousand-a-side battles. Strategy is different from other things in that if you mistake the Way even a little you will become bewildered and fall into bad ways. If you merely read this book you will not reach the Way of strategy. Absorb the things written in this book. Do not just read, memorize or imitate, but so that you realize the principle from within your own heart study hard to absorb these things into your body.

Spiritual Bearing in Strategy

In strategy your spiritual bearing must not be any different from normal. Both in fighting and in everyday life you should be determined though calm. Meet the situation without tenseness yet not recklessly, your spirit settled yet unbiased. Even when your spirit is calm do not let your body relax, and when your body is relaxed do not let your spirit slacken. Do not let your spirit be influenced by your body, or your body influenced by your spirit. Be neither insufficiently spirited nor over spirited. An elevated spirit is weak and a low spirit is weak. Do not let the enemy see your spirit. Small people must be completely familiar with the spirit of large people, and large people must be familiar with the spirit of small people. Whatever your size, do not be misled by the reactions of your own body. With your spirit open and unconstricted, look at things from a high point of view. You must cultivate your wisdom and spirit. Polish your wisdom: learn public justice, distinguish between good and evil, study the Ways of different arts one by one. When you cannot be deceived by men you will have realized the wisdom of strategy. The wisdom of strategy is different from other things. On the battlefield, even when you are hard-pressed, you should ceaselessly research the principles of strategy so that you can develop a steady spirit.

Stance in Strategy

Adopt a stance with the head erect, neither hanging down, nor looking up, nor twisted. Your forehead and the space between your eyes should not be wrinkled. Do not roll your eyes nor allow them to blink, but slightly narrow them. With your features composed, keep the line of your nose straight with a feeling of slightly flaring your nostrils. Hold the line of the rear of the neck straight: instill vigor into your hairline, and in the same way from the shoulders down through your entire body. Lower both shoulders and, without the buttocks jutting out, put strength into your legs from the knees to the tops of your toes. Brace your abdomen so that you do not bend at the hips. Wedge your companion sword in your belt against your abdomen, so that your belt is not slack — this is called "wedging in".
In all forms of strategy, it is necessary to maintain the combat stance in everyday life and to make your everyday stance your combat stance. You must research this well.

The Gaze in Strategy

The gaze should be large and broad. This is the twofold gaze "Perception and Sight". Perception is strong and sight weak.

In strategy it is important to see distant things as if they were close and to take a distanced view of close things. It is important in strategy to know the enemy's sword and not to be distracted by insignificant movements of his sword. You must study this. The gaze is the same for single combat and for large-scale combat. It is necessary in strategy to be able to look to both sides without moving the eyeballs. You cannot master this ability quickly. Learn what is written here: use this gaze in everyday life and do not vary it whatever happens.

Joseph Lumpkin

Holding the Long Sword

Grip the long sword with a rather floating feeling in your thumb and forefinger, with the middle finger neither tight nor slack, and with the last two fingers tight. It is bad to have play in your hands. When you take up a sword, you must feel intent on cutting the enemy. As you cut an enemy you must not change your grip, and your hands must not "cower". When you dash the enemy's sword aside, or ward it off, or force it down, you must slightly change the feeling in your thumb and forefinger. Above all, you must be intent on cutting the enemy in the way you grip the sword. The grip for combat and for sword-testing (footwork) is the same. There is no such thing as a "man-cutting grip". Generally, I dislike fixedness in both long swords and hands. Fixedness means a dead hand. Pliability is a living hand. You must bear this in mind.

With the tips of your toes somewhat floating, tread firmly with your heels. Whether you move fast or slow, with large or small steps, your feet must always move as in normal steps.

Sword Testing: Swords were tested by highly specialized professional testers. The sword would be fitted into a special mounting and test cuts made on bodies, bundles of straw, armor, sheets of metal, etc. Sometimes, appraisal marks of a sword testing inscribed on the tangs of old blades are found.

Footwork: Different methods of moving are used in different schools. Yin-Yang, or "In-Yo" in Japanese, is female-male, dark-light, right-left. Musashi advocates this "level mind" kind of walking, although he is emphatic about the significance of these parameters -issues of right and left foot arise in the Wind book of Go Rin No Sho.

Old Jujitsu schools advocate making the first attack with the left side forward. walking. I dislike the three walking methods known as "jumping-foot", "floating-foot" and "fixed-steps". So-called "Yin-Yang foot" is important to the Way. Yin-Yang foot means not moving only one foot. It means moving your feet left-right and right-left when cutting, withdrawing, or warding off a cut. You should not move one foot preferentially.

The Five Attitudes

The five attitudes are: Upper, Middle, Lower, Right Side, and Left Side. These are the five.

Although attitude has these five dimensions, the one purpose of all of them is to cut the enemy. There are none but these five attitudes. Whatever attitude you are in, do not be conscious of making the attitude; think only of cutting. Your attitude should be large or small according to the situation. Upper, Lower and Middle attitudes are decisive. Left Side and Right Side attitudes are fluid. Left and Right attitudes should be used if there is an obstruction overhead or to one side. The decision to use Left or Right depends on the place. The essence of the Way is this. To understand attitude you must thoroughly understand the Middle attitude. The Middle attitude is the heart of the attitudes. If we look at strategy on a broad scale, the Middle attitude is the seat of the commander, with the other four attitudes following the commander. You must appreciate this.

The Way of the Long Sword Knowing the Way of the long sword means we can wield with two fingers the sword that we usually carry.

If we know the path of the sword well, we can wield it easily. If you try to wield the long sword quickly you will mistake the Way. To wield the long sword well you must wield it calmly. If you try to wield it quickly, like a folding fan or a short sword, you will err by using "short sword chopping". You cannot cut a man with a long sword using this method. When you have cut downwards with the long sword, lift it straight upwards, when you cut sideways, return the sword along a sideways path. Return the sword in a reasonable way, always stretching the elbows broadly. Wield the sword strongly. This is the Way of the long sword. If you learn to use the five approaches of my strategy, you will be able to wield a sword well. You must train constantly.

The Five Approaches

• The first approach is the Middle attitude.
Confront the enemy with the point of your sword against his face.
When he attacks, dash his sword to the right and "ride" it. There is a
natural movement of the sword associated with a natural behavior
according to Kendo ethics.

Or, when the enemy attacks, deflect the point of his sword by hitting
downwards, keep your long sword where it is, and as the enemy
renews the attack cut his arms from below. This is the first method.
The five approaches are this kind of thing. You must train repeatedly
using a long sword in order to learn them. When you master my
Way of the long sword, you will be able to control any attack the
enemy makes. I assure you, there are no attitudes other than the five
attitudes of the long sword of NiTo.

The Way of the Long Sword: The Way as a way of life, and as the
natural path of a sword blade.

Folding fan: An item carried by men and women in the hot summer
months. Armored officers sometimes carried an iron war fan.

The Five Approaches: Who can understand Musashi's methods? It is
necessary to study traditional schools and basic cutting practice. Bear
in mind that fighting technique may start from a greater distance
than it seems to at a first glance. It is said that the man who has faced
death at the point of a sword has an elevated understanding.

• In the second approach with the long sword, from the Upper
attitude cut the enemy just as he attacks.

If the enemy evades the cut, keep your sword where it is and,
scooping from below, cut him as he renews the attack. It is possible
to repeat the cut from here. In this method there are various changes
in timing and spirit. You will be able to understand this by training
in the Ichi school. You will always win with the five long sword
methods. You must train repeatedly.

• In the third approach, adopt the Lower attitude, anticipating scooping up.

When the enemy attackes, hit his hands from below. As you do so, he may try to hit your sword down. If this is the case, cut his upper arm(s) horizontally with a feeling of "crossing". This means that from the Lower attitudes you hit the enemy at the instant that he attacks. You will encounter this method often, both as a beginner and in later strategy. You must train holding a long sword.

• In this fourth approach, adopt the Left Side attitude.

As the enemy attacks, hit his hands from below. If as you hit his hands he attempts to dash down your sword, with the feeling of hitting his hands, parry the path of his long sword and cut across from above your shoulder.

This is the Way of the long sword. Through this method you win by parrying the line of the enemy's attack. You must study this.

• In the fifth approach, the sword is in the Right Side attitude.

In accordance with the enemy's attack, cross your sword from below at the side to the Upper attitude. Then cut straight from above. This method is essential for knowing the Way of the long sword well. If you can use this method, you can freely wield a heavy long sword. I cannot describe in detail how to use these five approaches. You must become well acquainted with my "in harmony with the long sword" Way, learn large-scale timing, understand the enemy's long sword, and become used to the five approaches from the outset. You will always win by using these five methods, with various timing considerations discerning the enemy's spirit. You must consider all this carefully.

The "Attitude-No-Attitude" Teaching

"Attitude No-Attitude" means that there is no need for what are known as long sword attitudes. Even so, attitudes exist as the five ways of holding the long sword. However you hold the sword it must be in such a way that it is easy to cut the enemy well, in accordance with the situation, the place, and your relation to the enemy.

From the Upper attitude as your spirit lessens you can adopt the Middle attitude, and from the Middle attitude you can raise the sword a little in your technique and adopt the Upper attitude.

From the Lower attitude you can raise the sword a little and adopt the Middle attitudes as the occasion demands. According to the situation, if you turn your sword from either the Left Side or Right Side attitude towards the centre, the Middle or the Lower attitude results.

The principle of this is called "Existing Attitude — Non-existing Attitude". The primary thing when you take a sword in your hands is your intention to cut the enemy, whatever the means. Whenever you parry, hit, spring, strike or touch the enemy's cutting sword, you must cut the enemy in the same movement. It is essential to attain this. If you think only of hitting, springing, striking or touching the enemy, you will not be able actually to cut him. More than anything, you must be thinking of carrying your movement through to cutting him. You must thoroughly research this. Attitude in strategy on a larger scale is called "Battle Array". Such attitudes are all for winning battles. Fixed formation is bad. Study this well.

To Hit the Enemy "In One Timing"

"In One Timing" means, when you have closed with the enemy, to hit him as quickly and directly as possible, without moving your body or settling your spirit, while you see that he is still undecided. The timing of hitting before the enemy decides to withdraw, break or hit, is this "In One Timing". You must train to achieve this timing, to be able to hit in the timing of an instant.

The "Abdomen Timing of Two"

When you attack and the enemy quickly retreats, as you see him tense you must feint a cut. Then, as he relaxes, follow up and hit him. This is the "Abdomen Timing of Two". It is very difficult to attain this merely by reading this book, but you will soon understand with a little instruction.

No Design, No Conception

In this method, when the enemy attacks and you decide to attack, hit with your body, and hit with your spirit, and hit from the Void with your hands, accelerating strongly. This is the "No Design, No Conception" cut. This is the most important method of hitting. It is often used. You must train hard to understand it.

No Design, No Conception: "Munen Muso"-this means the ability to act calmly and naturally even in the face of danger. It is the highest accord with existence, when a man's word and his actions are spontaneously the same.

The Flowing Water Cut

The "Flowing Water Cut" is used when you are struggling blade to blade with the enemy. When he breaks and quickly withdraws trying to spring with his long sword, expand your body and spirit and cut him as slowly as possible with your long sword, following your body like stagnant water. You can cut with certainty if you learn this. You must discern the enemy's grade.

Continuous Cut

When you attack and the enemy also attacks, and your swords spring together, in one action cut his head, hands and legs. When you cut several places with one sweep of the long sword, it is the "Continuous Cut". You must practice this cut; it is often used. With detailed practice you should be able to understand it.

The Fire and Stones Cut

The Fires and Stones Cut means that when the enemy's long sword and your long sword clash together you cut as strongly as possible without raising the sword even a little. This means cutting quickly with the hands, body and legs — all three cutting strongly. If you train well enough you will be able to strike strongly.

The Red Leaves Cut

Red Leaves Cut Presumably Musashi is alluding here to falling, dying leaves.

The Red Leaves Cut means knocking down the enemy's long sword. The spirit should be getting control of his sword. When the enemy is in a long sword attitude in front of you and intent on cutting, hitting and parrying, you strongly hit the enemy's sword with the Fire and Stones Cut, perhaps in the design of the "No Design, No Conception" Cut. If you then beat down the point of his sword with a sticky feeling, he will necessarily drop the sword. If you practice this cut it becomes easy to make the enemy drop his sword. You must train repetitively.

The Body in Place of the Long Sword

Also "the long sword in place of the body". Usually we move the body and the sword at the same time to cut the enemy. However, according to the enemy's cutting method, you can dash against him with your body first, and afterwards cut with the sword. If his body is immoveable, you can cut first with the long sword, but generally you hit first with the body and then cut with the long sword. You must research this well and practice hitting.

Cut and Slash

To cut and slash are two different things. Cutting, whatever form of cutting it is, is decisive, with a resolute spirit. Slashing is nothing more than touching the enemy. Even if you slash strongly, and even if the enemy dies instantly, it is slashing. When you cut, your spirit is resolved. You must appreciate this. If you first slash the enemy's hands or legs, you must then cut strongly. Slashing is in spirit the same as touching. When you realize this, they become indistinguishable. Learn this well.

Chinese Monkey's Body

The Chinese Monkey's Body"*" is the spirit of not stretching out your arms. The spirit is to get in quickly, without in the least extending your arms, before the enemy cuts. If you are intent upon not stretching out your arms you are effectively far away, the spirit is to go in with your whole body. When you come to within arm's reach it becomes easy to move your body in. You must research this well.

Chinese Monkey's Body: A Chinese monkey here means a short-armed monkey.

Glue and Lacquer Emulsion Body

The spirit of "Glue and Lacquer Emulsion Body" is to stick to the enemy and not separate from him. When you approach the enemy, stick firmly with your head, body and legs. People tend to advance their head and legs quickly, but their body lags behind. You should stick firmly so that there is not the slightest gap between the enemy's body and your body. You must consider this carefully.

Glue and lacquer emulsion: The lacquer work which takes its name from Japan, used to coat furniture and home utensils, architecture, weapons and armor.

To Strive for Height

By "to strive for height" is meant, when you close with the enemy, to strive with him for superior height without cringing. Stretch your legs, stretch your hips, and stretch your neck face to face with him. When you think you have won, and you are the higher, thrust in strongly. You must learn this.

To Apply Stickiness

When the enemy attacks and you also attack with the long sword, you should go in with a sticky feeling and fix your long sword against the enemy's as you receive his cut. The spirit of stickiness is not hitting very strongly, but hitting so that the long swords do not separate easily. It is best to approach as calmly as possible when hitting the enemy's long sword with stickiness. The difference between "Stickiness" and "Entanglement" is that stickiness is firm and entanglement is weak. You must appreciate this.

The Body Strike

The Body Strike means to approach the enemy through a gap in his guard. The spirit is to strike him with your body. Turn your face a little aside and strike the enemy's breast with your left shoulder thrust out. Approach with a spirit of bouncing the enemy away, striking as strongly as possible in time with your breathing. If you achieve this method of closing with the enemy, you will be able to knock him ten or twenty feet away. It is possible to strike the enemy until he is dead. Train well.

Three Ways to Parry His Attack

There are three methods to parry a cut: First, by dashing the enemy's long sword to your right, as if thrusting at his eyes when he makes an attack. Or, to parry by thrusting the enemy's long sword towards his right eye with the feeling of snipping his neck. Or, when you have a short "long sword", without worrying about parrying the enemy's long sword, to close with him quickly, thrusting at his face with your left hand. These are the three ways of parrying. You must bear in mind that you can always clench your left hand and thrust at the enemy's face with your fist. For this it is necessary to train well.

To Stab at the Face

To stab at the face means, when you are in confrontation with the enemy, that your spirit is intent on stabbing at his face, following the line of the blades with the point of your long sword. If you are intent on stabbling at his face, his face and body will become rideable. When the enemy becomes rideable, there are various opportunities for winning. You must concentrate on this. When fighting and the enemy's body becomes as if rideable, you can win quickly, so you ought not to forget to stab at the face. You must persue the value of this technique through training.

To Stab at the Heart

To stab at the heart means, when fighting and there are obstructions above or to the sides, and whenever it is difficult to cut, to thrust at the enemy. You must stab the enemy's breast without letting the point of your long sword waver, showing the enemy the ridge of the blade square-on, and with the spirit of deflecting his long sword. The spirit of this principle is often useful when we become tired or for some reason our long sword will not cut. You must understand the application of this method.

To Scold "Tut-Tut!"

"Scold" means that, when the enemy tries to counter-cut as you attack, you counter-cut again from below as if thrusting at him, trying to hold him down. With very quick timing you cut, scolding the enemy. Thrust up, "Tut!", and cut "TUT!" This timing is encountered time and time again in exchanges of blows. The way to scold Tut-TUT is to time the cut simultaneously with raising your long sword as if to thrust the enemy. You must learn this through repetitive practice. (Accent on the second syllable tut-TUT indicates a speeding up and powerful cut on the last move, squeezing timing.)

The Smacking Parry

By "smacking parry" is meant that when you clash swords with the enemy, you meet his attacking cut on your long sword with a tee-dum, tee-dum rhythm, smacking his sword and cutting him. The spirit of the smacking parry is not parrying, or smacking strongly, but smacking the enemy's long sword in accordance with his attacking cut, primarily intent on quickly cutting him. If you understand the timing of smacking, however hard your long swords clash together, your swordpoint will not be knocked back even a little. You must research sufficiently to realize this.

There are Many Enemies

"There are many enemies" applies when you are fighting one against many. Draw both sword and companion sword and assume a wide-stretched left and right attitude. The spirit is to chase the enemies around from side to side, even though they come from all four directions. Observe their attacking order, and go to meet first those who attack first. Sweep your eyes around broadly, carefully examining the attacking order, and cut left and right alternately with your swords. Waiting is bad. Always quickly reassume your attitudes to both sides, cut the enemies down as they advance, crushing them in the direction from which they attack. Whatever you do, you must drive the enemy together, as if tying a line of fishes, and when they are seen to be piled up, cut them down strongly without giving them room to move.

The Advantage When Coming to Blows

You can know how to win through strategy with the long sword, but it cannot be clearly explained in writing. You must practice diligently in order to understand how to win. Oral tradition: The true Way of strategy is revealed in the long sword."

One Cut

You can win with certainty with the spirit of "one cut".^* It is difficult to attain this if you do not learn strategy well. If you train well in this Way, strategy will come from your heart and you will be able to win at will. You must train diligently.

Direct Communication

The spirit of "Direct Communication" is how the true Way of the NiTo Ichi school is received and handed down.

Oral tradition: "Teach your body strategy." Recorded in the above book is an outline of Ichi school sword fighting. To learn how to win with the long sword in strategy, first learn the five approaches and the five attitudes, and absorb the Way of the long sword naturally in your body. You must understand spirit and timing, handle the long sword naturally, and move body and legs in harmony with your spirit. Whether beating one man or two, you will then know values in strategy. Study the contents of this book, taking one item at a time, and through fighting with enemies you will gradually come to know the principle of the Way.

There are many enemies: Musashi is held to be the inventor of the Two Sword style. His school is sometimes called "Nito Ryu" (two sword school) and sometimes "Niten Ryu" (two heavens school). He writes that the use of two swords is for when there are many enemies, but people practice a style of fencing with a sword in each hand to give practical advantage in fencing. Musashi used the words "two swords" when meaning to use all one's resources in combat. He never used two swords when up against a skilled swordsman.

Oral tradition: Other Kendo schools also have oral traditions as opposed to teachings passed on in formal technique.

One cut Whatever this means, it is worthwhile to note the "Hitotsu Gachi" (One Victory), the Kiri Otoshi technique of the Itto Ryu school, where one cut provides attack and defense, cutting down the enemy's sword and spirit, and the related "Itchi no Tachi" (Long Sword of One) of the Shinto style.

Deliberately, with a patient spirit, absorb the virtue of all this, from time to time raising your hand in combat. Maintain this spirit whenever you cross swords with an enemy.

Step by step walk the thousand-mile road. Study strategy over the years and achieve the spirit of the warrior.

Today is victory over yourself of yesterday; tomorrow is your victory over lesser men.

Next, in order to beat more skilful men, train according to this book, not allowing your heart to be swayed along a side-track. Even if you kill an enemy, if it is not based on what you have learned it is not the true Way. If you attain this Way of victory, then you will be able to beat several tens of men. What remains is sword-fighting ability, which you can attain in battles and duels.

The Second Year of Shoho, the twelfth day of the fifth month (1645)

Teruo Magonojo SHINMEN MUSASHI

The Fire Book

In this the Fire Book of the NiTo Ichi school of strategy I describe fighting as fire. In the first place, people think narrowly about the benefit of strategy. By using only their fingertips, they only know the benefit of three of the five inches of the wrist. They let a contest be decided, as with the folding fan, merely by the span of their forearms.

They specialize in the small matter of dexterity, learning such trifles as hand and leg movements with the bamboo practice sword.

In my strategy, the training for killing enemies is by way of many contests, fighting for survival, discovering the meaning of life and death, learning the Way of the sword, judging the strength of attacks and understanding the Way of the "edge and ridge" of the sword.

You cannot profit from small techniques particularly when full armor is worn.

My Way of strategy is the sure method to win when fighting for your life one man against five or ten. There is nothing wrong with the principle "one man can beat ten, so a thousand men can beat ten thousand". You must research this. Of course you cannot assemble a thousand or ten thousand men for everyday training. But you can become a master of strategy by training alone with a sword, so that you can understand the enemy's strategies, his strength and resources, and come to appreciate how to apply strategy to beat ten thousand enemies.

Any man who wants to master the essence of my strategy must research diligently, training morning and evening. Thus can he polish his skill, become free from self, and realize extraordinary ability. He will come to possess miraculous power.
This is the practical result of strategy.

Bamboo practice sword: There have been practice swords of various kinds throughout the history of Kendo: some are made of spliced bamboo covered with cloth or hide.

Full armor: The words "Roku Gu" (six pieces) are used. This is a set of armor consisting of Cuiras, gauntlets, sleeves, apron and thigh pieces, or, according to another convention, body armor, helmet, mask, thigh pieces, gauntlets and leg pieces.

Depending on the Place

Examine your environment. Stand in the sun; that is, take up an attitude with the sun behind you. If the situation does not allow this, you must try to keep the sun on your right side. In buildings, you must stand with the entrance behind you or to your right. Make sure that your rear is unobstructed, and that there is free space on your left, your right side being occupied with your sword attitude. At night, if the enemy can be seen, keep the fire behind you and the entrance to your right, and otherwise take up your attitude as above. You must look down on the enemy, and take up your attitude on slightly higher places. For example, the Kamiza in a house is thought of as a high place. When the fight comes, always endeavor to chase the enemy around to your left side. Chase him towards awkward places, and try to keep him with his back to awkward places. When the enemy gets into an inconvenient position, do not let him look around, but conscientiously chase him around and pin him down.

In houses, chase the enemy into the thresholds, lintels, doors, verandas, pillars, and so on, again not letting him see his situation. Always chase the enemy into bad footholds, obstacles at the side, and so on, using the virtues of the place to establish predominant positions from which to fight. You must research and train diligently in this.

The Three Methods to Forestall the Enemy

The first is to forestall him by attacking. This is called Ken No Sen (to set him up). Another method is to forestall him as he attacks. This is called Tai No Sen (to wait for the initiative). The other method is when you and the enemy attack together. This is called Tai Tai No Sen (to accompany him and forestall him).

There are no methods of taking the lead other than these three. Because you can win quickly by taking the lead, it is one of the most important things in strategy. There are several things involved in taking the lead. You must make the best of the situation, see through the enemy's spirit so that you grasp his strategy and defeat him. It is impossible to write about this in detail.

The First-Ken No Sen

When you decide to attack, keep calm and dash in quickly, forestalling the enemy. Or you can advance seemingly strongly but with a reserved spirit, forestalling him with the reserve. Alternately, advance with as strong a spirit as possible, and when you reach the enemy move with your feet a little quicker than normal, unsettling him and overwhelming him sharply. Or, with your spirit calm, attack with a feeling of constantly crushing the enemy, from first to last. The spirit is to win in the depths of the enemy. These are all Ken No Sen.

The Second-Tai No Sen

When the enemy attacks, remain undisturbed but feign weakness. As the enemy reaches you, suddenly move away indicating that you intend to jump aside, then dash in attacking strongly as soon as you see the enemy relax. This is one way.

The three methods to forestall an enemy: A great swordsman or other artist will have mastered the ability to forestall the enemy. The great swordsman is always "before" his environment. This does not mean speed. You cannot beat a good swordsman, because he subconsciously sees the origin of every real action. One can still see in Kendo practice wonderful old gentlemen slowly hitting young champions on the head almost casually. It is the practiced ability to sum up a changing situation instantly. Or, as the enemy attacks, attack more strongly, taking advantage of the resulting disorder in his timing to win. This is the Tai No Sen principle.

Kazima: This is the residence of the ancestral spirit of a house; the head of the house sits nearest this place. It is often a slightly raised recess in a wall, sometimes containing a hanging scroll, armor, or other religious property.

The Third-Tai Tai No Sen

When the enemy makes a quick attack, you must attack strongly and calmly, aim for his weak point as he draws near, and strongly defeat him. Or, if the enemy attacks calmly, you must observe his movement and, with your body rather floating, join in with his movements as he draws near. Move quickly and cut him strongly. This is Tai Tai No Sen

These things cannot be clearly explained in words. You must research what is written here. In these three ways of forestalling, you must judge the situation. This does not mean that you always attack first; but if the enemy attacks first you can lead him around. In strategy, you have effectively won when you forestall the enemy, so you must train well to attain this.

To Hold Down a Pillow

"To Hold Down a Pillow" means not allowing the enemy's head to rise. In contests of strategy it is bad to be led about by the enemy. You must always be able to lead the enemy about. Obviously the enemy will also be thinking of doing this, but he cannot forestall you if you do not allow him to come out. In strategy, you must stop the enemy as he attempts to cut; you must push down his thrust, and throw off his hold when he tries to grapple. This is the meaning of "to hold down a pillow". When you have grasped this principle, whatever the enemy tries to bring about in the fight you will see in advance and suppress it.

The spirit is to check his attack at the syllable "at . . .", when he jumps check his advance at the syllable "ju . . .", and check his cut at "cu . . .". The important thing in strategy is to suppress the enemy's useful actions but allow his useless actions. However, doing this alone is defensive. First, you must act according to the Way, suppress the enemy's techniques, foiling his plans, and thence command him directly. When you can do this you will be a master of strategy. You must train well and research "holding down a pillow".

Note: To hold down a pillow: Note that samurai and Japanese ladies slept with heads on a small wooden pillow shaped to accommodate their hairstyle.

Crossing at a Ford

"Crossing at a ford" means, for example, crossing the sea at a strait, or crossing over a hundred miles of broad sea at a crossing place. I believe this "crossing at a ford" occurs often in a man's lifetime. It means setting sail even though your friends stay in harbour, knowing the route, knowing the soundness of your ship and the favour of the day. When all the conditions are met, and there is perhaps a favourable wind, or a tailwind, then set sail. If the wind changes within a few miles of your destination, you must row across the remaining distance without sailIf you attain this spirit, it applies to everyday life. You must always think of crossing at a ford. In strategy also it is important to "cross at a ford". Discern the enemy's capability and, knowing your own strong points, "cross the ford" at the advantageous place, as a good captain crosses a sea route. If you succeed in crossing at the best place, you may take your ease. To cross at a ford means to attack the enemy's weak point, and to put yourself in an advantageous position. This is how to win in large-scale strategy. The spirit of crossing at a ford is necessary in both large — and small-scale strategy. You must research this well.

To Know the Times

"To know the times" means to know the enemy's disposition in battle. Is it flourishing or waning? By observing the spirit of the enemy's men and getting the best position, you can work out the enemy's disposition and move your men accordingly. You can win through this principle of strategy, fighting from a position of advantage. When in a duel, you must forestall the enemy and attack when you have first recognised his school of strategy, perceived his quality and his strong and weak points. Attack in an unsuspected manner, knowing his meter and modulation and the appropriate timing. Knowing the times means, if your ability is high, seeing right into things. If you are thoroughly conversant with strategy, you will recognise the enemy's intentions and thus have many opportunities to win. You must sufficiently study this.

To Tread Down the Sword

"To tread down the sword" is a principle often used in strategy. First, in large-scale strategy, when the enemy first discharges bows and guns and then attacks, it is difficult for us to attack if we are busy loading powder into our guns or notching our arrows. The spirit is to attack quickly while the enemy is still shooting with bows or guns. The spirit is to win by "treading down" as we receive the enemy's attack. In single combat, we cannot get a decisive victory by cutting, with a "tee-dum tee- dum" feeling, in the wake of the enemy's attacking long sword. We must defeat him at the start of his attack, in the spirit of treading him down with the feet, so that he cannot rise again to the attack. "Treading" does not simply mean treading with the feet. Tread with the body, tread with the spirit, and, of course, tread and cut with the long sword. You must achieve the spirit of not allowing the enemy to attack a second time. This is the spirit of forestalling in every sense. Once at the enemy, you should not aspire just to strike him, but to cling after the attack. You must study this deeply.

To Know "Collapse"

Everything can collapse. Houses, bodies, and enemies collapse when their rhythm becomes deranged. In large-scale strategy, when the enemy starts to collapse you must persue him without letting the chance go. If you fail to take advantage of your enemies' collapse, they may recover. In single combat, the enemy sometimes loses timing and collapses. If you let this opportunity pass, he may recover and not be so negligent thereafter. Fix your eye on the enemy's collapse, and chase him, attacking so that you do not let him recover. You must do this. The chasing attack is with a strong spirit. You must utterly cut the enemy down so that he does not recover his position. You must understand how to utterly cut down the enemy.

To Become the Enemy

"To become the enemy" means to think yourself into the enemy's position. In the world people tend to think of a robber trapped in a house as a fortified enemy. However, if we think of "becoming the enemy", we feel that the whole world is against us and that there is no escape. He who is shut inside is a pheasant. He who enters to arrest is a hawk. You must appreciate this. In large-scale strategy, people are always under the impression that the enemy is strong, and so tend to become cautious. But if you have good soldiers, and if you understand the principles of strategy, and if you know how to beat the enemy, there is nothing to worry about. In single combat also you must put yourself in the enemy's position. If you think, "Here is a master of the Way, who knows the principles of strategy", then you will surely lose. You must consider this deeply.

To Release Four Hands

"To release four hands" is used when you and the enemy are

contending with the same spirit, and the issue cannot be decided. Abandon this spirit and win through an alternative resource. In large-scale strategy, when there is a "four hands" spirit, do not give up — it is man's existence. Immediately throw away this spirit and win with a technique the enemy does not expect. In single combat also, when we think we have fallen into the "four hands" situation, we must defeat the enemy by changing our mind and applying a suitable technique according to his condition. You must be able to judge this.

"To release four hands: "Yotsu te o hanasu"-the expression "Yotsu te" means the condition of grappling with both arms engaged with the opponent's arms, or "deadlock". It is also the name used to describe various articles with four corners joined, such as a fishing net, and was given to an article of ladies' clothing which consisted of a square of cloth tied from the back over each shoulder and under each arm, with a knot on the breast.

To Move the Shade

"To move the shade" is used when you cannot see the enemy's spirit. In large-scale strategy, when you cannot see the enemy's position, indicate that you are about to attack strongly, to discover his resources. It is easy then to defeat him with a different method once you see his resources. In single combat, if the enemy takes up a rear or side attitude of the long sword so that you cannot see his intention, make a feint attack, and the enemy will show his long sword, thinking he sees your spirit. Benefiting from what you are shown, you can win with certainty. If you are negligent you will miss the timing. Research this well.

To Hold Down a Shadow

"Holding down a shadow" is used when you can see the enemy's attacking spirit. In large-scale strategy, when the enemy embarks on an attack, if you make a show of strongly suppressing his technique, he will change his mind. Then, altering your spirit, defeat him by forestalling him with a Void spirit. Or, in single combat, hold down the enemy's strong intention with a suitable timing, and defeat him by forestalling him with this timing. You must study this well.

To Pass On

Many things are said to be passed on. Sleepiness can be passed on, and yawning can be passed on. Time can be passed on also.
In large-scale strategy, when the enemy is agitated and shows an inclination to rush, do not mind in the least. Make a show of complete calmness, and the enemy will be taken by this and will become relaxed. When you see that this spirit has been passed on, you can bring about the enemy's defeat by attacking strongly with a Void spirit. In single combat, you can win by relaxing your body and spirit and then, catching on the moment the enemy relaxes, attack strongly and quickly, forestalling him. What is known as "getting someone drunk" is similar to this. You can also infect the enemy with a bored, careless, or weak spirit. You must study this well.

To Cause Loss of Balance

Many things can cause a loss of balance. One cause is danger, another is hardship, and another is surprise. You must research this. In large-scale strategy it is important to cause loss of balance. Attack without warning where the enemy is not expecting it, and while his spirit is undecided follow up your advantage and, having the lead, defeat him. Or, in single combat, start by making a show of being slow, then suddenly attack strongly. Without allowing him space for breath to recover from the fluctuation of spirit, you must grasp the opportunity to win. Get the feel of this.

To Frighten

Fright often occurs, caused by the unexpected. In large-scale strategy you can frighten the enemy not by what you present to their eyes, but by shouting, making a small force seem large, or by threatening them from the flank without warning. These things all frighten. You can win by making best use of the enemy's frightened rhythm. In single combat, also, you must use the advantage of taking the enemy unawares by frightening him with your body, long sword, or voice, to defeat him. You should research this well.

To Soak In

When you have come to grips and are striving together with the enemy, and you realise that you cannot advance, you "soak in" and become one with the enemy. You can win by applying a suitable technique while you are mutually entangled. In battles involving large numbers as well as in fights with small numbers, you can often win decisively with the advantage of knowing how to "soak" into the enemy, whereas, were you to draw apart, you would lose the chance to win. Research this well.

To Injure the Corners

It is difficult to move strong things by pushing directly, so you should "injure the corners". In large-scale strategy, it is beneficial to strike at the corners of the enemy's force. If the corners are overthrown, the spirit of the whole body will be overthrown. To defeat the enemy you must follow up the attack when the corners have fallen. In single combat, it is easy to win once the enemy collapses. This happens when you injure the "corners" of his body, and this weakens him. It is important to know how to do this, so you must research this deeply.

To Throw into Confusion

This means making the enemy lose resolve. In large-scale strategy we can use your troops to confuse the enemy on the field. Observing the enemy's spirit, we can make him think, "Here? There? Like that? Like this? Slow? Fast?" Victory is certain when the enemy is caught up in a rhythm that confuses his spirit. In single combat, we can confuse the enemy by attacking with varied techniques when the chance arises. Feint a thrust or cut, or make the enemy think you are going close to him, and when he is confused you can easily win. This is the essence of fighting, and you must research it deeply.

The Three Shouts

The three shouts are divided thus: before, during and after. Shout according to the situation. The voice is a thing of life. We shout against fires and so on, against the wind and the waves. The voice shows energy. In large-scale strategy, at the start of battle we shout as loudly as possible. During the fight, the voice is low-pitched, shouting out as we attack. After the contest, we shout in the wake of our victory. These are the three shouts. In single combat, we make as if to cut and shout "Ei!" at the same time to disturb the enemy, then in the wake of our shout we cut with the long sword. We shout after we have cut down the enemy — this is to announce victory. This is called "sen go no koe" (before and after voice). We do not shout simultaneously with flourishing the long sword. We shout during the fight to get into rhythm. Research this deeply.

To Mingle

In battles, when the armies are in confrontation, attack the enemy's strong points and, when you see that they are beaten back, quickly separate and attack yet another strong point on the periphery of his force. The spirit of this is like a winding mountain path. This is an important fighting method for one man against many. Strike down the enemies in one quarter, or drive them back, then grasp the timing and attack further strong points to right and left, as if on a winding mountain path, weighing up the enemies' disposition. When you know the enemies' level, attack strongly with no trace of retreating spirit. In single combat, too, use this spirit with the enemy's strong points. What is meant by 'mingling' is the spirit of advancing and becoming engaged with the enemy, and not withdrawing even one step. You must understand this.

To Crush

This means to crush the enemy regarding him as being weak. In large-scale strategy, when we see that the enemy has few men, or if he has many men but his spirit is weak and disordered, we knock the hat over his eyes, crushing him utterly. If we crush lightly, he may recover. You must learn the spirit of crushing as if with a hand-grip. In single combat, if the enemy is less skilful than yourself, if his rhythm is disorganized, or if he has fallen into evasive or retreating attitudes, we must crush him straightaway, with no concern for his presence and without allowing him space for breath. It is essential to crush him all at once. The primary thing is not to let him recover his position even a little. You must research this deeply.

The Mountain-Sea Change

The "mountain-sea" spirit means that it is bad to repeat the same thing several times when fighting the enemy. There may be no help but to do something twice, but do not try it a third time. If you once make an attack and fail, there is little chance of success if you use the same approach again. If you attempt a technique, which you have previously tried unsuccessfully and fail yet again, then you must change your attacking method. If the enemy thinks of the mountains, attack like the sea; and if he thinks of the sea, attack like the mountains. You must research this deeply.

To Penetrate the Depths

When we are fighting with the enemy, even when it can be seen that we can win on the surface with the benefit of the Way, if his spirit is not extinguished, he may be beaten superficially yet undefeated in spirit deep inside. With this principle of "penetrating the depths" we can destroy the enemy's spirit in its depths, demoralizing him by quickly changing our spirit. This often occurs. Penetrating the depths means penetrating with the long sword, penetrating with the body, and penetrating with the spirit. This cannot be understood in a generalization. Once we have crushed the enemy in the depths, there is no need to remain spirited. But otherwise we must remain spirited. If the enemy remains spirited it is difficult to crush him. You must train in penetrating the depths for large-scale strategy and also single combat.

To Renew

"To renew" applies when we are fighting with the enemy, and an entangled spirit arises where there is no possible resolution. We must abandon our efforts, think of the situation in a fresh spirit then win in the new rhythm. To renew, when we are deadlocked with the enemy, means that without changing our circumstance we change our spirit and win through a different technique. It is necessary to consider how "to renew" also applies in large-scale strategy. Research this diligently.

Rat's Head, Ox's Neck

"Rat's head and ox's neck" means that, when we are fighting with the enemy and both he and we have become occupied with small points in an entangled spirit, we must always think of the Way of strategy as being both a rat's head and an ox's neck. Whenever we have become preoccupied with small details, we must suddenly change into a large spirit, interchanging large with small. This is one of the essences of strategy. It is necessary that the warrior think in this spirit in everyday life. You must not depart from this spirit in large-scale strategy nor in single combat.

The Commander Knows the Troops

"The commander knows the troops" applies everywhere in fights in my Way of strategy. Using the wisdom of strategy, think of the enemy as your own troops. When you think in this way you can move him at will and be able to chase him around. You become the general and the enemy becomes your troops. You must master this.

To Let Go the Hilt

There are various kinds of spirit involved in letting go the hilt. There is the spirit of winning without a sword. There is also the spirit of holding the long sword but not winning. The various methods cannot be expressed in writing. You must train well.

The Body of a Rock

When you have mastered the Way of strategy you can suddenly make your body like a rock, and ten thousand things cannot touch you. This is the body of a rock. You will not be moved. Oral tradition. What is recorded above is what has been constantly on my mind about Ichi school sword fencing, written down as it came to me. This is the first time I have written about my technique, and the order of things is a bit confused.

The body of the rock: This is recorded in the Terao Ka Ki, the chronicle of the house of Terao. Once, a lord asked Musashi "What is this 'Body of a rock'?" Musashi replied, "Please summon my pupil Terao Ryuma Suke". When Terao appeared, Musashi ordered him to kill himself by cutting his abdomen. Just as Terao was about to make the cut, Musashi restrained him and said to the lord, "This is the 'Body of the Rock'". It is difficult to express it clearly.

This book is a spiritual guide for the man who wishes to learn the Way. My heart has been inclined to the Way of strategy from my youth onwards. I have devoted myself to training my hand, tempering my body, and attaining the many spiritual attitudes of sword fencing. If we watch men of other schools discussing theory, and concentrating on techniques with the hands, even though they seem skilful to watch, they have not the slightest true spirit. Of course, men who study in this way think they are training the body and spirit, but it is an obstacle to the true Way, and its bad influence remains forever. Thus the true Way of strategy is becoming decadent and dying out. The true Way of sword fencing is the craft of defeating the enemy in a fight, and nothing other than this. If you attain and adhere to the wisdom of my strategy, you need never doubt that you will win.

The second year of Shoho, the fifth month, the twelfth day (1645)

Teruo Magonojo SHINMEN MUSASHI

The Wind Book

In strategy you must know the Ways of other schools, so I have written about various other traditions of strategy in this the Wind Book. Without knowledge of the Ways of other schools, it is difficult to understand the essence of my Ichi school. Looking at other schools we find some that specialize in techniques of strength using extra-long swords. Some schools study the Way of the short sword, known as kodachi. Some schools teach dexterity in large numbers of sword techniques, teaching attitudes of the sword as the "surface" and the Way as the "interior". That none of these are the true Way I show clearly in the interior of this book — all the vices and virtues and rights and wrongs.

My Ichi school is different. Other schools make accomplishments their means of livelihood, growing flowers and decoratively coloring articles in order to sell them. This is definitely not the Way of strategy. Some of the world's strategists are concerned only with sword fencing, and limit their training to flourishing the long sword and carriage of the body. But is dexterity alone sufficient to win? This is not the essence of the Way. I have recorded the unsatisfactory points of other schools one by one in this book. You must study these matters deeply to appreciate the benefit of my Ni To Ichi school.

Other Schools Using Extra-Long Swords

Some other schools have a liking for extra-long swords. From the point of view of my strategy these must be seen as weak schools. This is because they do not appreciate the principle of cutting the enemy by any means. Their preference is for the extra-long sword and, relying on the virtue of its length, they think to defeat the enemy from a distance. In this world it is said, "One inch gives the hand advantage", but these are the idle words of one who does not know strategy. It shows the inferior strategy of a weak sprit that men should be dependant on the length of their sword, fighting from a distance without the benefit of strategy. I expect there is a case for the school in question liking extra-long swords as part of it's doctrine, but if we compare this with real life it is unreasonable. Surely we need not necessarily be defeated if we are using a short sword, and have no long sword?

It is difficult for these people to cut the enemy when at close quarters because of the length of the long sword. The blade path is large so the long sword is an encumbrance, and they are at a disadvantage compared to the man armed with a short companion sword. From olden times it has been said: "Great and small go together." So do not unconditionally dislike extra-long swords. What I dislike is the inclination towards the long sword. If we consider large-scale strategy, we can think of large forces in terms of long swords, and small forces as short swords. Cannot few men give battle against many? There are many instances of few men overcoming many. Your strategy is of no account if when called on to fight in a confined space your heart is inclined to the long sword, or if you are in a house armed only with your companion sword. Besides, some men have not the strength of others. In my doctrine, I dislike preconceived, narrow spirit. You must study this well.

The Strong Long Sword Spirit in Other Schools

You should not speak of strong and weak long swords. If you just wield the long sword in a strong spirit your cutting will become coarse, and if you use the sword coarsely you will have difficulty in winning. If you are concerned with the strength of your sword, you will try to cut unreasonably strongly, and will not be able to cut at all. It is also bad to try to cut strongly when testing the sword. Whenever you cross swords with an enemy you must not think of cutting him either strongly or weakly; just think of cutting and killing him. Be intent solely on killing the enemy. Do not try to cut strongly and, of course, do not think of cutting weakly. You should only be concerned with killing the enemy. If you rely on strength, when you hit the enemy's sword you will inevitably hit too hard. If you do this, your own sword will be carried along as a result. Thus the saying, "The strongest hand wins", has no meaning. In large-scale strategy, if you have a strong army and are relying on strength to win, but the enemy also has a strong army, the battle will be fierce. This is the same for both sides. Without the correct principle the fight cannot be won.

The spirit of my school is to win through the wisdom of strategy, paying no attention to trifles. Study this well. Use of the Shorter Long Sword in Other Schools Using a shorter long sword is not the true Way to win. In ancient times, tachi and katana meant long and short swords. Men of superior strength in the world can wield even a long sword lightly, so there is no case for their liking the short sword. They also make use of the length of spears and halberds. Some men use a shorter long sword with the intention of jumping in and stabbing the enemy at the unguarded moment when he flourishes his sword. This inclination is bad. To aim for the enemy's unguarded moment is completely defensive, and undesirable at close quarters with the enemy. Furthermore, you cannot use the method of jumping inside his defense with a short sword if there are many enemies.

Some men think that if they go against many enemies with a shorter long sword they can unrestrictedly frisk around cutting in sweeps, but they have to parry cuts continuously, and eventually become entangled with the enemy. This is inconsistent with the true Way of strategy. The sure Way to win thus is to chase the enemy around in a confusing manner, causing him to jump aside, with your body held strongly and straight. The same principle applies to large-scale strategy. The essence of strategy is to fall upon the enemy in large numbers and to bring about his speedy downfall. By their study of strategy, people of the world get used to countering, evading and retreating as the normal thing. They become set in this habit, so can easily be paraded around by the enemy. The Way of strategy is straight and true. You must chase the enemy around and make him obey your spirit.

Other Schools with many Methods of using the Long Sword

I think it is held in other schools that there are many methods of using the long sword in order to gain the admiration of beginners. This is selling the Way. It is a vile spirit in strategy.

The reason for this is that to deliberate over many ways of cutting down a man is an error. To start with, killing is not the Way of mankind. Killing is the same for people who know about fighting and for those who do not. It is the same for women or children, and there are not many different methods. We can speak of different tactics such as stabbing and mowing down, but none other than these. Anyway, cutting down the enemy is the Way of strategy, and there is no need for many refinements of it. Even so, according to the place, your long sword may be obstructed above or to the sides, so you will need to hold your sword in such manner that it can be used. There are five methods in five directions. Methods apart from these five — hand twisting, body bending, jumping out, and so on, to cut the enemy — are not the true Way of strategy. In order to cut the enemy you must not make twisting or bending cuts. This is completely useless. In my strategy, I bear my spirit and body straight, and cause the enemy to twist and bend. The necessary spirit is to win by attacking the enemy when his spirit is warped. You must study this well.

Use of Attitudes of the Long Sword in Other Schools

Placing a great deal of importance on the attitudes of the long sword is a mistaken way of thinking. What is known in the world as "attitude" applies when there is no enemy. The reason is that this has been a precedent since ancient times, that there should be no such thing as "This is the modern way to do it" dueling. You must force the enemy into inconvenient situations.

Attitudes are for situations in which you are not to be moved. That is, for garrisoning castles, battle array, and so on, showing the spirit of not being moved even by a strong assault. In the Way of duelling, however, you must always be intent upon taking the lead and attacking. Attitude is the spirit of awaiting an attack. You must appreciate this. In duels of strategy you must move the opponent's attitude. Attack where his spirit is lax, throw him into confusion, irritate and terrify him. Take advantage of the enemy's rhythm when he is unsettled and you can win.

I dislike the defensive spirit known as "attitude". Therefore, in my Way, there is something called "Attitude-No Attitude". In large-scale strategy we deploy our troops for battle bearing in mind our strength, observing the enemy's numbers, and noting the details of the battlefield. This is at the start of the battle. The spirit of attacking is completely different from the spirit of being attacked. Bearing an attack well, with a strong attitude, and parrying the enemy's attack well, is like making a wall of spears and halberds. When you attack the enemy, your spirit must go to the extent of pulling the stakes out of a wall and using them as spears and halberds. You must examine this well.

Fixing the Eyes in Other Schools

Some schools maintain that the eyes should be fixed on the enemy's long sword. Some schools fix the eye on the hands. Some fix the eyes on the face, and some fix the eyes on the feet, and so on. If you fix the eyes on these places your spirit can become confused, and your strategy thwarted. I will explain this in detail.

Footballers do not fix their eyes on the ball, but by good play on the field they can perform well.

Footballers: Football was a court game in ancient Japan. There is a reference to it in Genji Monogatari.

When you become accustomed to something, you are not limited to the use of your eyes. People such as master musicians have the music score in front of their nose, or flourish the sword in several ways when they have mastered the Way, but this does not mean that they fix their eyes on these things specifically, or that they make pointless movements of the sword. It means that they can see naturally. In the Way of strategy, when you have fought many times you will easily be able to appraise the speed and position of the enemy's sword, and having mastery of the Way you will see the weight of his spirit. In strategy, fixing the eyes means gazing at the man's heart. In large-scale strategy the area to watch is the enemy's strength. "Perception" and "sight" are the two methods of seeing. Perception consists of concentrating strongly on the enemy's spirit, observing the condition of the battle field, fixing the gaze strongly, seeing the progress of the fight and the changes of advantage. This is the sure way to win. In single combat you must not fix the eyes on details. As I said before, if you fix your eyes on details and neglect important things, your spirit will become bewildered, and victory will escape you. Research this principle well and train diligently.

Use of the Feet in Other Schools

There are various methods of using the feet: floating foot, jumping foot, springing foot, treading foot, crow's foot, and such nimble walking methods. From the point of view of my strategy, these are all unsatisfactory. I dislike floating foot because the feet always tend to float during the fight. The Way must be trod firmly. Neither do I like jumping foot, because it encourages the habit of jumping, and a jumpy spirit. However much you jump, there is no real justification for it, so jumping is bad. Springing foot causes a springing spirit, which is indecisive. Treading foot is a "waiting" method, and I especially dislike it. Apart from these, there are various fast walking methods, such as crow's foot, and so on. Sometimes, however, you may encounter the enemy on marshland, swampy ground, river valleys, stony ground, or narrow roads, in which situations you cannot jump or move the feet quickly. In my strategy, the footwork does not change. I always walk as I usually do in the street. You must never lose control of your feet. According to the enemy's rhythm, move fast or slowly, adjusting your body not too much and not too little. Carrying the feet is important also in large-scale strategy. This is because, if you attack quickly and thoughtlessly without knowing the enemy's spirit, your rhythm will become deranged and you will not be able to win. Or, if you advance too slowly, you will not be able to take advantage of the enemy's disorder, the opportunity to win will escape, and you will not be able to finish the fight quickly. You must win by seizing upon the enemy's disorder and derangement,

Joseph Lumpkin

and by not according him even a little hope of recovery. Practice this well.

Speed in Other Schools

Speed is not part of the true Way of strategy. Speed implies that things seem fast or slow, according to whether or not they are in rhythm. Whatever the Way, the master of strategy does not appear fast. Some people can walk as fast as a hundred or a hundred and twenty miles in a day, but this does not mean that they run continuously from morning till night. Unpracticed runners may seem to have been running all day, but their performance is poor. In the Way of dance, accomplished performers can sing while dancing, but when beginners try this they slow down and their spirit becomes busy. The "old pine tree" melody beaten on a leather drum is tranquil, but when beginners try this they slow down and their spirit becomes busy. Very skilful people can manage a fast rhythm, but it is bad to beat hurriedly. If you try to beat too quickly you will get out of time. Of course, slowness is bad. Really skilful people never get out of time, and are always deliberate, and never appear busy. From this example, the principle can be seen. What is known as speed is especially bad in the Way of strategy. The reason for this is that depending on the place, marsh or swamp and so on, it may not be possible to move the body and legs together quickly. Still less will you be able to cut quickly if you have a long sword in this situation. If you try to cut quickly, as if using a fan or short sword, you will not actually cut even a little. You must appreciate this. In large-scale strategy also, a fast busy spirit is undesirable. The spirit must be that of holding down a pillow, then you will not be even a little late. When you opponent is hurrying recklessly, you must act contrarily, and keep calm. You must not be influenced by the opponent. Train diligently to attain this spirit.

Old pine tree: "KoMatsu Bushi", an old tune for flute or lyre.

"Interior" and "Surface" in Other Schools

There is no "interior" nor "surface" in strategy. The artistic accomplishments usually claim inner meaning and secret tradition, and "interior" and "gate","" but in combat there is no such thing as fighting on the surface, or cutting with the interior. When I teach my Way, I first teach by training in techniques which are easy for the pupil to understand, a doctrine which is easy to understand. I gradually endeavor to explain the deep principle, points which it is hardly possible to comprehend, according to the pupil's progress. In any event, because the way to understanding is through experience, I do not speak of "interior" and "gate". In this world, if you go into the mountains, and decide to go deeper and yet deeper, instead you will emerge at the gate. Whatever is the Way, it has an interior, and it is sometimes a good thing to point out the gate. In strategy, we cannot say what is concealed and what is revealed.

Accordingly I dislike passing on my Way through written pledges and regulations. Perceiving the ability of my pupils, I teach the direct Way, remove the bad influence of other schools, and gradually introduce them to the true Way of the warrior. The method of teaching my strategy is with a trustworthy spirit. You must train diligently. I have tried to record an outline of the strategy of other schools in the above nine sections. I could now continue by giving a specific account of these schools one by one, from the "gate" to the "interior", but I have intentionally not named the schools or their main points.

Gate: A student enrolling in a school would pass through the gate of the Dojo. To enter a teacher's gate means to take up a course of study.

The reason for this is that different branches of schools give different interpretations of the doctrines. In as much as men's opinions differ, so there must be differing ideas on the same matter. Thus no one man's conception is valid for any school. I have shown the general tendencies of other schools on nine points. If we look at them from an honest viewpoint, we see that people always tend to like long swords or short swords, and become concerned with strength in both large and small matters. You can see why I do not deal with the "gates" of other schools. In my Ichi school of the long sword there is neither gate nor interior. There is no inner meaning in sword attitudes. You must simply keep your spirit true to realize the virtue of strategy.

Twelfth day of the fifth month, the second year of Shoho (1645)

Teruo Magonojo SHINMEN MUSASHI

The Book of the Void

The Ni To Ichi Way of strategy is recorded in this the Book of the Void. What is called the spirit of the void is where there is nothing. It is not included in man's knowledge. Of course the void is nothingness. By knowing things that exist, you can know that which does not exist. That is the void. People in this world look at things mistakenly, and think that what they do not understand must be the void. This is not the true void. It is bewilderment. In the Way of strategy, also, those who study as warriors think that whatever they cannot understand in their craft is the void. This is not the true void. To attain the Way of strategy as a warrior you must study fully other martial arts and not deviate even a little from the Way of the warrior. With your spirit settled, accumulate practice day by day, and hour by hour. Polish the twofold spirit heart and mind, and sharpen the twofold gaze perception and sight.

When your spirit is not in the least clouded, when the clouds of bewilderment clear away, there is the true void. Until you realize the true Way, whether in Buddhism or in common sense, you may think that things are correct and in order. However, if we look at things objectively, from the viewpoint of laws of the world, we see various doctrines departing from the true Way. Know well this spirit, and with forthrightness as the foundation and the true spirit as the Way. Enact strategy broadly, correctly and openly. Then you will come to think of things in a wide sense and, taking the void as the Way, you will see the Way as void. In the void is virtue, and no evil. Wisdom has existence, principle has existence, the Way has existence, spirit is nothingness.

Twelfth day of the fifth month, second year of Shoho (1645)

Teruo Magonojo SHINMEN MUSASHI

Joseph Lumpkin

The Tao Te Ching

By Lao Tsu
Translation by Joseph Lumpkin

The Tao Te Ching was written by a man referred to as Lao Tzu. The unknown author's name means both "the old philosopher" and "the old philosophy." Hence Lao Tzu may also be the title for the book or the name or title of the author.

Lao Tzu lived in ancient China and was the keeper of the Imperial Library. Legends tell us he was famous for his wisdom. He was an advocate for personal inner growth, moral government, and the rights of the people. Perceiving the growing corruption of the government, he left for the countryside. On his way, the guard at the city gates asked Lao Tzu to write out his teachings for the benefit of future generations. Lao Tzu wrote the Tao Te Ching, and was never heard of again. The Tao Te Ching is the fundamental text of Taoism.

Taoism is a philosophy based upon the search for a middle path through life; avoiding extremes so that no act is followed by a reaction. This philosophy has come to influence many other aspects of Eastern life; including martial arts in such styles as Tai Chi, Aikido, Shinsei Hapkido, Jujutsu, and Judo. The concept of balancing the masculine and feminine or hard and soft applied also in Eastern medicines such as herbology and acupuncture.

The practice of Taoism is principally concerned with discovering balance and self-knowledge. All things, actions, and even intents, are broken into positive and negative, or masculine and feminine influences. Taoism advocates learning to sense the world directly, to "intuit" the flow of things, and to maintain a balance of opposing forces.

In doing so, one must contemplate impressions deeply as one attempts to become detached, without resorting to coloring intuitive

417

impressions with personal expectations. Taoism advises against relying on ideologies, because to do so will rob one's life of its meaning and personal intuition. By developing intuition, one acquires a deeper understanding of the world, one's place, and the future. Lao Tzu remarked that an excessive force tends to trigger an opposing force, and therefore the use of force cannot be the basis for establishing a strong and lasting social foundation of control, life, or government. The force used to lead others is said to be the "moral force," virtue, or wisdom of the Master.

There seem to be two theories disputing the dating of the Tao Te Ching. According to tradition, the work originates in the fourth century B.C., but recent discoveries confirm that the writings originate no earlier than the third or fourth century B.C. The oldest existing copy is from 206 or 195 B.C.

The second hypothesis concludes that the teaching may be old enough to pre-date the invention of paper. In fact, its form exhibits many of the features of an oral tradition, suggesting it may pre-date writing as well. Oral traditions are very difficult to trace and we may never know how old the verses are.

The Tao Te Ching as it exists today, consists of 81 short chapters among which 37 form the first part, the Book of the Way (Tao), and the next 44 form the Book of Te ("Te" is a word translated by James Legge as "virtue", pointing to the Tao of Heaven). Its division into chapters is considered to be the result of the remarks of Heschang Gong (Han dynasty). Other traditional interpretations conclude the name may be "The Book of the Way of Virtue" or "The Book of Flow and Harmony." It is in the association of intuition, insight, and the inward knowledge leading to God that the Gospel of Thomas and the Tao Te Ching come together to give the reader a most amazing experience.

Tao Te Ching

1

The Tao that can be explained is not the eternal Tao
The name that can be spoken is not the eternal Name.
The unspeakable is the beginning of everything.
Naming is the origin of every separate thing.
Free from desire, you realize the mystery.
Trapped in desire, you only see the manifestations.
Mystery and manifestations
Arise from the same source.
It is experienced as darkness.
Darkness within darkness.
The gateway to all understanding.

2

When beautiful things appear as beautiful,
It is because other things are called ugly.
We can only know Good because there is Evil.
Having and not having are born together
Being and not being arise from each other.
Difficult and easy support each other.
Long and short define each other.
High and low depend on each other.
Sound and silence harmonize each other.
Before and after follow each other.
Therefore the Master acts without doing anything
And teaches without talking.
He allows things to come and go naturally.
He does not hold on.
He has but doesn't possess, works but takes no credit,
He has no expectations.
It is done and then forgotten, therefore, it lasts forever.

3

If you give some men power, others become powerless.
If you assign value, people begin to steal.
Desire begets confusion of the heart.
The Master leads by emptying people's minds
And filling their hearts,
By weakening their ambition
And toughening their resolve.
He helps people lose everything they know and desire.
Craftiness, ambition, and expectations
Cause things to go badly.
Practice not-doing, and everything will be well.

4

The Tao is like an empty well; used but never emptied.
It is the eternal void and source of all possibilities.
It blunts the sharpness, untangles the knot;
Softens the glare, unifies the dust.
It is hidden deeply but is ever-present.
I don't know from where it came.
It is older than God.

5

The Tao is impartial;
It gives birth to both good and evil.
The Master is impartial
He sees people as both good and evil.
The Tao is like a bellows:
It is empty, changeable, potential in form.
The more it is moved, the more it yields.
The more you talk of it, the less your words count.
Hold on to the stillness of the center.

6

The Tao is called the Great Mother:
Empty yet inexhaustible,
It gives birth to all things.
It is always present within you.
If you use it; it will never fail.

7

Tao is eternal.
Why is it eternal?
It was never born;
Thus, it can never die.
The Master stays behind;
That is why he is ahead.
He is detached from all things;
That is why he is one with all things.
Because he has let go of himself,
He attains fulfillment.

8

The highest good is like water,
Which nourishes all things without trying.
It is content with the low places that people disdain.
Water, goodness and the Tao flow in the same way.

In dwelling, be close to the land.
In thinking, go deep within the heart.
In conflict, be fair and generous.
When speaking, be true.
When ruling, be fair
In work, be competent.
In life, be completely present in each moment.

If there is no fight, there will be no blame.

9

Better to stop short than to fill your bowl to overflowing.
The sharpened knife dulls easily.
Collect money, jewels, and possessions
And you cannot protect them.
Gather wealth and position
And your heart will be their captive.

Do your work, then let it go.
This is the only path to inner peace.

10

Can you control your mind and
Keep it from wandering?
Can you keep your oneness and focus?
Can you let your body become
Supple as a newborn child's?

Can you cleanse your inner vision?
Can you love people and lead them
Without guile or will?
Can you deal with the most vital matters
By letting events take their course?

Can you be lead by Tao,
Keeping an empty mind
And thus being open to all things?

Are you able to be still,
Give birth and nourish,
Create and bear without possessing,
Act without expectations,
Lead without controlling?

This is the highest virtue.

11

Thirty spokes join at the hub of a wheel.
It is the center hole that has use.

We shape clay into a pot.
It is the emptiness inside
That makes it useful.

Shape doors and windows for a house.
It is the space that makes it useful.

We derive benefit from things that are there
And usefulness from what is not there.

12

Five colors blind the eye.
Five sounds deafen the ear.
Five flavors dull the taste.
Racing thoughts confuse the mind.
Desires lead the heart astray.

The Master observes the world
But is guided by his intuition.
He allows things to come and go
As they will.

13

Disgrace and success are the same.
Misfortune is a condition of life.

What does it mean to accept disgrace willingly?
Accept not being important.
Do not be concerned with loss or gain.

What does it mean to accept misfortune
As a condition of life?
Misfortune comes to all who are born.
If you had no body, what misery would you have?

Submit to the Tao and you will be trusted with everything.
Love the world as yourself and you can care for all beings.

14

Look for it, and it can't be seen.
Listen for it, and it can't be heard.
Grasp for it, and it can't be held.
These three are beyond understanding,
Thus they as joined as one.

Above, it isn't bright.
Below, it isn't dark.
It is an unbroken, unnamable thread
That returns to the nothingness.
The form is void.
Image without visage,
Beyond definition and imagination.

Examine it. You will see no beginning;
Follow it and there is no end.
With the ancient Tao you will be in the eternal present.
Knowing the beginning is the essence of wisdom.

15

The ancient Masters were profound and subtle.
The depth of their wisdom cannot be measured.
Because they cannot be recognized
We can only describe their appearance.

They were careful

As someone crossing an ice covered stream,
Alert like a warrior in hostile territory,
Courteous as a guest,
Yielding as melting ice,
Simple as a block of wood,
Receptive as a valley,
Clear as a pool of water.

Can you be still inside while the mud settles?
Can you wait quietly until the moment is right?

The Master doesn't seek fulfillment.
Without seeking and without expectations
He is not confused by desires.

16

Empty your mind of every thought.
Let your mind and heart still.
Watch things come and go without attachment.

Everything grows, matures,
And returns to the same source.
Returning to the source is stillness and peace.
This is the way of nature.
It is the way of Tao and it is unchanging.

Not knowing the consistency of Tao
Leads to confusion and disaster.
When you realize the common source
You become tolerant and charitable.
Giving to others, you will be as a king.
Being as a king, you will become divine.
Being divine, you will become one with the eternal Tao.

When death comes, you will be ready
For you will know that the Tao will never pass away.

17

When the Master governs,
The people are hardly aware that he exists.
An ineffective leader is known and loved.
A poor leader is feared.
The worst is one who is despised.

If you don't trust the people, you will not be trusted.
If you do not trust, you make others untrustworthy.

The Master says nothing.
When work is done
The people say, We did it all!

18

When the great Tao is forgotten,
Goodness and piety arise.

When the cleverness and knowledge begin
Falseness and pretension are born.

When there is no peace in the family,
Filial piety and devotion begin.
When the country falls into disarray,
Patriotism is born.

19

Give up trying to be holy or wise,
And people will be a hundred times happier.
Throw away ideas of kindness, morality, and justice,
And people will rediscover love and family.
Give up industriousness and profit,
And there won't be any robbers or thieves.
These three are outward forms and are useless.

Cultivate simplicity.
Realize your true nature.
Renounce selfishness.
Do not desire.

20

Stop thinking, and end your troubles.
What is the difference between yes and no?
What is the difference between success and failure?
Why must I fear what others fear? How ridiculous!

Other people are content with a feast or party
Others are content with a park in spring or a beautiful view.
I alone wonder, uncaring, expressionless as a newborn child.
I have no home.
Other people have what they need;

I alone have nothing.
Others are clear and witty.
I am nothing. My mind is empty.

Other people are bright;
I alone am dim.
Other people are clever;
I alone am dull.
Other people have a purpose;
I am aimless and depressed,
Drifting like a wave on the ocean,
Blown by the wind.

I am different from the others.
I am nourished from the Great Mother.

21

The Master keeps his own mind
And is alone, always at one with the Tao.

Joseph Lumpkin

The Tao cannot be imagined, yet within is image.
It is elusive and formless, yet within is form.

The Tao is dark and void, yet within is radiance.
The essence is real, yet within is faith.

Since before time was and until now, the Tao is.
How do I know the way of creation is true?
Because of this; I look inside myself and it is there.

22

Yield and overcome;
Bend and be straight;
If you want to become full,
Let yourself be empty.
If you want to be reborn,
Let yourself die.
If you want to be fulfilled,
Give up everything.

Therefore, the Master, resides in the Tao,
And thus sets an example for all.
Because he doesn't put on a display,
People can see his true light.
Because he does not boast,
People can trust his words.
Because he doesn't know who he is,
People recognize themselves in him.
Not bragging or boasting,
There is no quarrel or dissention.

The ancient Masters said,
If you want to receive all,
Give up everything,
Be complete and all things will be yours.

23

To talk a little is natural.
High winds and heavy rains soon exhaust themselves.
If Heaven and Earth cannot sustain then how can man?

He who lives the Tao,
Is in unity with the Tao
He who lives virtuously experiences virtue.
If you lose the Way,
You are willingly lost.

Trust your natural responses;
And everything will be as it should.

24

He who stands on tiptoe is not steady.
He who strives cannot maintain the pace.
He who wishes to be known is not enlightened.
He who defines himself cannot know his true nature.
He who lords over others cannot empower himself.
He who brags will not be remembered.
According to followers of Tao,
These are extra food and unnecessary baggage.
They weigh you down, slow you down and impede your joy.
The followers of Tao reject them.

25

There was something formless and mysterious
Born before anything existed.
It is silent, peaceful, and empty.
Solitary. Unchanging.
Eternally present – peace within motion.
It is the mother of the all things.
It is unnamable.

Joseph Lumpkin

I call it the Tao.

It flows through all things,
And then returns.

The Tao is great.
The sky is great.
Earth is great.
Man is great.
These are the four great powers.

Man follows the Earth.
Earth follows the sky.
The sky follows the Tao.
The Tao follows only its own natural way.

26

The heavy is the root of the light.
The still is the master of all that moves.

Thus the Master travels all day
Without leaving "home."
However splendid the views,
He stays unattached and calm.

Why should the lord of a country
Flit about like a fool?
If you let yourself be quixotic,
You lose your root.
To be restless or anxious,
Is to lose control and move too soon.

27

A good traveler leaves no tracks.
A good speaker does not stutter.
A good mathematician needs no paper.

A good door needs no lock yet cannot be opened.
A good binding needs no knots yet cannot be untied.

Thus the Master cares for all people
And doesn't reject anyone.
He nurtures all and abandons nothing.
This is called embodying the light.

What is a good man but a bad man's teacher?
What is a bad man but a good man's responsibility?
The teacher is to be respected,
And the student is to be nurtured.
However intelligent you are, if you do not follow
This way there will be confusion.
This is the secret.

28

Know the strength of the masculine,
Yet keep the heart of the feminine:
Let all things flow through you,
Like a true and constant stream.
If you do this, the Tao will never forsake you
And you will be like a little child once more.

Know the white,
Yet keep the black:
Be an example for the world.
If you are an example for the world,
Like a true and constant stream.
You will return, flowing to the All.

Know honor,
Yet have no care for it:
Be the valley of the world.
Being the valley of the world,
All things will flow into you.

Return to the simple state of a block of wood.

Then you will be useful and full of potential.
When the Master uses the Tao he rises above the rest.
Thus a master tailor cuts little.

29

Do you want to improve the world?
I don't think it can be done.

The world is sacred.
It can't be improved.
If you tamper with it, you will ruin it.
If you treat it like a procession, you will lose it.
There are some that are ahead,
And some behind;
Some that are difficult;
And some that are easy;
Some that are weak,
And some that are strong;
Some that will endure,
And some that will be overthrown.

Seek the center path.
Avoid extremes and excess.
Seek balance.
The Master sees and accepts
Without trying to control.

30

If you counsel a warrior about the Tao
Tell him he should not use force.
For every force there is a reaction.
Briars grow where armies tread.
Famine follows in the wake of war.
Do only what needs to be done.
Never take advantage of power.

Achieve results but the results are not your own.
Do not let pride interfere.
Thus, there is nothing to brag about.
Nothing to be proud of.
Nothing to fight about.

Force is followed by weakness.
This is not the Way of the Tao.
Any other way will lead to premature destruction.

31

Weapons are the instruments of violence;
All decent men hate them.

Weapons are the instruments of fear;
A decent man will avoid them until there is no choice.
Peace and serenity are his highest desires.
Victory is no cause for celebration.
For, it you celebrate victory you celebrate death and defeat.
He cannot delight in the slaughter of men.

On happy occasions the underdog is celebrated.
On sad occasions people look to their leaders.
Generals stand to the left, kings and presidents to the right.
Therefore, war is conducted like a funeral
He enters a battle gravely, and has compassion for those killed,
As with a funeral.

32

The Tao cannot be defined.
It is smaller than anything formed,
And cannot be grasped.

If powerful men could use it all things would flow naturally.
Men would do as they should
And the rain would come in its season.

When the whole is divided, all parts must have a name.
Knowing when to stop you can avoid troubles.

All things end in the Tao
As rivers end at the sea.

33

Knowing others is intelligence;
Knowing yourself is wisdom.
Mastering others requires force;
Mastering yourself requires inner strength.

If you realize that you have enough,
You are truly rich.
Tenacity and a perfect finish require willpower.
He who stays centered will endure.
To live in the eternal present one will never die.

34

The great Tao flows everywhere.
All things are born from it and it holds nothing in reserve.
It creates naturally and is not possessive.
It pours itself into its work,
Yet it makes no claim.

It nourishes infinite worlds,
Yet it doesn't hold on to them.

Since it is merged with all things
And is hidden in their hearts,
It can be called humble.
Since all things vanish into it,
It alone endures.
It can be called great.
It isn't aware of its greatness;

Thus it is truly great.

35

All men seek he who is centered in the Tao.
There they find rest, joy, and peace of mind.

Music or the smell of good food
May entice people to stop and enjoy.
But conversations about Tao
Seem boring and bland.
When you look for it, it cannot be seen.
When you listen for it, it cannot be heard.
When you use it, it cannot be exhausted.

36

If you wish to diminish it, allow it to expand.
If you wish to end it, allow it to mature.
If you wish to bash it against the ground, first raise it up.
If you wish to take something, it must first be given.
This is called understanding the nature of things.

The soft overcomes hard.
Weak overcomes strong.
Stay in your element.
Hide your strength until it is needed.

37

The Tao does not strive, yet it leaves nothing undone.

If powerful men observe this Way
The whole world would be as it should be.
If they wanted to act, they would resume
Their simple, everyday lives,
In harmony and free of desire.
Then, there would be peace.

38

A good man does not try to be good. It comes from his heart.
A fool tries to be good but he it is not his natural way.
Therefore, a good man does not strive yet good comes from him.
A fool rushes about trying to act good but no good comes of it.

The kind man acts, kindness leaves nothing lacking.
When a just man acts,
There is judgment and things are left to do.
When the man of discipline follows, no one responds
Until he begins his enforcement.

When the Tao is lost, there is kindness.
When kindness is lost, there is justice.
When justice is lost, there is ritual.
Ritual is the corpse of true faith,
And the beginning of foolishness.

Therefore the Master looks deeply
And is not concerned with how things appear on the surface,

He examines the fruit and not the flower.
He dwells on what is real and not on appearances.

39

From Ancient times, all things arise from the One:
The sky is clear and complete.
The earth is solid and complete.
The spirit is strong and complete.
The valley is full and complete.
All things are alive, complete, and content.

When the Way is not followed,
The sky becomes tarnished,

The earth becomes wasteland,
The spirit is depleted,
Creatures become extinct.

The Master humbly follows the Way, yet seems noble.
He acts as a lowly servant and is thus raised in stature.
Rulers and men of authority feel orphaned, widowed, and alone.
This is humility.
Too much success draws attention like sounding chimes and Rattling
jade stones.

40

Returning is the movement of the Tao.
Yielding is the way of the Tao.

All things are born of that which is and was.
Being is born of nothingness.

41

When the seeker hears of the Tao,
He immediately begins to practice it.
When an average man hears of the Tao,
He thinks about it but does not practice it.
When a foolish man hears of the Tao, he laughs out loud.
If he didn't laugh, it wouldn't be the Tao.

Thus it is said:
The path into the light seems dark,
The path forward seems to go back,
The direct path seems prolonged,
The highest good seems empty.
Great purity seems sullied.
Depth of spirit seems inadequate.
True stability seems changeable,
Great talent matures in time.

The highest notes are beyond hearing.
The Tao is obscured and nameless.

It nourishes and completes all things.

42

The Tao created One.
One created Two.
Two created Three.
Three gives birth to all things.

All things carry the feminine
But demonstrate the masculine.
When masculine and feminine are balanced,
Harmony is achieved.

Ordinary men hate being deserted, abandoned, or alone.
But these are how the Master is described.
Embracing his solitude
He becomes aware that he is one with All.

43

The softest thing in the world overcomes the hardest thing.
That which has no substance enters where there is no space.
This shows the value of non-action.

Teaching without words, and acting without movement is the
Master's way.

44

Fame or integrity: which is more important?
Integrity or wealth: which do you desire more?
Success or failure: which is more damaging?

If you are attached to processions you will suffer.
If you hoard you will lose.
Contentment assuages disappointment.
Know when to stop, avoid troubles, and remain forever satisfied.

45

Great accomplishments seem imperfect,
But, they continue to be useful.
True fullness seems lacking,
Yet, it is never emptied.

True straightness seems crooked.
True wisdom seems foolish.
True grace seems awkward.

If you are cold, move.
If you are warm, be still.
In stillness and peace, the order of the universe is established.

46

When Tao is present within a country,
Fine horses are free to fertilize the fields.
When Tao is missing in a country,
War horses are bred in the county side.

There is no greater fault than desire,
No worse goad than discontentment.
No greater shame than selfishness
He who realizes enough is enough will be fulfilled.

47

Without traveling you can know the world.
Without looking out of a window,
You can see the ways of the Tao.
The more you go outside yourself the less you understand.

The Master understands without traveling,
Sees the Way without looking,
Achieves without action.

48

In pursuit of knowledge, every day something is added.
In the pursuit of the Tao, every day something is dropped.
Less and less is needed until stillness is achieved.
When nothing is done, nothing is left undone.
If things are left alone they will take their own course.
If you interfere - chaos.

49

The Master has no mind of his own.
He serves the needs of others.

He is good to people who are good.
He is also good to people who are not good.
This is true goodness.

He has faith in people who are trustworthy.
He also has faith in people who are not trustworthy.
This is true trust.

The Master is quiet, timid and does not consider himself.
People do not understand him.
They look and listen even though
They think he behaves like a child

50

In the space of a lifetime,

One third follows the way of life,
One third follows the way of death
And one third follows nothing
And drift through life,
Having no purpose.

He who knows how to live in the Tao
Moves without fear or thought of his actions.
He will not be harmed because there is no place
For weapons or beasts to enter him.
If there is no fear of death,
The mind is clear and death has no place.

51

Everything in existence is an expression of the Tao.
The Tao nurtures them.
Unconsciously and spontaneously, they take on form.
They allow circumstances to shape them.
That is why everything honors the Tao.

The Tao gives birth to all things.
Its goodness nourishes them, protects them, and comforts them.
The Tao does not possess them.
The Tao does not boast of them.
The Tao has no expectations of them.
The Tao guides without interfering.
This is the highest good.

52

Tao is the mother and beginning of all things.
All things issue from it like children from their mother;
And all things return to it.
To know the maker is to know the creation.
Recognize the children and know the mother,
And free yourself from fear and sorrow.

Stay quiet, focus the mind and life will be peaceful.
Speak without thought, entertain desires,
Rush about and your heart will be troubled.

Seeing into darkness and detail is insight.
Yielding is strength.

Use your own light.
Inwardly lighting your own path is wisdom.
Consistence leads to perfection.

53

The center path is the main road.
If I have any sense I will walk the clear path.
But, people prefer the side roads,
Even though the main path is easiest.
Step left or right and things are out of balance.
Stay centered within the Tao.

When rulers demand too much
Common folks lose their land;
When rulers spend money, buy weapons, and distribute wealth
Some wear expensive clothing
Eat fine food, and have many possessions;
While others go hungry.
These rulers are robbers and thieves.
They do not follow the Tao.

54

Whoever is firmly planted in the Tao
Will not be uprooted.
Whoever embraces the Tao
Will not slip away.
His name will be honored
From generation to generation.

Cultivate goodness in your life
And you will become genuine.
Encourage goodness in your family
And your family will flourish.
Spread it throughout your country
And your country will be an example
To the rest of the world.

Let it be present in the universe
And it will be omnipresent.

First, there must be virtue in family, then village,
Then nation, then the world.

How do I know this is true?
By looking.

55

He who is in harmony with the Tao
Is like a newborn child.
Whose bones are soft, its muscles are weak,
Yet its grip is strong.
It doesn't know about sex
Yet its penis can stand erect.
It screams and cries all day long,
Yet never becomes hoarse.
It is in harmony with the Tao.

Harmony brings consistency.
Consistency brings enlightenment.

The Master's power is in his timing.
He lets all things happen without rush or desire;
So he does not exhaust his energies.
He never expects results;
So, he is never disappointed.
What does not follow the Tao will not endure.

56

Those who know don't speak of it.
Those who speak of it do not know it.

Close your mouth,
Guard your senses,
Blunt your sharpness,
Reduce complex problems to basic issues,
Soften your glare,
Let your dust settle.
This is the primal Oneness.

When there is oneness
We will not distinguish between friend or enemy,
Gain or harm, honor or failure.
We will give ourselves continually.

This is the highest state of being.

57

If you want to be a great leader,
You must learn to rule justly.
Do not try to control, allow plans to change on their own.
Rule without striving.

The more laws enacted,
The more cunning people will become.
The more violent the weapons,
The less secure people will be.
The more cunning the people are
The more difficult it is to solve the crimes.

Therefore the Master says:
I let go of the law and ritual,
And people become honest and peaceful.

I do not wage war, and people become prosperous,
I let go of all desire to be a ruler
And the people become good and peaceful.

58

If a country is governed with tolerance,
The people are down to earth and honest.
If people are repressed, they become cunning and crafty.
When happiness is contrived, it is meaningless
Try to make people happy,
And the result is misery.
Try to legislate morality,
And you sow immorality and vice.

Thus the Master is intelligent but not cunning,
To the point but not hurtful.
Straightforward, but not rude.
Radiant, but not blinding.

59

For governing a country and serving the common good,
There is nothing better than moderation.

The moderation comes from laying aside your own ideas.
It depends on wisdom gathered through maturity.
If wisdom is acquired, nothing is impossible.
When possibilities are seen as limitless,
Then, a man is equipped to rule.

Nothing is impossible for him.
Because he has become the mother of all people.
He has deep and solid roots into the Tao.
He will nourish others and have a long, happy life.
He will be able to see the outcome from the beginning.

60

Governing a large country
Is like frying a small fish.
It breaks apart if poked too much.
Too much disturbance leads to damage.

Let Tao be the center
And evil will have no power over you.
Evil exists but can be avoided and does not propagate.

The Master avoids evil.
And thus protects both himself and others.

61

When a country obtains great power,
It becomes like the sea:
All streams run downward into it.
The more powerful it grows,
The greater the need for humility.
Yielding to a smaller country,
The greater country will absorb it.
If the smaller country yields to the greater country
It remains whole and can conquer from within.
Humility means trusting the Tao,
Thus, never needing to be defensive.
It is natural for the lesser to serve and the greater to lead.
It is Virtue for the greater to yield.

62

The Tao is the source of all things.
It is the good man's treasure,
And the bad man's refuge.

Honors can be bought with flattering words,

Respect can be won with good deeds;
The Tao does not choose,
So, do not abandon the bad man.
One day he could be king.
On that day, do not send gifts,
But instead, offer the Tao.

Why did the ancient ones esteem the Tao?
Because, when you seek, you find;
And when you make a mistake, you are forgiven.

Therefore, it is the greatest gift of all.

63

Act without striving;
Work without effort.
See the small as large
And the few as many.
Confront the bitter.
Simplify the complicated.
Attend to details.
Achieve greatness by
Accomplishing small feats.

By doing these things
You will do great things as if they were easy.

Great acts are made of small deeds.
The Master never attempts greatness;
Thus, he achieves greatness.

When there is difficulty, he is not concerned.
He has no preconceived ideas of how things should be.

64

What is rooted is easy to nourish.

Joseph Lumpkin

What is first beginning is easily stopped.
What is brittle is easy to break.
What is small is easy to scatter.

Prevent trouble before it arises.
Put things in order before there is confusion.

The giant tree the size of a man's arm span
Grows from a small seedling.
A skyscraper begins with a pile of dirt.
The journey of a thousand miles
Starts from beneath your feet.

Rushing into action, you are defeated from the start.
Trying to grasp things, they will slip through your fingers.

Therefore the Master takes no action.
Letting things unfold, he is not defeated.
He does not try to hold anything,
Thus by claiming nothing, he has nothing to lose.

People fail when nearing the end.
The finish should be as strong as the beginning.
Then there will be no failure.

What he desires is non-desire;
He owns nothing but collects nothing.
He has nothing but gives men All.
He helps others find their own nature.
He can care for all things by doing nothing but showing forth the
Tao.

65

The ancient Masters
Didn't try to educate the people.
They kept knowledge to themselves.

When people think they know the answers,

They are difficult to guide.
Rulers who use deceit cheat the country.
If you want to learn how to govern,
Avoid being clever or deceitful.
The simplest pattern is the clearest
And easiest to follow.

Cleverness or simplicity; these are the two options.
Understanding this leads to goodness.
The highest good leads all men back to Tao.

66

All streams flow to the sea.
Because it is lower
It receives and rules ten thousand streams.

If the Master would lead,
He must place himself humbly below them.
If you want to lead,
You must learn how to follow.

If a ruler serves the people
No one feels oppressed.
If he stands before the people to guide them,
No one feels manipulated.
The whole world is grateful to him and will not tire of him.
Because he does not compete
He meets with no resistance.

67

Some say that my teaching of the Tao cannot be understood.
Others call it lofty but impractical.
It is different and thus has endured.

I have only three treasures to teach;

Simplicity, patience, and compassion.

Being simple in actions and in thoughts,
You return to the source of being.

Being patient with all,
You are in harmony with the Tao
And care not whether you are ahead or behind others.
In compassion you reconcile all beings and are yourself, Reconciled.

Lack of compassion is not bravery.
Lack of patience is not spontaneity.
Lack of simplicity is not cleverness
To lack any of these things is certain death.

Compassion brings victory in battle and a steadfast defense.
Heaven saves and guards its own when there is compassion.

68

The best soldier is controlled and thoughtful.
The best general patiently searches the mind of his enemy.
The best businessman serves the good of his clients.
The best victor is merciful.

These embody the virtue of not striving.
They have the gift of knowing others
And thus knowing themselves.
This is in harmony with the Tao.

69

The generals have a saying:
He who moves first loses.
It is better to wait and see.
Better to retreat an inch
Than to take a foot by force.
But, where the enemy leaves an inch, I fill the void.

This is called going forward without seeming to advance;
Attacking without using weapons.

There is no greater misfortune
Than underestimating your enemy.

Underestimating your enemy means vilifying
And lessening him in your mind.

Thus you destroy your three treasures
And become your own worst enemy.

When the battle is joined,
He who is patient and yields, wins.

70

My teachings are easy to understand
And easy to perform.
Yet if you try to practice them you will fail.
If you grasp them they will slip away from you.

My teachings are older than the world.
My actions are from self-knowledge.
Men cannot understand these things
Because they do not seek them within.
Thus, I am abused and they are honored.

The Master wears simple clothes
And holds the treasure within him.

71

Not-knowing is true knowledge.
Ignoring this is sickness.
If you realize that you are sick;
You can start to become healthy.

The Master is sick of being sick
Thus, he has become well.

72

When they lose their sense of awe and mystery,
People turn to religion and law.

When they no longer trust,
If you visit, they are suspicious.
If you offer them work,
They will be wary.
Self-confidence will fail,
And they will become dependant on authority.

Therefore the Master steps back,
So that people won't be confused.
He is not arrogant and does not need to rule.
He lets go and gets out of the way.

73

A brave and driven man will place life on the line.
A brave and patient man will value life.
Of these two, one is good and one is injurious.

Heaven favors certain attributes.
Even the Master does not know why.

The Tao does not try but it covers the whole world.
It asks no questions but is answered to by all.
It is not petitioned but supplies every need.
It has no observable goal but fulfills all that is required.

The Tao casts its net wider than the world.
Though the mesh is large, it holds and keeps all things.

74

If men are not afraid to die,
There is nothing that can stop them.

Law is upheld by fear of punishment.
But who wants to be the executioner?
If you take his place you will harm yourself.
It is like trying to take the place of a master carpenter.

If you use his tools,
You may cut your hand.

75

When taxes are too high, people go hungry.
When the government is too restrictive, people soon rebel.

When the price of life becomes too high,
People think of death more often.

Having little to live on, or few things to live for,

The value of life falls.

76

Men are born soft and supple;
Dead, they are stiff and hard.
Green plants are tender and pliant;
When they die they become brittle, brown, and dry.

Thus whoever is stiff and inflexible
Is a disciple of death.
Whoever is soft and yielding

Is a disciple of life.

A tree that will not bend will be broken.
Tactics of life should be fluid to meet changing circumstance.
Flow with change or meet defeat.
For the hard and unyielding will be broken.
The soft and yielding will prevail.

77

As the Tao acts, it is like the bending of a bow.
The high is bent downward;
The low is raised up.
It adjusts excess and deficiency of strength,
Measure, and status
And blends all into harmony and balance.

The Tao takes from those with more and gives
To those with less.
Man's ways are the opposite.
Man esteems those who are wealthy and shuns
Those with little.
Only the Tao gives what it has.

The Master produces without owning,
Works without credit, succeeds without plaudits.
He is not proud or arrogant.

78

Nothing in the world is as soft and yielding as water.
Yet it wears down the hard and inflexible.
Nothing can withstand it.

The weak overcomes strong;
Gentle overcomes the rigid.
Everyone knows this but few practice it.

Therefore the Master knows
That only he who takes on the hardship of the people
Is fit to rule them.

And he who does not shield himself from common disaster
Is fit to rule a country.

True words seem paradoxical.

79

After a fight, resentment remains.
It cannot be helped.

Therefore, the Master fulfills his promises,
Corrects his own mistakes,
And has no expectations of others.

When there is honor, one does his part.
When there is no honor, one makes demands.
The Tao is everywhere,
But it rests on the good man.

80

If a country has few inhabitants,
They enjoy their work,
They don't need complicated machinery,
They love their homes,
They aren't interested in travel.

There are wagons and boats left unused,
Armor that is never worn.
People enjoy their simple ways, food, and clothes.

They live in peace with their neighbor.
Dogs bark and roosters crow and all is heard for miles away.

They are content to grow old and die without
Straying from their ways.

81

True words are not pleasing;
Pleasing words are not true.
Wise men don't need to prove their point;
Men who push their point are fools.
Those who know they could be wrong are learned.
Those who are certain they know are ignorant.
The Master has no possessions.
He hoards nothing.
The more he gives, the more he has.
The more he serves, the happier he is.
The Tao pierces the heart but does not harm.
The Tao nourishes by letting go.

By not dominating, the Master's obligation is done.

BEHOLD THE SECOND HORSEMAN

The Four Horsemen of the Apocalypse

Albrecht Durer, 1498

Compiled by Joseph B. Lumpkin

Prelude to war

THE UNIVERSAL SOLDIER
Donovan, 1968

He's five foot-two, and he's six feet-four,
He fights with missiles and with spears.

He's all of thirty-one, and he's only seventeen,
Been a soldier for a thousand years.

He's a Catholic, a Hindu, an Atheist, a Jain,
A Buddhist and a Baptist and a Jew.
And he knows he shouldn't kill,
And he knows he always will,
Kill you for me, my friend, and me for you.

And he's fighting for Democracy,
He's fighting for the Reds,
He says it's for the peace of all.
He's the one who must decide,
Who's to live and who's to die,
And he never sees the writing on the wall.

He's the Universal Soldier and he really is to blame,
His orders come from far away no more,
They come from here and there and you and me,
And brothers can't you see,
This is not the way we put the end to war.

Since the dawn of his existence, this creature, this homosapien, this fierce and violent beast has been obsessed with conflict. Like an incurable disease – an addictive drug, war has terrified him, teased and taunted him; repelled yet fascinated him. The power of conquest, the lure of sex, and the craving for wealth have fashioned the ultimate Beast of War – Man.

In his own words throughout the centuries, Man speaks to us of War – its allure and its tragedy. Come with us now and listen to the words of our own kind. Words that continue to haunt us in our nightmares.
The Editors

And I saw, and beheld a white horse: and he that sat on him had a bow; and a crown was given unto him and he went forth conquering, and to conquer. And when he had opened the second seal, I heard the second beast say, Come and see. And there went out another horse that was red: and power was given to him that sat thereon to take peace from the earth, and that they should kill one another: and

there was given unto him a great sword. And when he had opened the third seal, I heard the third beast say, Come and see. And I beheld, and lo a black horse; and he that sat on him had a pair of balances in his hand. And when he had opened the fourth seal, I heard the voice of the fourth beast say, Come and see. And I looked, and behold a pale horse: and his name that sat on him was Death, and Hell followed with him. And power was given unto them over the fourth part of the earth, to kill with sword, and with hunger, and with death, and with the beasts of the earth.
Holy Bible, Book of Revelation

I hate war as only a soldier who has lived it can, only as one who has seen its brutality, its stupidity.
Dwight D. Eisenhower

Power corrupts, and absolute power corrupts absolutely.
Lord Acton

Victor and vanquished never unite in substantial agreement.
Tacitus

War is delightful to those who have no experience of it.
Erasmus

A bad peace is even worse than war.
Tacitus

Where force is necessary, there it must be applied boldly, decisively and completely. But one must know the limitations of force; one must know when to blend force with a maneuver, a blow with an agreement.
Leon Trotsky

You can't say civilization don't advance...in every war they kill you in a new way.
Will Rogers

The human race's prospects of survival were considerably better when we were defenseless against tigers than they are today when we have become defenseless against ourselves.

Arnold J. Toynbee

We are not interested in the possibilities of defeat. They do not exist.
Queen Victoria

The body of a dead enemy always smells sweet.
Titus Flavius Vespasian

The strongest of all warriors are these two - Time and Patience.
Leo Tolstoy.

Any man may make a mistake; none but a fool will persist in it.
Cicero

Forewarned, forearmed; being prepared is half the victory.
Miguel de Cervantes

The foolish and the dead never change their opinions.
Jack Russell Lowell

All warfare is based on deception.
Sun Tzu

The power of commitment is wondrous and can transcend all other
forces.
Tak Kubota

He that is slow to anger is better than the mighty; and he that ruleth
his spirit than he that taketh a city.
Holy Bible, Book of Proverbs

Pretend to be weak, that he may grow arrogant.
Sun Tzu

Trouble is easily overcome before it starts...Deal with it before it
happens. Set things in order before there is confusion.
Lao Tzu

There is a tide in the affairs of men which, taken at the flood, leads on
to fortune; omitted, all voyage in their life is bound in shallows and

miseries. We must take the current when it serves, or lose our ventures.
William Shakespeare

Men are your castles. Men are your walls. Sympathy is your ally. Enmity is your foe.
Takeda Shingen

The warrior is not the brute. War makes them look alike. Life separates them fully.
Joseph Lumpkin

I feel I have this great creativity and spiritual force within me that is greater than faith, greater than ambition, greater than confidence, greater then determination, greater than vision. It is all of these combined. ... It is like a strong emotion mixed with faith, but a lot stronger.
Bruce Lee

In all aspects of life, relationships form the basis of everything. In all things, think with one's starting point in man.
Nabeshima Naoshige

Always forgive your enemies: nothing annoys them so much.
Oscar Wilde

No dictator, no invader, can hold an imprisoned population by force of arms forever. There is no greater power in the universe than the need for freedom. Against that power, governments and tyrants and armies cannot stand. The Centauri learned this lesson once. We will teach it to them again. Though it takes a thousand years, we will be free.
J. Michael Straczynski

I would rather die fighting than fight dying.
Kevin Grant

We will either find a way or make one.
Hannibal

It is not because things are difficult that we do not dare, it is because we do not dare that things are difficult.
Seneca

Human beings, who are almost unique in having the ability to learn from the experience of others, are also remarkable for their apparent disinclination to do so.
Douglas Noel Adams

The possession of unlimited power will make a despot of almost any man. There is a possible Nero in the gentlest human creature that walks.
Thomas Bailey

Nobody can give you freedom.
Malcolm X

Sometime they'll give a war and nobody will come.
Carl Sandburg

An army made up of creatures of impulse would be only a mob.
Ralph W. Sockman

I hate war as only a soldier who has lived it can, only as one who has seen its brutality, its futility, its stupidity.
Dwight Eisenhower

The Samurai Code:
I have no parents; I make the Heavens and the Earth my parents.
I have no home; I make the Tan T'ien my home.
I have no divine power; I make honesty my Divine Power.
I have no means; I make Docility my means.
I have no magic power; I make personality my Magic Power.
I have neither life nor death; I make A Um my Life and Death.

I have no body; I make Stoicism my Body.
I have no eyes; I make The Flash of Lightning my eyes.
I have no ears; I make Sensibility my Ears.
I have no limbs; I make Promptitude my Limbs.
I have no laws; I make Self-Protection my Laws.

I have no strategy; I make the Right to Kill and the Right to Restore Life my Strategy.

I have no designs; I make Seizing the Opportunity by the Forelock my Designs.

I have no miracles; I make Righteous Laws my Miracle.

I have no principles; I make Adaptability to all circumstances my Principle.

I have no tactics; I make Emptiness and Fullness my Tactics.

I have no talent; I make Ready Wit my Talent.

I have no friends; I make my Mind my Friend.

I have no enemy; I make Incautiousness my Enemy.

I have no armour; I make Benevolence my Armour.

I have no castle; I make Immovable Mind my Castle.

I have no sword; I make No Mind my Sword.

REVEILLE, n. A signal to sleeping soldiers to dream of battlefields no more, but get up and have their blue noses counted.
Ambrose Bierce

DRAGOON, n. A soldier who combines dash and steadiness in so equal measure that he makes his advances on foot and his retreats on horseback.
Ambrose

War is an instrument entirely inefficient toward redressing wrong; and multiplies, instead of indemnifying losses.
Thomas Jefferson

For it has been said so truthfully that it is the soldier, not the reporter, who has given us the freedom of the press. It is the soldier, not the poet, who has given us freedom of speech. It is the soldier, not the agitator, who has given us the freedom to protest. It is the soldier who salutes the flag, serves beneath the flag, whose coffin is draped by the flag, who gives that protester the freedom to abuse and burn that flag.
Zell Miller

The conquest of the earth, which mostly means the taking it away from those who have a different complexion or slightly flatter noses than ourselves, is not a pretty thing when you look into it too much.
Joseph Conrad

Great men rejoice in adversity, just as brave soldiers triumph in war
Seneca

If you will not fight for the right when you can easily win without bloodshed, if you will not fight when victory will be sure and not so costly, you may come to the moment when you will have to fight with all the odds against you and only a precarious chance of survival. There may be a worse case. You may have to fight when there is no chance of victory, because it is better to perish than to live as slaves.
Sir Winston Churchill

Right is more precious than peace.
Woodrow Wilson

War is an ugly thing, but not the ugliest of things. The decayed and degraded state of moral and patriotic feeling which thinks that nothing is worth war is much worse. The person who has nothing for which he is willing to fight, nothing which is more important than his own personal safety, is a miserable creature and has no chance of being free unless made and kept so by the exertions of better men than himself.
John Stuart Mill

All that is essential for the triumph of evil is that good men do nothing.
Edmund Burke

War is delightful to those who have had no experience of it.
Desiderius Erasmus

The God of War hates those who hesitate.
Euripides

Politics is war without bloodshed while war is politics with bloodshed.
Mao Tze-tung

Either war is obsolete or men are.
R. Buckminster Fuller

Never, never, never believe any war will be smooth and easy, or that anyone who embarks on the strange voyage can measure the tides and hurricanes he will encounter. The statesman who yields to war fever must realize that once the signal is given, he is no longer the master of policy but the slave of unforeseeable and uncontrollable events.
Sir Winston Churchill

The life of a modern soldier is ill represented by heroic fiction. War has means of destruction more formidable than the cannon and the sword.
Samuel Johnson

Politics and War

OVER THERE
World War I Ballad by George M. Cohan, 1917

Over there, over there
Send the word, send the word over there
That the Yanks are coming, the Yanks are coming,

The drums rum-tumming ev'rywhere
So prepare say a pray'r
Send the word, send the word to beware
We'll be over, we're coming over,

And we won't come back till it's over over there!

If we choose, we may hear the drums and trumpets sounding their martial calls...the jingling of bridles on the massive warhorses...the deep rumblings of mechanized vehicles...the roar of lethal weapons from the sky. Dare we ask who sent them forward? Dare we question the commands of the Imperial They? In the deepness of our souls we have known Them and fear Them; They terrify us – the ones in the palaces, in the senates, in the parliaments...because we made Them.
The Editors

We have met the enemy... and he is us.
Pogo

It belongs to human nature to hate those you have injured. It is only necessary to make war with five things: with the maladies of the body, with the ignorances of the mind, with the passions of the body, with the seditions of the city, with the discords of families.
Tacitus

Diplomats are just as essential to starting a war as soldiers are for finishing it. You take diplomacy out of war, and the thing would fall flat in a week.
Will Rogers

The time comes upon every public man when it is best for him to keep his lips closed.
Abraham Lincoln

A world without nuclear weapons would be less stable and more dangerous for all of us.
Margaret Thatcher

Of course it's the same old story. Truth usually is the same old story.
Margaret Thatcher

Vietnam presumably taught us that the United States could not serve as the world's policeman; it should also have taught us the dangers

of trying to be the world's midwife to democracy when the birth is scheduled to take place under conditions of guerrilla war.
Jeanne Kirkpatrick

You can make a throne of bayonets, but you can't sit on it for long.
Boris Yeltsin

Some of the critics viewed Vietnam as a morality play in which the wicked must be punished before the final curtain and where any attempt to salvage self-respect from the outcome compounded the wrong. I viewed it as a genuine tragedy. No one had a monopoly on anguish.
Henry Kissinger

Above all, Vietnam was a war that asked everything of a few and nothing of most in America.
Myra MacPherson

No event in American history is more misunderstood than the Vietnam War. It was misreported then, and it is mis-remembered now.
Richard M. Nixon

All the wrong people remember Vietnam. I think all the people who remember it should forget it, and all the people who forgot it should remember it.
Michael Herr

Television brought the brutality of war into the comfort of the living room. Vietnam was lost in the living rooms of America--not on the battlefields of Vietnam.
Marshall McLuhan

Power is like being a lady... if you have to tell people you are, you aren't.
Margaret Thatcher

Let us understand: North Vietnam cannot defeat or humiliate the United States. Only Americans can do that.
Richard M. Nixon

It was sheer professionalism and inspiration and the fact that you really cannot have people marching into other people's territory and staying there.
Margaret Thatcher

This war has already stretched the generation gap so wide that it threatens to pull the country apart.
Frank Church

If the Americans do not want to support us anymore, let them go, get out! Let them forget their humanitarian promises!
Nguyen Van Thieu

You have a row of dominoes set up; you knock over the first one, and what will happen to the last one is that it will go over very quickly.
Dwight D. Eisenhower

This is not a jungle war, but a struggle for freedom on every front of human activity.
Lyndon B. Johnson

We are at war with the most dangerous enemy that has ever faced mankind in his long climb from the swamp to the stars, and it has been said if we lose that war, and in so doing lose this way of freedom of ours, history will record with the greatest astonishment that those who had the most to lose did the least to prevent its happening.
Ronald Reagan

'Resource-constrained environment' are fancy Pentagon words that mean there isn't enough money to go around.
John W. Vessey, Jr.

You don't have a peaceful revolution. You don't have a turn-the-other-cheek revolution. There's no such thing as a nonviolent revolution. Revolution is bloody. Revolution is hostile. Revolution knows no compromise. Revolution overturns and destroys everything that gets in its way.
Malcom X

I have witnessed the tremendous energy of the masses. On this foundation it is possible to accomplish any task whatsoever.
Mao Tze-tung

Standing in the middle of the road is very dangerous; you get knocked down by the traffic from both sides.
Margaret Thatcher

If you want to know the taste of a pear, you must change the pear by eating it yourself. If you want to know the theory and methods of revolution, you must take part in revolution. All genuine knowledge originates in direct experience.
Mao Tze-tung

All that is necessary for evil to triumph is for good men to do nothing.
Edmund Burke

It is a melancholy fact that many of the worst laws put upon the statute-books have been put there with the best of intentions by thoroughly well-meaning people. Mere desire to do right can no more by itself make a good statesman than it can make a good general.
Theodore Roosevelt

I do not want the best to be any more the deadly enemy of the good. We climb through degrees of comparison.
Archbishop Edward White Benson

A fanatic is one who can't change his mind and won't change the subject.
Sir Winston Churchill

Generally speaking, people do not care who is in charge as long as things run well. The charisma and expertise of the leader should be used to build the organization, but things should not be tied so closely to the leader that his downfall would adversely impact the organization.
Joseph Lumpkin

You may have to fight a battle more than once to win it.
Margaret Thatcher

When you engage in actual fighting, if victory is long in coming, then men's weapons will grow dull and their ardor will be damped. If you lay siege to a town, you will exhaust your strength. Again, if the campaign is protracted, the resources of the State will not be equal to the strain. Thus, though we have heard of stupid haste in war, cleverness has never been seen associated with long delays. There is no instance of a country having benefited from prolonged warfare. In war, then, let your great object be victory, not lengthy campaigns. Thus it may be known that the leader of armies is the arbiter of the people's fate, the man on whom it depends whether the nation shall be in peace or in peril.
Sun Tzu

Every gun that is fired, every warship launched, every rocket fired, signifies a theft from those who hunger and are not fed, those who are cold but not clothed. The world in arms is not spending money alone. It is spending the sweat of its laborers, the genius of its scientists, the hopes of its children.
Dwight D. Eisenhower

I have come to the conclusion that politics are too serious a matter to be left to the politicians.
General Charles De Gaulle

Politics is the gentle art of getting votes from the poor and campaign funds from the rich by promising to protect each from the other.
Oscar Ameringer

Patriotism is your conviction that this country is superior to all others because you were born in it.
George Bernhard Shaw

I'm extraordinarily patient provided I get my own way in the end.
Margaret Thatcher

A diplomat is someone who can tell you to go to hell in such a way that you look forward to the trip.
Cashie Stinnett

The real problem is what to do with the problem solvers after the problems are solved.
G. Talese

If you can't convince them, confuse them.
Harry S. Truman

The empires of the future will be the empires of the mind.
Sir Winston Churchill

In politics there is no honour.
Benjamin Disraeli

Few things are as immutable as the addiction of political groups to the ideas by which they have once won office.
John Kenneth Galbraith

Since a politician never believes what he says, he is surprised when others believe him.
Charles de Gaulle

When a politician changes his position it's sometimes hard to tell whether he has seen the light or felt the heat.
Robert Fuoss

As a general rule the most successful man in life is the man who has the best information.
Benjamin Disraeli

In war, truth is so precious she must always be escorted by a bodyguard of lies.
Sir Winston Churchill

Among the calamities of war may be justly numbered the diminution of the love of truth, by the falsehoods which interest dictates and credulity encourages.

Joseph Lumpkin

Samuel Johnson

Justice is what you get when you run out of money.
H.L. Mencken

Never in the field of human suffering, was so much made, by so few, from so many.
Sir Murray Rivers QC (Bryan Dawe)

An appeaser is one who feeds a crocodile - hoping that it will eat him last.
Sir Winston Spencer Churchill

It is better to have a lion at the head of an army of sheep than a sheep at the head of an army of lions.
Daniel Defoe

You cannot shake hands with a clenched fist.
Indira Gandhi

If error is corrected whenever it is recognized as such, the path of error is the path of truth.
Hans Reichenbach

I know that my unity with all people cannot be destroyed by national boundaries and government orders.
Leo Tolstoy

Where the willingness is great, the difficulties cannot be great.
Niccolò Machiavelli

In the councils of government, we must guard against the acquisition of unwarranted influence, whether sought or unsought, by the military-industrial complex. The potential for the disastrous rise of misplaced power exists and will persist.
Dwight D. Eisenhower

Because just as good morals, if they are to be maintained, have need of the laws, so the laws, if they are to be observed, have need of good morals.

Niccolò Machiavelli

The greatest blunders, like the thickest ropes, are often compounded of a multitude of strands. Take the rope apart, separate it into the small threads that compose it, and you can break them one by one. You think, 'That is all there was!' But twist them all together and you have something tremendous.
Victor Hugo

I never did give anybody hell. I just told the truth, and they thought it was hell.
Harry S Truman

How can you govern a country which has 246 varieties of cheese?
Charles De Gaulle

We live in a Newtonian world of Einsteinian physics ruled by Frankenstein logic.
David Russell

A single death is a tragedy; a million deaths is a statistic.
Joseph Stalin

I am not bound to win, but I am bound to be true. I am not bound to succeed, but I am bound to live by the light that I have. I must stand with anybody that stands right, and stand with him while he is right, and part with him when he goes wrong.
Abraham Lincoln

I can make more generals, but horses cost money.
Abraham Lincoln

It is a melancholy fact that many of the worst laws put upon the statute-books have been put there with the best of intentions by thoroughly well-meaning people. Mere desire to do right can no more by itself make a good statesman than it can make a good general.
Theodore Roosevelt

It is often easier to fight for one's principles that to live up to them.

Joseph Lumpkin

Adlai Stevenson

If you do not tell the truth about yourself, you cannot tell it about other people.
Virginia Woolf

The foolish and the dead never change their opinions.
Jack Russell Lowell

To err is human, but when the eraser wears out ahead of the pencil, you're overdoing it.
Josh Jenkins

Life can only be understood backwards; but it must be lived forwards.
Soren Kiekegaad

A diplomat is someone who can tell you to go to hell in such a way that you look forward to the trip.
Cashie Stinnett

The real problem is what to do with the problem solvers after the problems are solved.
G. Talese

If you can't convince them, confuse them.
Harry S. Truman

If it ain't broke, break it.
Richard Pascale

Oppose, adapt, adopt.
Benjamin Disraeli

There can be no justice so long as rules are absolute.
Patrick Stewart

The great masses of people will more easily fall victims to a big lie than to a small one, especially if it is repeated over and over.
Adolph Hitler

The lie can be maintained only for such time as the State can shield the people from the political, economic and/or military consequences of the lie. It thus becomes vitally important for the State to use all of its powers to repress dissent, for the truth is the mortal enemy of the lie, and thus by extension, the truth becomes the greatest enemy of the State.
Dr. Joseph M. Goebbels

When a politician changes his position it's sometimes hard to tell whether he has seen the light or felt the heat.
Robert Fuoss

In our time, political speech and writing are largely the defense of the indefensible.
George Orwell

I know what a statesman is. He's a dead politician. We need more statesmen.
Robert C. Edwards

Few people think more than two or three times a year; I have made an international reputation for myself by thinking once or twice a week.
George Bernard Shaw

Bad administration, to be sure, can destroy good policy; but good administration can never save bad policy.
Adlai Stevenson

HISTORY, An account mostly false, of events mostly unimportant, which are brought about by rulers, mostly knaves, and soldiers, mostly fools.
Ambrose Bierce

I like to believe that people in the long run are going to do more to promote peace than our governments. Indeed, I think that people want peace so much that one of these days governments had better get out of their way and let them have it.
Dwight D. Eisenhower

Joseph Lumpkin

The more corrupt the republic, the more numerous the laws.
Tacitus

In the councils of government, we must guard against the acquisition of unwarranted influence, whether sought or unsought, by the military-industrial complex.
Dwight D. Eisenhower

Diplomats are just as essential in starting a war as soldiers are in finishing it. . . . You take diplomacy out of war, and the thing would fall flat in a week.
Will Rogers

There is nothing that surpasses ruling with benevolence. However, to put into practice enough benevolent governing to rule the country is difficult. To do this lukewarmly will result in neglect. If governing with benevolence is difficult, then it is best to govern strictly. To govern strictly means to be strict before things have arisen, and to do things in such a way that evil will not arise. To be strict after the evil has arisen is like laying a snare. There are few people who will make mistakes with fire after having once been burned. Of people who regard water lightly, many have been drowned.
Tzu Ch'an

If you are able to vote, then do so. There may be no candidates or issues you want to vote for... but there will certainly be someone or something to vote against. In case of doubt, vote against. By this rule you will rarely go wrong.
Lazarus Long

Patriotism means to stand by the country. It does not mean to stand by the President or any other public official save exactly to the degree in which he himself stands by the country. It is patriotic to support him insofar as he efficiently serves the country. It is unpatriotic not to oppose him to the exact extent that by inefficiency or otherwise he fails in his duty to stand by the country.
Theodore Roosevelt

That kind of patriotism which consists in hating all other nations.

Elizabeth Gaskell

Are you a politician who says to himself: I will use my country for my own benefit? Or are you a devoted patriot, who whispers in the ear of his inner self: I love to serve my country as a faithful servant.?
Kahlil Gibran

Each man must for himself alone decide what is right and what is wrong, which course is patriotic and which isn't. You cannot shirk this and be a man.
Thomas Tusser

In time of war the first casualty is truth.
Boake Carter

War is mainly a catalogue of blunders.
Sir Winston Churchill

In war, as in life, it is often necessary, when some cherished scheme has failed, to take up the best alternative open, and if so, it is folly not to work for it with all your might.
Sir Winston Churchill

People never lie so much as after a hunt, during a war or before an election.
Otto von Bismarck

The whole aim of practical politics is to keep the populace alarmed (and hence clamorous to be led to safety) by menacing it with an endless series of hobgoblins, all of them imaginary.
H.L. Mencken

Conceit, arrogance and egotism are the essentials of patriotism. Patriotism assumes that our globe is divided into little spots, each one surrounded by an iron gate. Those who had the fortune of being born on some particular spot, consider themselves better, nobler, grander, more intelligent than the living beings inhabiting any other spot. It is, therefore, the duty of everyone living on that chosen spot to fight, kill, and die in the attempt to impose his superiority upon all others.

Joseph Lumpkin

Emma Goldman

When a whole nation is roaring patriotism at the top of its voice, I am fain to explore the cleanness of its hands and the purity of its heart.
Ralph Waldo Emerson

Herein lies a riddle: How can a people so gifted by God become so seduced by naked power, so greedy for money, so addicted to violence, so slavish before mediocre and treacherous leadership, so paranoid, deluded, lunatic?
Philip Berrigan

If the Nuremberg laws were applied, then every post-war American president would have been hanged.
Noam Chomsky

Praising our leaders, we're getting in tune with the music played by the madmen.
Alphaville

Authoritarian government required to speak, is silent. Representative government required to speak, lies with impunity.
Napoleon Bonaparte

Few of us can easily surrender our belief that society must somehow make sense. The thought that The State has lost its mind and is punishing so many innocent people is intolerable. And so the evidence has to be internally denied.
Arthur Miller

There exists a shadowy Government with its own Air Force, its own Navy, its own fundraising mechanism, and the ability to pursue its own ideas of national interest, free from all checks and balances, and free from the law itself.
Senator Daniel K. Inouye

The people can have anything they want. The trouble is, they do not want anything. At least they vote that way on election day.
Eugene Debs

Why of course the people don't want war. Why should some poor slob on a farm want to risk his life in a war when the best he can get out of it is to come back to his farm in one piece? Naturally, the common people don't want war: neither in Russia, nor in England, nor for that matter in Germany. That is understood. But after all it is the leaders of the country who determine the policy, and it is always a simple matter to drag the people along, whether it is a democracy, or a fascist dictatorship, or a parliament, or a communist dictatorship. Voice or no voice, the people can always be brought to the bidding of the leaders. That is easy. All you have to do is to tell them they are being attacked, and denounce the pacifists for lack of patriotism and exposing the country to danger.
Hermann Goering

It is the duty of the patriot to protect his country from its government.
Thomas Paine

Truth will do well enough if left to shift for herself. She has no need of force to procure entrance into the minds of men.
Thomas Jefferson

It does not require a majority to prevail, but rather an irate, tireless minority keen to set brush fires in people's minds.
Samuel Adams

Never give in, never, never, never, never, in nothing great or small, large or petty, never give in except to convictions of honour and good sense. Never yield to force ... never yield to the apparently overwhelming might of the enemy.
Sir Winston Churchill

We shall fight on the beaches. We shall fight on the landing grounds. We shall fight in the fields, and in the streets, we shall fight in the hills. We shall never surrender!
Sir Winston Churchill

History does not long entrust the care of freedom to the weak or timid.
Dwight Eisenhower

No man is entitled to the blessings of freedom unless he be vigilant in its preservation.
General Douglas MacArthur

The political object is the goal, war is the means of reaching it, and the means can never be considered in isolation from their purposes.
Karl von Clausewitz

Diplomats are just as essential in starting a war as soldiers are in finishing it.
Will Rogers

The only power tyrants have, is the power relinquished to them by their victims.
Ettiene de la Boetie

With reasonable men I will reason; with humane men I will plead; but to tyrants I will give no quarter, nor waste arguments, where they will certainly be lost.
William Lloyd Garrison

War is a game that is played with a smile. If you can't smile, grin. If you can't grin, keep out of the way till you can.
Sir Winston Churchill

One should never allow chaos to develop in order to avoid going to war, because one does not avoid a war but instead puts it off to his disadvantage.
Machiavelli

Laws are inoperative in war
Cicero

War is the continuation of politics by other means.
Karl von Clausewitz

Controlled, universal disarmament is the imperative of our time. The demand for it by the hundreds of millions whose chief concern is the long future of themselves and their children will, I hope, become so

universal and so insistent that no man, no government anywhere, can withstand it.
Dwight D. Eisenhower

The ballot is stronger than the bullet.
Abraham Lincoln

Business and War

JOHNNY HAS GONE FOR A SOLDIER
A Revolutionary War Song

O Johnny dear has gone away
He has gone afar across the bay,

O my heart is sad and weary today,
Johnny has gone for a soldier.

In the boardrooms and the battlefields, in flying machines at 30,000 feet...the art of war has always been the same.

The weapons are equally deadly. The rule of no rule applies. Strength, power, poise, deception, and the luck of the moment determine the outcome of the battle...and the victorious always writes the history.
The Editors

But war is not the whole business of life; it happens but seldom, and every man, either good or wise, wishes that its frequency were still

less. That conduct which betrays designs of future hostility, if it does not excite violence, will always generate malignity; it must forever exclude confidence and friendship, and continue a cold and sluggish rivalry, by a sly reciprocation of indirect injuries, without the bravery of war or the security of peace.
Samuel Johnson

Motivation is everything. You can do the work of two people, but you can't be two people. Instead, you have to inspire the next guy down the line and get him to inspire his people.
Lee Iacocca

If you want to succeed, you should strike out on new paths rather than travel the worn paths of accepted success.
John D. Rockefeller, Jr.

Eagles don't flock -- You have to find them one at a time.
H Ross Perot

I'm a deeply superficial person.
Andy Warhol

If you are going to try to go to war, or to prepare for war, in a capitalist country, you have got to let business make money out of the process or business won't work.
Henry Lewis Stimson

If you have ideas, you have the main asset you need, and there isn't any limit to what you can do with your business and your life. They are any man's greatest asset -- IDEAS.
Harvey S. Firestone

Nothing is illegal if a hundred businessmen decide to do it, and that's true anywhere in the world.
Andrew Young

So long as war is the main business of nations, temporary despotism --despotism during the campaign --is indispensable.
Walter Bagehot

When two men in business always agree, one of them is unnecessary.
W. Wrigley Jr.

Making a killing – the business of war.
Phillip Van Niekerk

The only competition worthy of a wise man is with himself.
Washington Allston

You can get much farther with a kind word and a gun than you can with a kind word alone.
Al Capone

The secret of success is to know something nobody else knows.
Aristotle Onassis

You can't do business sitting on your arse.
Lord MacLaurin

If it ain't broke, break it.
Richard Pascale

It is better to be defeated on principle than to win on lies.
Arthur Calwell

The rule of business is how fast you can get your idea to market. Those whose systems do not allow them to move quickly are doomed.
Ken Tuchman

The art of giving advice is to make the recipient believe he thought of it himself.
Frank Tyger

Dreaming is zero value. I mean, anyone can dream.
Bill Gates

He who never fell, never climbed.
Anon

I dream of a company where people come to work every day in a rush to try something they woke up thinking about the night before. We want them to go home from work wanting to talk about what they did that day, rather than trying to forget it. We want factories where the whistle blows and everybody wonders where the time went, and then somebody suddenly wonders aloud why we need a whistle. We want a company where people find a better way, everyday, of doing things, and where by shaping their own work experience, they make their lives better and their company the best.
Jack Welch

Make three correct guesses consecutively and you will establish a reputation as an expert.
Laurence J Peter

Success depends on your backbone, not your wishbone.
Anon

Obstacles are things a person sees when he takes his eyes off his goal.
E. Joseph Cossman

When there is an original sound in the world, it makes a hundred echoes.
John Shedd

Leaders get out in front and stay there by raising the standards by which they judge themselves - and by which they are willing to be judged.
Fredrick Smith

In any great organization it is far, far safer to be wrong with the majority than to be right alone.
John Kenneth Galbraith

There are two kinds of people, those who do the work and those who take the credit. Try to be in the first group; there is less competition there.
Indira Gandhi

Education's purpose is to replace an empty mind with an open one.

Malcolm S. Forbes

It is not the employer who pays the wages. Employers only handle the money. It is the customer who pays the wages.
Henry Ford

The test of a first-rate intelligence is the ability to hold two opposing ideas in mind at the same time and still retain the ability to function.
F. Scott Fitzgerald

Risk-taking is the essence of innovation.
Herman Kahn

Better ask questions twice than lose your way once.
Danish Proverb

People who are just in it for the money - they usually fail.
Robert Holmes A Court.

Thinking is the hardest work there is, which is probably the reason why so few engage in it.
Henry Ford

If you're not confused about the current state of the economy then you clearly do not understand what is going on.
Dr Chris Caton

Deciding what to do is easy, deciding what not to do is hard.
Michael Dell

To every man, every day, will come one valuable thought.
Thomas Edison

Systems allow us to apply the best thinking and give us a benchmark against which to measure and evaluate future ideas.
Alan Patching
Imagination is more important than knowledge.
Albert Einstein

Remarkable people in all fields of endeavor move the world forward
- they never give up.
Kevin Gosper

Once you have passionately sold the core values of your organization
to your people, they have a very simple choice, and that is to be an
Ambassador or an Assassin of those values.
Geoff Burch

Only a radically new kind of creativity will keep you and your
organization up there with the best.
Dr Kobus Neethling

Success is the maximum utilization of the ability that you have.
Zig Zilglar

Everyone is a genius at least once a year; a real genius has his
original ideas closer together.
Georg Lichtenberg

I have not failed. I've just found 10,000 ways that won't work.
Thomas Alva Edison

Not everything that can be counted counts, and not everything that
counts can be counted.
Albert Einstein

The race for market share is a race against time, not against
competitors. Good ideas sooner always beat good ideas eventually.
Today's marketplace is more concerned about when than who.
Bruce Haddon

Successful is the person who has lived well, laughed often and loved
much, who has gained the respect of children, who leaves the world
better than they found it, who has never lacked appreciation for the
earth's beauty, who never fails to look for the best in others or give
the best of themselves.
Ralph Waldo Emerson

Age is only a number, a cipher for the records. A man can't retire his experience. He must use it. Experience achieves more with less energy and time.
Bernard Baruch

Whenever you see a successful business, someone once made a courageous decision.
Peter F. Drucker

When you reduce a complex message to something customers can understand, you're not dumbing it down. You're smartening it up.
Bruce Haddon

When two men in business always agree, one of them is unnecessary.
W. Wrigley

The only competition worthy of a wise man is with himself.
Washington Allston

If you haven't had a failure in this business, you haven't been around long enough.
Sandra Levy

As a general rule the most successful man in life is the man who has the best information.
Benjamin Disraeli

In a competitive world so much importance is put on winning at any cost that we sometimes forget that honesty, decency and integrity are the ultimate victors in both business and life.
Bryce Courtenay

To manage a business successfully requires as much courage as that possessed by the soldier who goes to war. Business courage is the more natural because all the benefits which the public has in material wealth come from it.
Charles F. Abbott

I believe that banking institutions are more dangerous to our liberties than standing armies if the American people ever allow private

banks to control the issue of currency...the banks and corporations that will grow up around them will deprive the people of all property until their children will wake up homeless on the continent that their fathers conquered.
Thomas Jefferson

Such as it is, the press has become the greatest power within the Western World, more powerful than the legislature, the executive and judiciary. One would like to ask: by whom has it been elected, and to whom is it responsible?
Alexander Solzhenitsyn

Do not fear the enemy, for your enemy can only take your life. It is far better that you fear the media, for they will steal your HONOR. That awful power, the public opinion of a nation, is created in America by a horde of ignorant, self-complacent simpletons who failed at ditching and shoemaking and fetched up in journalism on their way to the poorhouse.
MarkTwain

Business is war. It is a battle to reach your objectives. It is a battle to keep your key talent from leaving and "upgrading" your competitors. Above all, it is a battle to dominate your competition.
Greg Langston

Business is war...war is business.
Carl von Clausewitz

The Philosophy of War

WE ARE TENTING TONIGHT
Civil War Ballad by Walter Kittredge, 1864

Many are the hearts that are weary tonight,

Wishing for the war to cease;

Many are the hearts looking for the right
To see the dawn of peace.

Tenting tonight, Tenting tonight,.

Tenting on the old Campground.

Great warrior! Wars not make one great!
Yoda, The Empire Strikes Back

Four things come not back:
The spoken word,
The sped arrow,
The past life,
The neglected opportunity.
Arabian Proverb

I have a dream that one day this nation will rise up and live out the true meaning of its creed: 'We hold these truths to be self-evident that all men are created equal'.
Martin Luther King Jr.

A house divided against itself cannot stand. I believe this government cannot endure permanently half slave and half free.
Abraham Lincoln

Europe was created by history. America was created by philosophy.
Margaret Thatcher

The end may justify the means as long as there is something that justifies the end.
Leon Trotsky

The historic ascent of humanity, taken as a whole, may be summarized as a succession of victories of consciousness over blind forces - in nature, in society, in man himself.
Leon Trotsky

There are no absolute rules of conduct, either in peace or war. Everything depends on circumstances.
Leon Trotsky

There is a limit to the application of democratic methods. You can inquire of all the passengers as to what type of car they like to ride in, but it is impossible to question them as to whether to apply the brakes when the train is at full speed and accident threatens.
Leon Trotsky

Believe that we too love freedom and desire it. To us it is more desirable than anything in the world. If you strike us down now, we shall rise again and renew the fight. You cannot conquer Ireland; you cannot extinguish the Irish passion for freedom; if our deed has not been sufficient to win freedom then our children will win it with a better deed.
Padraig Pearse

The fierce pulsation of resurgent pride that disclaims servitude may one day cease to throb in the heart of Ireland - but the heart of Ireland will that day be dead. While Ireland lives, the brain and the brawn of her manhood will strive to destroy the last vestige of British rule in her territory.
Thomas MacDonagh

This is the beginning, our fight has saved Ireland. The soldiers of tomorrow will finish the task. our deed has not been sufficient to win freedom then our children will win it with a better deed.
Thomas Clarke

I leave for the guidance of other revolutionaries, who may tread the path which I have trod, this advice; never treat with the enemy, never to surrender to his mercy, but to fight to a finish.
Eamonn Ceannt

A handful of men, inured to war, proceed to certain victory, while on the contrary, numerous armies of raw and undisciplined troops are but multitudes of men dragged to the slaughter.
Flavius Renatus Vegetius

He who puts out his hand to stop the wheel of history will have his fingers crushed.
Lech Walesa

History is made at night. Character is what you are in the dark.
Lord John Whorfin

Classes struggle, some classes triumph, others are eliminated. Such is history; such is the history of civilization for thousands of years.
Mao Tze-tung

A man who has committed a mistake and doesn't correct it is committing another mistake
Confucius

When valor preys on reason, it eats the sword it fights with.
William Shakespeare

He who will not apply new remedies must expect old evils.
Sir Francis Bacon

The control of a large force is the same principle as the control of a few men: it is merely a question of dividing up their numbers. Fighting with a large army under your command is nowise different from fighting with a small one: it is merely a question of instituting signs and signals.
Sun Tzu

If the teacher is not respected and the students not cared for, confusion will arise, however clever one is. This is the crux of mystery.
Lao Tzu

The principle on which to manage an army is to set up one standard of courage which all must reach.

Joseph Lumpkin

Sun Tzu

All men come to he who keeps unity. For there lie rest, happiness, and peace. Lao Tzu

The humble is the root of nobility. Low is the foundation of high. Princes and lords consider themselves orphaned, widowed, and worthless. Do they not depend on being humble? Too much success is not an advantage. Do not tinkle like jade or clatter like a stone chime.
Lao Tzu

And if we are able thus to attack an inferior force with a superior one, our opponents will be in dire straits.
Sun Tzu

When invading hostile territory, the general principle is, that penetrating deeply brings cohesion; penetrating but a short way means dispersion.
Sun Tzu

That the impact of your army may be like a grindstone dashed against an egg--this is effected by the science of weak points and strong. In all fighting, the direct method may be used for joining battle, but indirect methods will be needed in order to secure victory.
Sun Tzu

You may advance and be absolutely irresistible, if you make for the enemy's weak points; you may retire and be safe from pursuit if your movements are more rapid than those of the enemy.
Sun Tzu

Rapidity is the essence of war: take advantage of the enemy's unreadiness, make your way by unexpected routes, and attack unguarded spots.
Sun Tzu

Disciplined and calm, to await the appearance of disorder and hubbub amongst the enemy - this is the art of retaining self-possession.

Sun Tzu

Whoever is first in the field and awaits the coming of the enemy, will be fresh for the fight; whoever is second in the field and has to hasten to battle will arrive exhausted.
Sun Tzu

Patience is not passive. Patience is concentrated strength.
Bruce Lee

One must not be negligent of learning. Lun Yu says, to study and not to think is darkness. To think without study is dangerous.
Takeda Nobushige

If you sit, sit. If you stand, stand but never wobble.
Master Ummon

The art of war teaches us to rely not on the likelihood of the enemy's not coming, but on our own readiness to receive him; not on the chance of his not attacking, but rather on the fact that we have made our position unassailable.
Sun Tzu

Conquering evil, not the opponent, is the essence of swordsmanship.
Yagyu Munenori

In connection with military matters, one must never say what can absolutely not be done. By this, the limitations of one's heart will be exposed.
Asakura Norikage

Conquer the self and you will conquer the opponent.
Takuan Soho

Argue for your limitations, and sure enough, they are yours.
Richard Bach

It's a difficult thing to truly know your own limits and points of weakness.
Hagakure

He who is aware of his own weakness will remain master of himself in any situation.
Gichin Funakoshi

The mind of the warrior remains focused on his own mortality. The sacredness and brevity of life is always in his thoughts. Life is lived to the fullness, moment by moment, when the possibility of death is realized. The warrior's heart is not reserved for those who do battle with others, but is kept secret for those who battle themselves and their own limitations.
Joseph Lumpkin

Take the arrow in your forehead, but never in your back.
HwaRang maxim

There is no such thing as an effective segment of totality.
Bruce Lee

Whenever you meet difficult situations dash forward bravely and joyfully.
Tsunetomo Yamamoto

DEU 20:3 - 4 And shall say unto them, Hear, O Israel, ye approach this day unto battle against your enemies: let not your hearts faint, fear not, and do not tremble, neither be ye terrified because of them; For the LORD your God is he that goeth with you, to fight for you against your enemies, to save you.

Knowing is not enough; we must apply. Willing is not enough; we must do.
Bruce Lee

The man whose profession is arms (fighting) should calm his own spirit and look into the depths of others. Doing so is likely the best of the martial arts.
Shiba Yoshimasa

Yield and overcome. Bend and be straight. Empty and be filled. Wear out and become new. Have little and gain. Have much and be confused.
Lao Tsu

If a man becomes alienated from his friends, he should make endeavors in the way of humanity.... One should not turn his back on reproof.
Takeda Nobushige

Knowing your ignorance is strength. Ignoring knowledge is sickness. When one becomes sick of sickness he is no longer sick.
Lao Tzu

It is hardly necessary to record that both learning and the military arts are the Way of the Warrior, for it is an ancient law that one should have Learning on the left and martial arts on the right.
Hojo Nagauji

The martial arts consider intelligence most important because intelligence involves the ability to plan and to know when to change effectively.
Sun Tzu

The sage takes care of all men, and abandons no one. He takes care of all things and abandons nothing. This is called following the light.
Lao Tzu

I have heard that when a man has literary business, he will always take military preparations; and when he has military business, he will always take literary preparations.
Confucius

Without knowledge of Learning, one will have no military victories.
Imagawa Sadayo

Unhappy is the fate of one who tries to win his battles and succeed in his attacks without cultivating the spirit of enterprise; for the result is a waste of time and general stagnation. Hence the saying: The

enlightened ruler lays his plans well ahead; the good general cultivates his resources.
Sun Tzu

A goal is not always meant to be reached, it often serves simply as something to aim at.
Bruce Lee

Achieve results, but never glory in them. Achieve results, but do not boast. Achieve results, but do not be proud. Achieve results, because it is the natural way. Achieve results, but not through violence.
Lao Tzu

What others teach, I also teach; that is, A violent man will die a violent death. This is the essence of my teaching.
Lao Tsu

Train. An unpolished crystal does not shine; an undisciplined Samurai does not have brilliance. A Samurai therefore should cultivate his mind.
Anonymous

To invalidate the opponent's expectation, a person must know others and know the self.
Sun Tzu

The victory of a military force is determined by the opponent and his reactions.
Sun Tzu

Water shapes its course according to the nature of the ground over which it flows; the soldier works out his victory in relation to the foe who he is facing. Therefore, just as water retains no constant shape, so in warfare there are no constant conditions. He who can modify his tactics in relation to his opponent and thereby succeed in winning, may be called a heaven-born captain.
Sun Tzu

The art of war is to avoid big battles.
Sun Tzu

Experience is the name that everyone gives to his mistakes.
Oscar Wilde.

Being defeated is often a temporary condition. Giving up is what makes it permanent.
Marilyn Vos Savant

Education is when you read the fine print. Experience is what you get if you don't.
Pete Seeger

The breakfast of champions is not cereal, it's the opposition.
Nick Seitz

It is the eternal struggle between these two principles - right and wrong. They are the two principles that have stood face to face from the beginning of time and will ever continue to struggle. It is the same spirit that says, You work and toil and earn bread, and I'll eat it.
Abraham Lincoln

Do not do unto others as you would they should do unto you. Their tastes may not be the same.
George Bernard Shaw

We are most nearly ourselves when we achieve the seriousness of the child at play.
Heraclitus

The brighter you are, the more you have to learn.
Don Herold

I am free of all prejudices. I hate everyone equally.
W.C. Fields

In the midst of great joy do not promise anyone anything. In the midst of great anger do not answer anyone's letter.
Chinese Proverb

A smooth sea never made a skilful mariner.

English Proverb

Consistency is the hobgoblin of small minds.
J Frank Dobie

Insanity in individuals is something rare - but in groups, parties, nations and epochs, it is the rule.
Friedrich Nietzsche

Everybody sets out to do something, and everybody does something, but no one does what he sets out to do.
George Moore

To achieve great things, we must live as though we are never going to die.
Luc de Clarnes Vauvenargues

Optimism is the faith that leads to achievement. Nothing can be done without hope.
Helen Keller

Happiness is not a possession to be prized. It is a quality of thought, a state of mind.
Daphne du Maurier

It is better to deserve honours and not have them, than to have them and not deserve them.
Mark Twain

Cherish your visions and your dreams as they are the children of your soul; the blueprints of your ultimate achievements.
Napoleon Hill

Glory is fleeting, but obscurity is forever.
Napoleon Bonaparte

Don't be so humble - you are not that great.
Golda Meir

People demand freedom of speech to make up for the freedom of thought which they avoid.
Soren Kierkegaard

Jokes of the proper kind, properly told, can do more to enlighten questions of politics, philosophy, and literature than any number of dull arguments.
Isaac Asimov

Doing easily what others find difficult is talent; doing what is impossible for talent is genius.
Henri-Frédéric Amiel

It is the mark of an educated mind to be able to entertain a thought without accepting it.
Aristotle

All things that are truly great are at first thought impossible.
Friedrich Nietzsche

Dare to believe only in yourself.
Friedrich Nietzsche

It is hard enough to remember my opinions, without also remembering my reasons for them!
Friedrich Nietzsche

Man is the cruelest animal.
Friedrich Nietzsche

I would never die for my beliefs because I might be wrong.
Bertrand Russell

Science is what you know, philosophy is what you don't know.
Bertrand Russell

All knowledge, we feel, must be built up upon our instinctive beliefs; and if these are rejected, nothing is left.
Bertrand Russell

We think in generalities, but we live in detail.
Alfred North Whitehead

A man who has committed a mistake and doesn't correct it is committing another mistake.
Confucius

All that is necessary for evil to triumph is for good men to do nothing.
Edmund Burke

Any man may make a mistake; none but a fool will persist in it.
Cicero

Each morning puts a man on trial and each evening passes judgment.
Ray L. Smith

It is better to be defeated on principle than to win on lies.
Arthur Caldwell

When there is an original sound in the world, it makes a hundred echoes.
John Shedd

If error is corrected whenever it is recognized as such, the path of error is the path of truth.
Hans Reichenbach

How shall we rank thee upon glory's page, Thou more than soldier, and just less than sage?
Thomas Moore

The first who was king was a fortunate soldier: Who serves his country well has no need of ancestors.
Voltaire

What can they see in the longest kingly line in Europe, save that it runs back to a successful soldier?
Sir Walter Scott

Shall I ask the brave soldier who fights by my side in the cause of mankind, if our creeds agree?
Charles Lamb

The schoolmaster is abroad, and I trust to him, armed with his primer, against the soldier in full military array.
Henry Peter Brougham

Although too much of a soldier among sovereigns, no one could claim with better right to be a sovereign among soldiers.
Sir Walter Scott

In war, there are no unwounded soldiers.
Jose Narosky

The Way lies at hand yet it is sought afar off; the thing lies in the easy yet it is sought in the difficult.
Mencius

The dance of battle is always played to the same impatient rhythm. What begins in a surge of violent motion is always reduced to the perfectly still.
Sun Tzu

The undisturbed mind is like the calm body water reflecting the brilliance of the moon. Empty the mind and you will realize the undisturbed mind.
Yagyu Jubei

You might as well stand and fight because if you run, you will only die tired.
Vern Jocque - Sei Shin Kan.

Am I not destroying my enemies when I make friends of them?
Abraham Lincoln

It is easy to kill someone with a slash of a sword. It is hard to be impossible for others to cut down.
Yagyu Munenori

Mental bearing (calmness), not skill, is the sign of a matured samurai. A Samurai therefore should neither be pompous nor arrogant.
Sukahara Bokuden.

One finds life through conquering the fear of death within one's mind. Empty the mind of all forms of attachment, make a go-for-broke charge and conquer the opponent with one decisive slash.
Togo Shigekata.

It does not matter how slowly you go so long as you do not stop.
Confucius

I have a high art, I hurt with cruelty those who would damage me.
Archilocus

You must concentrate upon and consecrate yourself wholly to each day, as though a fire were raging in your hair.
Taisen Deshimaru

Given enough time, any man may master the physical. With enough knowledge, any man may become wise. It is the true warrior who can master both....and surpass the result.
Tien T'ai

Act like a man of thought - Think like a man of action.
Thomas Mann

Civilize the mind but make savage the body.
Mao Tze-tung

The belief in the possibility of a short decisive war appears to be one of the most ancient and dangerous of human illusions.
Robert Lynd

The violence of war admits no distinction; the lance, that is lifted at guilt and power, will sometimes fall on innocence and gentleness.
Samuel Johnson

I think that, as life is action and passion, it is required of a man that he should share the passion and action of his time at peril of being judged not to have lived.
Oliver Wendell Holmes Jr.

True knowledge is to experience the inner self, but since the inner being is unique to every individual, knowledge cannot be assimilated by talking about it.
Theun Mares

Knowledge gained from someone else lacks the confidence necessary to implement that knowledge. Confidence is cultivated only through practice.
Theun Mares

No man can trade upon the knowledge of another person, simply because the only knowledge which we can use with certainty is that knowledge which we have acquired through personal experience. Any person can read the theory entailed in walking a tightrope, but such theory will not keep him from falling off should he try to walk a rope. Only through repeated practice, and through trial and error, does he finally manage to walk on the rope without falling off.
Theun Mares

As we have already seen, at the end of the day it does not matter how many battles we have won or how many we have lost, as the only thing of importance is whether or not we fought and, if we did, how well we fought. Did we run from a battle because of fear, or did we fight bravely, giving it our all?
Theun Mares

Explanations are not reality – only a makeshift arrangement of the world.
Theun Mares

Confusion is a willfully induced state of mind. We can enter or exit it at will. Man deliberately confuses himself in order to plead ignorance.
Theun Mares

Confusion is a most convenient escapism used by man whenever he has to face something that frightens him, or that he does not like. However, we are always fully aware of what we are doing, even though we may choose not to acknowledge our true motives.
Theun Mares

The warrior, knowing that there is nothing to understand, acknowledges a barrier when he comes to it, and then jumps over it.
Theun Mares

When the warrior encounters a problem in his life he puts his mind at rest by acknowledging it for the obstacle it is, but instead of getting caught up in rationalizations in an effort to understand the problem, he simply tackles it immediately. Problems in themselves have no value other than to make us emotionally stronger, mentally more agile and spiritually wiser.
Theun Mares

Should a warrior feel the need to be comforted, he simply chooses anyone or anything, be it a friend, dog, or mountain, to whom he expresses his innermost feelings. It does not matter to the warrior if he is not answered, or if he is not heard, because the warrior not seeking to be understood or helped – by verbalizing his feelings, he is merely releasing the pressure of his battle.

The hunter is intimately familiar with his world, yet remains detached from it.
Theun Mares

Ultimately, you must forget about technique. The further you progress, the fewer teachings there are. The Great Path is really NO PATH.
Ueshiba Morihei

In the beginners mind there are many possibilities, but in the expert's mind there are few.
Suzuki

To practice Zen or the Martial Arts, you must live intensely, wholeheartedly, without reserve - as if you might die in the next instant.
Taisen Deshimaru

Don't think dishonestly
The Way is in training
Become acquainted with every art
Know the ways of all professions
Distinguish between gain and loss
Develop intuitive judgment and understanding for everything
Perceive those things which cannot be seen
Pay attention even to trifles
Do nothing which is of no use.
Mayomoto Musashi

Empty your mind,
Be formless, shapeless, like water.
Now you put water into a cup, it becomes the cup.
You put water into a bottle, it becomes the bottle.
You put water into a teapot, it becomes the teapot.
Now water can flow, or it can crash,
Be water my friend.
Bruce Lee

The consciousness of self is the greatest hindrance to the proper execution of all physical action.
Bruce Lee

Victory goes to the one who has no thought of himself.
Shinkage School of Swordsmanship

It is truly regrettable that a person will treat a man who is valuable to him well, and a man who is worthless to him poorly.
Samurai Quotation

It is a principle of the art of war that one should simply lay down his life and strike. If one's opponent also does the same, it is an even match. Defeating one's opponent is then a matter of faith and destiny.

Joseph Lumpkin

Yamamoto Tsunetomo

If a warrior is not unattached to life and death, he will be of no use whatsoever. The saying that All abilities come from one mind sounds as though it has to do with sentient matters, but it is in fact a matter of being unattached to life and death. With such non-attachment one can accomplish any feat. Martial arts and the like are related to this insofar as they can lead to the Way.
Yamamoto Tsunetomo

This is essentially a people's contest... whose leading object is to elevate the condition of men - to lift artificial weights from all shoulders - to clear the paths of laudable pursuit for all - to afford all, an unfettered start and a fair chance, in the race of life.
Abraham Lincoln

If a man does not keep pace with his companions, perhaps it is because he hears a different drummer. Let him step to the music which he hears, however measured or far away.
Henry Thoreau

At the time of the attack on the castle at Shimabara, Tazaki Geki was wearing very resplendent armor. Lord Katsushige was not pleased by this, and after that every time he saw something showy he would say, That's just like Geki's armor." In the light of this story, military armor and equipment that are showy can be seen as being weak and having no strength. By them one can see through the wearer's heart.
From the Hagakure

As long as people believe in absurdities, they will continue to commit atrocities.
Voltaire

Feeling deeply the difference between oneself and others, bearing ill will and falling out with people--these things come from a heart that lacks compassion. If one wraps up everything with a heart of compassion, there will be no coming into conflict with people.
From the Hagakure

Violence is the last refuge of the incompetent.

Issac Asimov

There is surely nothing other than the single purpose of the present moment. A man's whole life is a succession of moment after moment. If one fully understands the present moment, there will be nothing else to do, and nothing else to pursue. Live being true to the single purpose of the moment.
From the Hagakure

Live Free or Die.
New Hampshire State Motto

It is better to die on your feet than to live on your knees.
Dolores Ibarruri

These are the levels in general; But there is one transcending level, and this is the most excellent of all. This person is aware of the endlessness of entering deeply into a certain Way and never thinks of himself as having finished. He truly knows his own insufficiencies and never in his whole life thinks that he has succeeded. He has no thoughts of pride but with self-abasement knows the Way to the end. It is said that Master Yagyu once remarked, I do not know the way to defeat others, but the way to defeat myself.
From the Hagakure

Men of high position, low position, deep wisdom and artfulness all feel that they are the ones who are working righteously, but when it comes to the point of throwing away one's life for his lord, all get weak in the knees. This is rather disgraceful. The fact that a useless person often becomes a matchless warrior at such times is because he has already given up his life and has become one with his master.
From the Hagakure

True patriotism hates injustice in its own land more than anywhere else.
Clarence Darrow

The only thing necessary for the triumph of evil is for good men to do nothing.
Edmund Burke

Patriotism is not short, frenzied outbursts of emotion, but the tranquil and steady dedication of a lifetime.
Adlai Stevenson

One of the great attractions of patriotism - it fulfils our worst wishes. In the person of our nation we are able, vicariously, to bully and cheat. Bully and cheat, what's more, with a feeling that we are profoundly virtuous.
Aldous Huxley

Speaking the Truth in times of universal deceit is a revolutionary act.
George Orwell

The more individuals capable of watching the world theater calmly and critically, the less danger of monumental mass stupidities in first of all, wars.
Hermann Hesse

My instinct as an individualist and artist has always warned me most urgently against this capacity of men for becoming drunk on collective suffering, collective pride, collective hatred, and collective honor. When this morbid exaltation becomes perceptible in a room, a hall, a village, a city, or a country, I grow cold and distrustful; a shudder comes over me, for already, while most of my fellow men are still weeping with rapture and enthusiasm, still cheering and venting protestations of brotherhood, I see blood flowing and cities going up in flames.
Hermann Hesse

There is one tactical principal which is not subject to change. It is to use the means at hand to inflict the maximum amount of wounds, death and destruction on the enemy in the minimum amount of time.
General George S. Patton Jr.

Nothing focuses the mind and cleanses the soul so well as facing your own execution in the morning.
Joseph Lumpkin

It is sweet and honorable to die for your country.

Horace

If ever there was a holy war, it was that which saved our liberties and gave us independence.
Thomas Jefferson

In the long run luck is given only to the efficient.
Helmuth von Moltke

The concentration of troops can be done fast and easy, on paper.
Field Marshal Radomir Putnik

Victory in war does not depend entirely on numbers or courage; only skill and discipline will ensure it.
Flavius Vegetius

Wars may be fought with weapons, but they are won by men. It is the spirit of men who follow and of the man who leads that gains the victory.
General George S. Patton

What is our aim? Victory, victory at all costs, victory in spite of all terror; Victory how ever long and hard the road may be.
Sir Winston Churchill

In war there is no substitute for victory.
Douglas MacArthur

There is one source, O Athenians, of all your defeats. It is that your citizens have ceased to be soldiers.
Demosthenes

Once we have a war there is only one thing to do. It must be won. For defeat brings worse things than any that can ever happen in war.
Ernest Hemingway

Older men declare war. But it is the youth that must fight and die.
Herbert Hoover

The sinews of war are not gold, but good soldiers; for gold alone will not procure good soldiers, but good soldiers will always procure gold.
Machiavelli

So in war, the way is to avoid what is strong and to strike at what is weak.
Sun Tzu,

The art of war is, in the last result, the art of keeping one's freedom of action
Xenophon

When bad men combine, the good must associate else they will fall one by one, an unpitied sacrifice in a contemptible struggle.
Edmund Burke

The belief in the possibility of a short decisive war appears to be one of the most ancient and dangerous of human illusions.
Robert Lynd

Wars are not acts of God. They are caused by man, by man-made institutions, by the way in which man has organized his society. What man has made, man can change.
Frederick Moore Vinson

The character of a soldier is high. They who stand forth the foremost in danger, for the community, have the respect of mankind. An officer is much more respected than any other man who has as little money. In a commercial country, money will always purchase respect. But you find, an officer, who has, properly speaking, no money, is every where well received and treated with attention. The character of a soldier always stands him in good stead.
Samuel Johnson

A soldier's time is passed in distress and danger, or in idleness and corruption.
Samuel Johnson

I suppose every man is shocked when he hears how frequently soldiers are wishing for war. The wish is not always sincere; the greater part are content with sleep and lace, and counterfeit an ardor which they do not feel; but those who desire it most are neither prompted by malevolence nor patriotism; they neither pant for laurels, nor delight in blood; but long to be delivered from the tyranny of idleness, and restored to the dignity of active beings.
Samuel Johnson

The Character of War

WHITE CLIFFS OF DOVER
World War II Song by Nat Burton and
Walter Kent, 1941

There'll be bluebirds over,
The white cliffs of Dover,
Tomorrow, just you wait and see.

There'll be love and laughter, And peace ever after,
Tomorrow when the world is free.

Courage, fearlessness, intelligence, cunning, bravery, integrity, discipline, humility, faithfulness, confidence, valor....all describe the indescribable.

The character of war is all of these yet more; all of these yet less. One single man or one infinite army, one hero or one despicable tyrant...alike yet completely disparate...imbued with the call to conquer, the mission to dominate.   The Editors

There are three essentials to leadership: humility, clarity and courage.
Fuchan Yean

A leader is a dealer in hope.
Napoleon Bonaparte

I don't know what effect these men will have on the enemy, but, by God, they terrify me.
Duke of Wellington

Those in supreme power always suspect and hate their next heir.
Tacitus

Reason and judgment are the qualities of a leader.
Tacitus

Life is hard. Life is harder if you're stupid.
John Wayne

A little man often cast a long shadow.
G. M. Trevelyan

Action springs not from thought, but from a readiness for responsibility.
G. M. Trevelyan

Anger is a momentary madness, so control your passion or it will control you.
G. M. Trevelyan

It is not those who can inflict the most, but those that can suffer the most who will conquer.
Terence MacSwiney

Don't get small units caught in between the forces of history.
John W. Vessey, Jr.

More has been screwed up on the battlefield and misunderstood in the Pentagon because of a lack of understanding of the English language than any other single factor.
John W. Vessey, Jr.

Our strategy is one of preventing war by making it self-evident to our enemies that they're going to get their clocks cleaned if they start one.
John W. Vessey, Jr.

The courage of a soldier is heightened by his knowledge of his profession.
Flavius Renatus Vegetius

We find that the Romans owed the conquest of the world to no other cause than continual military training, exact observance of discipline in their camps, and unwearied cultivation of the other arts of war.
Flavius Renatus Vegetius

If God wanted us to be brave, why did he give us legs?
Marvin Kitman

There will be times when we think hard work and training are of no use. We will blame our lot in life on fate alone. In our despair we may say: Valor is of no service, chance rules all, and the bravest often fall by the hands of cowards.
Tacitus

I think with the Romans, that the general of today should be the soldier of tomorrow if necessary.
Thomas Jefferson

Never forget the fact that all leaders have strengths. In addition to that, all leaders have weaknesses. The problem for the leader is to avoid pride. If not, a leader will see his strengths become his weaknesses.
Joseph Lumpkin

To justify a fault is to argue for your own downfall.
Joseph Lumpkin

After I, a man of little rank, unexpectedly took control of the province, I have put forth great effort both day and night, at one time to gather together famous men of all kinds, listened to what they and to say, and have continued in such a way up to this time.

Joseph Lumpkin

Asakura Toshikage

The clever combatant looks to the effect of combined energy, and does not require too much from individuals. Hence his ability to pick out the right men and utilize combined energy.
Sun Tzu

One matures into leadership. It overtakes him as he learns and expands. Those who seek leadership usually do so prematurely. When people see those things in you they desire in themselves, they will follow. Only then are you a leader.
Joseph Lumpkin

It is the business of a general to be quiet and thus ensure secrecy; and to be upright and just, and thus maintain order.
Sun Tzu

The supreme quality for leadership is unquestionably integrity. Without it, no real success is possible, no matter whether it is on a section gang, a football field, in an army, or in an office.
Dwight D. Eisenhower

Those who desire to govern their states should first put their families in order. And those who desire to put their families in order would first discipline themselves.
Confucius

The consummate leader cultivates the moral law, and strictly adheres to method and discipline; thus it is in his power to control success.
Sun Tzu

A functional military does not rely on the enemy not coming, but relies on the fact that he himself is waiting; one does not rely on the enemy not attacking, but relies on the fact that he himself is unassailable.
Sun Tzu

Character is like a tree and reputation like a shadow. The shadow is what we think of it; the tree is the real thing.
Abraham Lincoln

The quality of decision is like the well-timed swoop of a falcon, which enables it to strike and destroy its victim. Therefore the good fighter will be terrible in his onset, and prompt in his decision. Energy may be likened to the bending of a crossbow; decision, to the releasing of a trigger.
Sun Tzu

Training is the education of instinct.
Anonymous

In this uncertain world, ours should be the path of discipline.
Hiba Yoshimasa

One should exert himself in martial arts absolutely. There are no weak soldiers under a strong general.
Takeda Nobushige

There is no deadlier weapon than the will! The sharpest sword is not equal to it. There is no robber so dangerous as nature. Yet, it is not nature that does the damage: it is man's own will! Chuang Tzu

The fatal flaw of one promoted to a position of authority is to forget from where he came. If he remembers his previous low position he will see that every fool has a chance to advance. He will understand it is by grace and chance that he is there. If he understands there are many as good as he, the position will not seem so high and he will know all men are replaceable. This will keep him humble. In his humility he will treat others well and they will follow him willingly.
Joseph Lumpkin

Like everyone else, you want to learn the way to win, but never to accept the way to lose. To accept defeat - to learn to die - is to be liberated from it. Once you accept, you are free to flow and to harmonize.
Bruce Lee

I will stand off the forces of the entire county here, and die a glorious death.
Torii Mototada

Not being tense, but ready. Not thinking yet not dreaming, not being set, but flexible - it is being wholly and quietly alive, aware and alert, ready for whatever may come.
Bruce Lee

A person's character and depth of mind is seen by his behavior. Thus, one should understand that even the walls and fences have eyes... one should not take a single step in vain, or speak a word in a way that others may speak of him as shallow.
Shiba Yoshimasa

The wise adapt themselves to circumstances, as water molds itself to the pitcher.
Chinese proverb

Beware of the leader who bangs the drums of war to whip the citizenry into a patriotic fever. For patriotism is indeed a double edged sword. It both emboldens the blood, just as it narrows the mind. When the drums of war have reached a fever pitch, and the blood boils with hate and the mind is closed, the leader will have no need in seizing the rights of the citizenry. Rather the citizenry, infused with fear and blinded by patriotism, will offer up all of their rights unto the leader and do it gladly so. How do I know? I know for this is what I have done. And I am Caesar.
William Shakespeare

I cannot hear what you say for the thunder of what you are.
Zulu Proverb

Good leaders must first become good servants. Prosperity is a great teacher; adversity a greater.
William Hazlitt

Duty is what one expects from others.
Oscar Wilde

Courage is doing what you're afraid to do. There can be no courage unless you're scared.
Eddie Rickenbacker

Only those who dare to fail greatly can ever achieve greatly.
Robert F. Kennedy

As for courage and will - we cannot measure how much of each lies within us, we can only trust there will be sufficient to carry through trials which may lie ahead.
Andre Norton

Disobedience, the rarest and most courageous of the virtues, is seldom distinguished from neglect, the laziest and commonest of the vices.
George Bernard Shaw

You may be deceived if you trust too much, but you will live in torment if you don't trust enough.
Frank H. Crane

If you can talk brilliantly about a problem, it can create the consoling illusion that it has been mastered.
Stanley Kubrick

Courage is grace under pressure.
Ernest Hemingway

The world breaks everyone, and afterward, some are stronger at the broken places.
Ernest Hemingway

It is hard to fight an enemy who has outposts in your head.
Sally Kempton

Life is what happens to us while we're making other plans.
Thomas LaMance

Leaders get out in front and stay there by raising the standards by which they judge themselves - and by which they are willing to be judged.
Fredrick Smith

A successful man is one who can lay a firm foundation with the bricks that others throw at him.
Sidney Greenberg

Good leaders must first become good servants.
Robert Greenleaf

One does not discover new lands without consenting to lose sight of the shore for a very long time.
André Gide

If you want to piss with the big dogs, you'd better learn to lift your leg first; otherwise you just might get pissed on.
E.M. Glenn

It's there within us all. It costs nothing, takes almost no time and is powerful beyond measure. Unleash the power of praise and reap the rewards.
Susan Mitchell

Better to remain silent and be thought a fool than to speak out and remove all doubt.
Abraham Lincoln

I'm a slow walker, but I never walk back.
Abraham Lincoln

It's the process of striving that makes us grow - not necessarily the result. Running against the wind makes us better than running with it.
Herb Elliott

Humility must always be the portion of any man who receives acclaim earned in blood of his followers and sacrifices of his friends.
Dwight D. Eisenhower

Courage is the finest of human qualities because it is the quality which guarantees all others.
Sir Winston Churchill

What, courage, man! What though care killed a cat, thou hast mettle enough in thee to kill care.
William Shakespeare

Courage - a perfect sensibility of the measure of danger, and a mental willingness to endure it.
William T. Sherman

Courage, hard work, self-mastery, and intelligent effort are all essential to successful life.
Theodore Roosevelt

What we need in this country today is more courage and more belief in the things that we have.
Thomas Watson

A lot of people do not muster the courage to live their dreams because they are afraid to die.
Les Brown

It is curious that physical courage should be so common in the world and moral courage so rare.   Mark Twain

Hope, like faith, is nothing if it is not courageous; it is nothing if it is not ridiculous.
Thornton Wilder

Courage consists not in hazarding without fear, but being resolutely minded in a just cause.
Plutarch

It is stupidity rather than courage to refuse to recognize danger when it is close upon you.
Sir Arthur Conan Doyle

Support the strong, give courage to the timid, remind the indifferent, and warn the opposed.
Whitney M. Young

It's not the maker of the sword, but the courage and skill of the swordsman that wins the day.
Robert M. Irwin

The only kind of courage that matters is the kind that gets you from one minute to the next.
Mignon McLaughlin

Pain nourishes courage. You can't be brave if you've only had wonderful things happen to you.
Mary Tyler Moore

You cannot build character and courage by taking away men's initiative and independence.
William J. H. Boetcker

It takes as much courage to have tried and failed as it does to have tried and succeeded.
Anne Lindbergh

The courage we desire and prize is not the courage to die decently, but to live manfully.   Thomas Carlyle

Hope awakens courage. He who can implant courage in the human soul is the best physician.
Karl Ludwig von Knebel

Courage is doing without witnesses that which we would be capable of doing before everyone.
Duc de La Rochefoucauld

To call war the soil of courage and virtue is like calling debauchery the soil of love.
George Santayana

No man in the world has more courage than the man who can stop after eating one peanut.
Channing Pollock

A good man will certainly also possess courage; but a brave man is not necessarily good.
Confucius

Never ask the Gods for life set free from grief, but ask for courage that endureth long.
Menander

Courage consists not in blindly overlooking danger, but in seeing it, and conquering it.
Jean Paul Richter

It takes vision and courage to create - it takes faith and courage to prove.
Owen D. Young

The greatest test of courage on earth is to bear defeat without losing heart.
Robert Ingersoll

He who is not courageous enough to take risks will accomplish nothing in life.   Muhammad Ali

Good ideas and innovations must be driven into existence by courage and patience.
Hyman Rickover

We must have courage to bet on our ideas, to take the calculated risk, and to act.
Maxwell Maltz

You cannot discover new oceans unless you have the courage to lose sight of the shore.
Unknown

All you need is the plan, the road map, and the courage to press on to your destination
Earl Nightingale

To see what is right, and not do it, is want of courage, or of principle.
Confucius

We must constantly build dikes of courage to hold back the flood of fear.
Martin Luther King Jr.

Success is never final and failure never fatal. It's courage that counts.
George F. Tilton

True courage is a result of reasoning. A brave mind is always impregnable.
Jeremy Collier

Courage is the strong desire to live taking the form of a readiness to die.
Gilbert Chesterton

Courage is resistance to fear, mastery of fear - not absence of fear.
Mark Twain

Some temptations are so great it takes great courage to yield to them.
Oscar Wilde

A great part of courage is the courage of having done the thing before.
Ralph Waldo Emerson

The secret of Happiness is Freedom, and the secret of Freedom, Courage.
Thucydides

Courage conquers all things; it even gives strength to the body.
Ovid

Efforts and courage are not enough without purpose and direction.
John F. Kennedy

Courage is the capacity to confirm what you can be imagined. Leo Calvin Rosten

Confidence is directness and courage in meeting the facts of life.
John Dewey

Failure is unimportant. It takes courage to make a fool of oneself.
Charlie Chaplin

If we survive danger it steels our courage more than anything else.
Reinhold Niebuhr

Few persons have courage enough to appear as good as they really are.
Augustus Hare

Most men have more courage than even they themselves think they have.
Lord Brook Fulke Greville

Courage is like love - it must have hope to nourish it.
Napoleon Bonaparte

Courage is being scared to death-and saddling up anyway.
John Wayne

To see what is right and not to do it, is want of courage.
Confucius

It takes courage to grow up and become who you really are.
e.e. cummings

Courage is the ladder on which all the other virtues mount.
Clare Boothe Luce

Nothing gives a fearful man more courage than another's fear.
Umberto Eco

Courage is simply the willingness to be afraid and act anyway.
Dr. Robert Anthony

Until the day of his death, no man can be sure of his courage.
Jean Anouilh

True courage is like a kite; a contrary wind raises it higher.
John Petit-Senn

One man with courage makes a majority.
Andrew Jackson

Without courage, wisdom bears no fruit.
Baltasar Gracian

Courage is fire, and bullying is smoke.
Benjamin Disraeli

Courage without conscience is a wild beast.
Robert Ingersoll

Courage is fear holding on a minute longer.
George Patton

Courage leads to heaven; fear leads to death.
Seneca

Fortune can take away riches, but not courage.
Seneca

It requires more courage to suffer than to die.
Napoleon Bonaparte

It is in great dangers that we see great courage.
Jean Francois Regnard

Life shrinks or expands in proportion to one's courage.
Anais Nin

These are the times that try men's souls. The summer soldier and the sunshine patriot will, in this crisis, shrink from the service of his country; but he that stands it now, deserves the love and thanks of

man and woman. Tyranny, like hell, is not easily conquered; yet we have this consolation with us, that the harder the conflict, the more glorious the triumph. What we may obtain too cheap, we esteem too lightly.
Thomas Paine

Courage follows action.
Mack R. Douglas

Have the courage to be wise.
Horatius

Courage is grace under pressure.
Ernest Hemingway

Without justice, courage is weak.
Benjamin Franklin

You can't test courage cautiously.
Annie Dillard

Necessity does the work of courage.
George Eliot

One man with courage is a majority.
Thomas Jefferson

Courage in danger is half the battle.
Titus Maccius Plautus

To be a successful soldier you must know history. What you must know is how man reacts. Weapons change but the man who uses them changes not at all. To win battles you do not beat weapons - you beat the soul of man of the enemy man.
George Patton

Leadership is solving problems. The day soldiers stop bringing you their problems is the day you have stopped leading them. They have either lost confidence that you can help or concluded you do not care. Either case is a failure of leadership.

Karl Popper

I have seen soldiers panic at the first sight of battle, and a wounded squire pulling arrows out from his wound to fight and save his dying horse. Nobility is not a birth right but is defined by one's action.
Robin Hood, Prince of Thieves

Valor, glory, firmness, skill, generosity, steadiness in battle and ability to rule - these constitute the duty of a soldier. They flow from his own nature.
Bhagavad Gita

Every man who expresses an honest thought is a soldier in the army of intellectual liberty.
Robert Ingersoll

The most vital quality a soldier can possess is self-confidence, utter, complete and bumptious.
George Patton

The dignity of man is vindicated as much by the thinker and poet as by the statesman and soldier.
James Bryant Conant

No matter whether a person belongs to the upper or lower ranks, if he has not put his life on the line at least once he has cause for shame.
Nabeshima Naoshige

Being affected by the avarice for office and rank, or wanting to become a daimyo and being eager for such things ... will not one then begin to value his life? And how can a man commit acts of martial valor if he values his life? A man who has been born into the house of a warrior and yet places no loyalty in his heart and thinks only of the fortune of his position will be flattering on the surface and construct schemes in his heart, will forsake righteousness and not reflect on his shame, and will stain the warrior's name of his household to later generations. This is truly regrettable.
Torii Mototada

Life is like unto a long journey with a heavy burden. Let thy step be slow and steady, that thou stumble not. Persuade thyself that imperfection and inconvenience are the natural lot of mortals, and there will be no room for discontent, neither for despair. When ambitious desires arise in thy heart, recall the days of extremity thou has passed through. Forbearance is the root of quietness and assurance forever. Look upon the wrath of the enemy. If thou knowest only what it is to conquer, and knowest not what it is to be defeated, woe unto thee; it will fare ill with thee. Find fault with thyself rather than with others. Tokugawa

In strategy your spiritual bearing must not be any different from normal. Both in fighting and in everyday life you should be determined though calm. Meet the situation without tenseness yet not recklessly, your spirit settled yet unbiased. If the enemy thinks of the mountains, attack like the sea; and if he thinks of the sea, attack like the mountains.
Miyamoto Musashi

The combining of these three virtues may seem unobtainable to the ordinary person, but it is easy. Intelligence is nothing more than discussing things with others. Wisdom comes from this. Humanity is something done for the sake of others, simply comparing oneself with them and putting them in the fore. Courage is gritting one's teeth; it is simply doing that and pushing ahead, paying no attention to the circumstances. Anything that seems above these three is not necessary to be known.
From the Hagakure

When one is attending to matters, there is one thing that comes forth from his heart. That is, in terms of one's lord, loyalty; in terms of one's parents, filial piety; in martial affairs, bravery; and apart from that, something that can be used by all the world.
From the Hagakure

The essential American character is hard, isolated, stoic, and a killer.
D.H. Lawrence

In a civil war, a general must know ... exactly when to move over to the other side.

Henry Reed

Patriotism is the virtue of the vicious.
Oscar Wilde

It is curious that physical courage should be so common in the world and moral courage so rare.
Mark Twain

Courage is resistance to fear, master of fear -- not absence of fear. Except a creature be part coward, it is not a compliment to say it is brave.
Mark Twain

A really great people, proud and high-spirited, would face all the disasters of war rather than purchase that base prosperity which is bought at the price of national honor.
Theodore Roosevelt

No man can sit down and withhold his hands from the warfare against wrong and get peace from his acquiescence.
Woodrow Wilson

No person was ever honored for what he received. Honor has been the reward for what he gave.
Calvin Coolidge

Never in the field of human conflict was so much owed by so many to so few.
Sir Winston Churchill

Let us solemnly remember the sacrifices of all those who fought so valiantly, on the seas, in the air, and on foreign shores, to preserve our heritage of freedom, and let us re-consecrate ourselves to the task of promoting an enduring peace so that their efforts shall not have been in vain.
Dwight Eisenhower

Cowards die many times before their deaths; the valiant never taste of death but once.

William Shakespeare

I believe that military service in the Armed Forces of the United States is a profound form of service to all humankind. You stand engaged in an effort to keep America safe at home, to protect our allies and interests abroad, to keep the seas and the skies free of threat. Just as America stands as an example to the world of the inestimable benefits of freedom and democracy, so too an America with the capacity to project her power for the purpose of protecting and expanding freedom and democracy abroad benefits the suffering people of the world.
Ronald Reagan

To lead uninstructed people to war is to throw them away.
Confucius

The secret of all victory lies in the organization of the non-obvious.
Marcus Aurelius

Time is a sort of river of passing events, and strong is its current; no sooner is a thing brought to sight than it is swept by and another takes its place, and this too will be swept away.
Marcus Aurelius

Leadership is the art of getting someone else to do something you want done because he wants to do it.
Dwight D. Eisenhower

Nearly all men can stand adversity, but if you want to test a man's character, give him power.
Abraham Lincoln

Joseph Lumpkin

LOVE AND WAR

DANNY BOY
Lyrics by Fred Weatherly, 1911

Oh, Danny boy, the pipes, the pipes are calling
From glen to glen, and down the mountain side.
The summer's gone, and all the roses falling,
It's you, it's you must go and I must bide.
But come ye back when summer's in the meadow,
Or when the valley's hushed and white with snow,
It's I'll be here in sunshine or in shadow,
Oh, Danny boy, O Danny boy, I love you so!

But when ye come, and all the flowers are dying,
If I am dead, as dead I well may be.
Ye'll come and fine the place where I am lying,
And kneel and say an Ave' there for me.
And I shall hear, though soft you tread above me,
And all my grave will warmer, sweeter be,
For you will bend and tell me that you love me,
And I shall sleep in peace until you come to me.

The rush of submission blends with the power of victory. Yet in the final analysis there are winners and there are losers and the survivors are forever damaged and scarred.

Love and war...war and love; inextricably joined; painfully entwined, one with the other, for all eternity.
        The Editors

Love is like war: easy to begin but very hard to stop.
H. L. Mencken

530

Love does not begin and end the way we seem to think it does. Love is a battle, love is a war; love is a growing up.
James A. Baldwin

What a cruel thing war is... to fill our hearts with hatred instead of love for our neighbors.
Robert E. Lee

The Wedding March always reminds me of the music played when soldiers go into battle.
Heinrich Heine

Nothing is miserable unless you think it is so.
Boethius

War is like love; it always finds a way.
Bertold Brecht

For in all adversity of fortune the worst sort of misery is to have been happy.
Boethius

Love is an ocean of emotions entirely surrounded by expenses.
Lord Thomas Dewar

The art of love is largely the art of persistence.
Albert Ellis

Fortune and love favor the brave.
Ovid

Love conquers all.
Virgil

Love is a canvas furnished by nature and embroidered by imagination.
Voltaire

You don't have to have fought in a war to love peace.
Geraldine A. Ferraro

Don't tell me peace has broken out.
Bertoltd Brecht

Love is a net that catches hearts like a fish.
Muhammad Ali

Love is the only force capable of transforming an enemy into friend.
Martin Luther King, Jr.

Man must evolve for all human conflict a method which rejects
revenge, aggression and retaliation. The foundation of such a method
is love.
Martin Luther King, Jr.

All brave men love; for he only is brave who has affections to fight
for, whether in the daily battle of life, or in physical contests.
Nathaniel Hawthorne

All married couples should learn the art of battle as they should learn
the art of making love. Good battle is objective and honest - never
vicious or cruel. Good battle is healthy and constructive, and brings
to a marriage the principles of equal partnership.
Ann Landers

A woman watches her body uneasily, as though it were an unreliable
ally in the battle for love.
Leonard Cohen

All the passions make us commit faults; love makes us commit the
most ridiculous ones.
Francois de La Rochefoucauld

And yet a little tumult, now and then, is an agreeable quickener of
sensation; such as a revolution, a battle, or an adventure of any lively
description.
Lord Byron

Marriage is an adventure, like going to war.
Gilbert K. Chesterton

Men like war: they do not hold much sway over birth, so they make up for it with death. Unlike women, men menstruate by shedding other people's blood.
Lucy Ellman

Power is my mistress. I have worked too hard at her conquest to allow anyone to take her away from me.
Napoleon Bonaparte

War has always been the grand sagacity of every spirit which has grown too inward and too profound; its curative power lies even in the wounds one receives.
Friedrich Nietzsche

You say that love is nonsense....I tell you it is no such thing. For weeks and months it is a steady physical pain, an ache about the heart, never leaving one, by night or by day; a long strain on one's nerves like toothache or rheumatism, not intolerable at any one instant, but exhausting by its steady drain on the strength.
Henry Brooks Adams

Love of country is like love of woman--he loves her best who seeks to bestow on her the highest good.
Felix Adler

In dreams and in love there are no impossibilities.
Janos Arany

Love does not begin and end the way we seem to think it does. Love is a battle, love is a war; love is a growing up.
James Baldwin

But, O Sarah! if the dead can come back to this earth and flit unseen around those they loved, I shall always be near you; In the gladdest days and in the darkest nights . . . always, always, and if there be a soft breeze upon your cheek, it shall be my breath, as the cool air fans your throbbing temple, it shall be my spirit passing by. Sarah do not mourn me dead; think I am gone and wait for thee, for we shall meet again.

Major Sullivan Ballou

And in the end, the love you take is equal to the love you make.
The Beatles

The only way of knowing a person is to love them without hope.
Walter Benjamin

Love is very patient, Love is very kind, Love is never envious. Or vaunted up with pride.
1 Corinthians 13:4-7

Love is always either increasing or decreasing.
Andreas Capellanus

We always deceive ourselves twice about the people we love - first to their advantage, then to their disadvantage.
Albert Camus

Give me more love or more disdain; The torrid or the frozen zone; Bring equal ease unto my pain; The temperate affords me none.
Thomas Carew

Of all the pain, the greatest pain, Is to love, but to love in vain.
Abraham Cowley

Love is a power too strong to be overcome by anything but flight.
Miguel de Cervantes

There are people who would have never fallen in love if they never heard of love.
François, Duc de La Rochefoucauld

Love never dies of starvation, but often of indigestion.
Ninon de Lenclos

Love does not consist in gazing at each other but in looking together in the same direction.
Antoine de Saint-Exupery

Love is an irresistible desire to be irresistibly desired.
Robert Frost

It is the special quality of love not to be able to remain stationary, to be obliged to increase under pain of diminishing.
Andre Gide

Love is a perky elf dancing a merry little jig and then suddenly he turns on you with a miniature machine-gun.
Matt Groening

Never judge someone by who he's in love with; judge him by his friends. People fall in love with the most appalling people.
Cynthia Heimel

Hatred paralyzes life; love releases it. Hatred confuses life; love harmonizes it. Hatred darkens life; love illuminates it.
Martin Luther King, Jr.

Love is the delusion that one woman differs from another.
H. L. Mencken

Love is the triumph of imagination over intelligence.
H. L. Mencken

To be in love is merely to be in a state of perpetual anesthesia - to mistake an ordinary young woman for a goddess.
H. L. Mencken

Alas! how light a cause may move
Dissension between hearts that love!
Hearts that the world in vain had tried,
And sorrow but more closely tied;
That stood the storm when waves were rough,
Yet in a sunny hour fall off.
Thomas Moore

Love is much like a wild rose, beautiful and calm, but willing to draw blood in its defense.
Mark A. Overby

Love and dignity cannot share the same abode.
Ovid

Love is like quicksilver in the hand. Leave the fingers open and it stays. Clutch it, and it darts away.
Dorothy Parker

Love is a reciprocal torture.
Marcel Proust

Love is like the moon; when it does not increase it decreases.
Segur

Love is a smoke made with the fume of sighs,
Being purged, a fire sparkling in lovers' eyes,
Being vexed, a sea nourished with lovers' tears.
What is it else? A madness most discreet,
A choking gall and a preserving sweet.
William Shakespeare
Men have died from time to time, and the worms have eaten 'em, but not for love.

William Shakespeare

First love is only a little foolishness and a lot of curiosity.
George Bernard Shaw

All is fair in love and war.
Francis Edward Smedley

The joy of late love is like green firewood when set aflame, for the longer the wait in lighting, the greater heat it yields and the longer its force lasts.
Chrétien de Troyes

Love begets love, love knows no rules, this is the same for all.
Virgil

It has ever been since time began, And ever will be, till time lose breath,
That love is a mood - no more - to man, And love to a woman is life or death.
Ella Wheeler Wilcox

Yet each man kills the thing he loves,
By each let this be heard,
Some do it with a bitter look
Some with a flattering word,
The coward does it with a kiss,
The brave man with a sword!
Oscar Wilde

Warriors on War

THE BATTLE CRY OF FREEDOM
Civil War Song, George F. Root, 1861

Yes, we'll rally round the flag, boys,
we'll rally once again,
Shouting the battle-cry of Freedom;
We will rally from the hillside,
we will gather from the plain,
Shouting the battle-cry of Freedom.

Onward in infinite legions they march, wave after wave…the centurians, the magyars, the mongols, the samurai, the chevaliers, the hwa rang, the green berets…all breathing fire, all eyes forward; to glory or to death, into the teeth of battle. What do they think and feel, these human machines of war – do we detect fear? Regret? Joy? Rage? The overpowering rush of adrenalin, and then it begins.

Joseph Lumpkin

The Editors

Cry 'Havoc', and let slip the dogs of war.
William Shakespeare

Minds are like parachutes. They function only when they are open!
Lord Dewar

One 'Oh shit' wipes out 30 'Atta boys'!
U.S. Marine Corps

If you want a decision, go to the point of danger.
Gen James M. Gavin

When we jumped into Sicily, the units became separated, and I couldn't find anyone. Eventually I stumbled across two colonels, a major, three captains, two lieutenants, and one rifleman, and we secured the bridge. Never in the history of war have so few been led by so many.
Gen James M. Gavin

Onward we stagger, and if the tanks come, may God help the tanks.
Col. William O. Darby

The essence of a general's job is to assist in developing a clear sense of purpose to keep the junk from getting in the way of important things.
Lt Gen Walter F. Ulmer

You can kill ten of my men for every one I kill of yours, but even at those odds, you will lose and I will win.
Ho Chi Minh

Intuitive decision-making and mastering this profession are one in the same.
Lt Gen Van Riper

The purpose of studying the new sciences is simple... We want to learn to understand war through the most powerful means

available... to encompass the ideas contained in quantum mechanics, nonlinear systems, and chaos and complexity theories.
Lt Gen Van Riper

This is an era of violent peace.
Adm. James D. Watkins

All the business of war, and indeed all the business of life, is to endeavor to find out what you don't know by what you do; that's what I called 'guess what was at the other side of the hill'.
Duke of Wellington

An extraordinary affair. I gave them their orders and they wanted to stay and discuss them.
Duke of Wellington

War is fear cloaked in courage.
William C. Westmoreland

War can only be abolished through war, and in order to get rid of the gun it is necessary to take up the gun.
Mao Tze-tung

It is a fact that under equal conditions, large-scale battles and whole wars are won by troops which have a strong will for victory, clear goals before them, high moral standards, and devotion to the banner under which they go into battle.
Marshal Georgi Zhukov

The nature of encounter operations required of the commanders limitless initiative and constant readiness to take the responsibility for military actions.
Marshal Georgi Zhukov

It is better to die on your feet than to live on your knees.
Emiliano Zapata

An army without culture is a dull-witted army, and a dull-witted army cannot defeat the enemy.
Mao Tze-tung

A man who has been shot at is a new realist, and what do you say to a realist when the war is a war of ideals?
Michael Shaara

It's the idea that we all have value, you and me, and we're worth something more than the dirt. I never saw dirt I'd die for, but I'm not asking you to come join us and fight for dirt. What we're all fighting for, in the end, is each other.
Michael Shaara

They would fight again, and when they came he would be behind another stone wall waiting for them, and he would stay there until he died or until it ended, and he was looking forward to it with an incredible eagerness, as you wait for the great music to begin after the silence.
Michael Shaara

It's all components and it's combat service support, combat support and combat arms that have to be fielded because the enemy doesn't sign up for this Department of the Army Master Priority List business, and they don't sign up for combat arms first. They don't understand that.
Gen. Kevin P. Byrnes

Humility must always be the portion of any man who receives acclaim earned in the blood of his fellows and the sacrifice of his friends.
Gen Dwight D. Eisenhower

Comrades, you have lost a good captain to make a bad general.
Saturninus

Leadership is a combination of strategy and character. If you must be without one, be without the strategy.
General H. Norman Schwarzkopf

The transition from the defensive to the offensive is one of the most delicate operations of war.
Napoleon Bonaparte

When an archer is shooting for nothing he has all of his skill. If he shoots for a brass buckle he is already nervous. If he shoots for a prize of gold he goes blind... his skill has not changed, but the prize divides him.
Chuang Tzu

As far as Saddam Hussein being a great military strategist, he is neither a strategist, nor is he schooled in the operational arts, nor is he a tactician, nor is he a general, nor is he a soldier. Other than that, he's a great military man. I want you to know that.
Norman Schwarzkopf

Do what is right, not what you think the high headquarters wants or what you think will make you look good.
Norman Schwarzkopf

It doesn't take a hero to order men into battle. It takes a hero to be one of those men who goes into battle.
Norman Schwarzkopf

Leadership is a potent combination of strategy and character. But if you must be without one, be without the strategy.
Norman Schwarzkopf

The day we executed the air campaign, I said, we gotcha!
Norman Schwarzkopf

When placed in command - take charge.
Norman Schwarzkopf

A good plan violently executed now is better than a perfect plan executed next week.
George S. Patton

A piece of spaghetti or a military unit can only be led from the front end.
George S. Patton

A pint of sweat saves a gallon of blood.

George S. Patton

Accept the challenges so that you can feel the exhilaration of victory.
George S. Patton

All very successful commanders are prima donnas and must be so treated.
George S. Patton

Always do everything you ask of those you command.
George S. Patton

Americans love to fight. All real Americans love the sting of battle.
George S. Patton

Americans play to win at all times. I wouldn't give a hoot and hell for a man who lost and laughed. That's why Americans have never lost nor ever lose a war.
George S. Patton

Battle is an orgy of disorder.
George S. Patton

Battle is the most magnificent competition in which a human being can indulge. It brings out all that is best; it removes all that is base. All men are afraid in battle. The coward is the one who lets his fear overcome his sense of duty. Duty is the essence of manhood.
George S. Patton

Better to fight for something than live for nothing.
George S. Patton

Courage is fear holding on a minute longer.
George S. Patton

Do your damnedest in an ostentatious manner all the time.
George S. Patton

I do not fear failure. I only fear the slowing up of the engine inside of me which is pounding, saying, Keep going, someone must be on top, why not you?

George S. Patton

I don't measure a man's success by how high he climbs but how high he bounces when he hits bottom.
George S. Patton

If a man does his best, what else is there?
George S. Patton

If everyone is thinking alike, then somebody isn't thinking.
George S. Patton

If we take the generally accepted definition of bravery as a quality which knows no fear, I have never seen a brave man. All men are frightened. The more intelligent they are, the more they are frightened.
George S. Patton

If you tell people where to go, but not how to get there, you'll be amazed at the results.
George S. Patton

It is foolish and wrong to mourn the men who died. Rather we should thank God that such men lived.
George S. Patton

Never tell people how to do things. Tell them what to do and they will surprise you with their ingenuity.
George S. Patton

The object of war is not to die for your country, but to make the other bastard die for his.
George S. Patton

Nobody ever defended anything successfully, there is only attack and attack and attack some more.

George S. Patton

Prepare for the unknown by studying how others in the past have coped with the unforeseeable and the unpredictable.
George S. Patton

Take calculated risks. That is quite different from being rash.
George S. Patton

The leader must be an actor. But with him as with his bewigged counterpart he is unconvincing unless he lives his part.
George S. Patton

The time to take counsel of your fears is before you make an important battle decision. That's the time to listen to every fear you can imagine! When you have collected all the facts and fears and made your decision, turn off all your fears and go ahead!
George S. Patton

There is a time to take counsel of your fears, and there is a time to never listen to any fear.
George S. Patton

There is only one sort of discipline, perfect discipline.
George S. Patton

Untutored courage is useless in the face of educated bullets.
George S. Patton

Wars may be fought with weapons, but they are won by men. It is the spirit of men who follow and of the man who leads that gains the victory.
George S. Patton

War is an art and as such is not susceptible of explanation by fixed formula
General George Patton

Watch what people are cynical about, and one can often discover what they lack.

George S. Patton

We herd sheep, we drive cattle, we lead people. Lead me, follow me, or get out of my way.
George S. Patton

War is an art and as such is not susceptible of explanation by fixed formula
General George Patton

You need to overcome the tug of people against you as you reach for high goals.
George S. Patton

A general is just as good or just as bad as the troops under his command make him.
Douglas Macarthur

Americans never quit.
Douglas MacArthur

And like the old soldier in that ballad, I now close my military career and just fade away, an old soldier who tried to do his duty as God gave him the sight to see that duty.
Douglas MacArthur

Build me a son, O Lord, who will be strong enough to know when he is weak, and brave enough to face himself when he is afraid, one who will be proud and unbending in honest defeat, and humble and gentle in victory.
Douglas MacArthur

Could I have but a line a century hence crediting a contribution to the advance of peace, I would yield every honor which has been accorded by war.
Douglas MacArthur

I am concerned for the security of our great Nation; not so much because of any threat from without, but because of the insidious forces working from within.

Douglas MacArthur

I can recall no parallel in history where a great nation recently at war has so distinguished its former enemy commander.
Douglas MacArthur

I suppose, in a way, this has become part of my soul. It is a symbol of my life. Whatever I have done that really matters, I've done wearing it. When the time comes, it will be in this that I journey forth. What greater honor could come to an American, and a soldier?
Douglas MacArthur

I've looked that old scoundrel death in the eye many times but this time I think he has me on the ropes.
Douglas MacArthur

In my dreams I hear again the crash of guns, the rattle of musketry, the strange, mournful mutter of the battlefield.
Douglas MacArthur

In war there is no substitute for victory.
Douglas MacArthur

In war, you win or lose, live or die-and the difference is just an eyelash.
Douglas MacArthur

It is fatal to enter any war without the will to win it.
Douglas MacArthur

Last, but by no means least, courage - moral courage, the courage of one's convictions, the courage to see things through. The world is in a constant conspiracy against the brave. It's the age-old struggle-the roar of the crowd on one side and the voice of your conscience on the other.
Douglas Macarthur

Life is a lively process of becoming.
Douglas MacArthur

My first recollection is that of a bugle call.
Douglas MacArthur

Never give an order that can't be obeyed.
Douglas MacArthur

No man is entitled to the blessings of freedom unless he be vigilant in its preservation.
Douglas MacArthur

Old soldiers never die; they just fade away. And like the old soldier in that ballad, I now close my military career and just fade away, an old soldier who tried to do his duty as God gave him the sight to see that duty.
Douglas MacArthur

One cannot wage war under present conditions without the support of public opinion, which is tremendously molded by the press and other forms of propaganda.
Douglas MacArthur

Part of the American dream is to live long and die young. Only those Americans who are willing to die for their country are fit to live.
Douglas Macarthur

The best luck of all is the luck you make for yourself.
Douglas Macarthur

The outfit soon took on color, dash and a unique flavor which is the essence of that elusive and deathless thing called soldiering.
Douglas MacArthur

The soldier above all others prays for peace, for it is the soldier who must suffer and bear the deepest wounds and scars of war.
Douglas MacArthur

The world is in a constant conspiracy against the brave. It's the age-old struggle: the roar of the crowd on the one side, and the voice of your conscience on the other.
Douglas MacArthur

There is no security on this earth; there is only opportunity.
Douglas MacArthur

They died hard, those savage men - like wounded wolves at bay.
They were filthy, and they were lousy, and they stunk. And I loved
them.
Douglas MacArthur

We are bound no longer by the straitjacket of the past and nowhere is
the change greater than in our profession of arms. What, you may
well ask, will be the end of all of this? I would not know! But I would
hope that our beloved country will drink deep from the chalice of
courage.
Douglas MacArthur

We are not retreating - we are advancing in another direction.
Douglas MacArthur

Whether in chains or in laurels, liberty knows nothing but victories.
Douglas MacArthur

You are remembered for the rules you break.
Douglas MacArthur

A true man of honor feels humbled himself when he cannot help
humbling others.
Robert E. Lee

Duty is the sublimest word in the language. You can never do more
than your duty. You should never wish to do less.
Robert E. Lee

I have been up to see the Congress and they do not seem to be able to
do anything except to eat peanuts and chew tobacco, while my army
is starving.
Robert E. Lee

I like whiskey. I always did, and that is why I never drink it.
Robert E. Lee

In all my perplexities and distresses, the Bible has never failed to give me light and strength.
Robert E. Lee

It is well that war is so terrible. We should grow too fond of it.
Robert E. Lee

Let the tent be struck.
Robert E. Lee

My experience through life has convinced me that, while moderation and temperance in all things are commendable and beneficial, abstinence from spirituous liquors is the best safeguard of morals and health.
Robert E. Lee

Never do a wrong thing to make a friend or to keep one.
Robert E. Lee

The devil's name is dullness.
Robert E. Lee

The education of a man is never completed until he dies.
Robert E. Lee

We failed, but in the good providence of God apparent failure often proves a blessing.
Robert E. Lee

We have fought this fight as long, and as well as we know how. We have been defeated. For us as a Christian people, there is now but one course to pursue. We must accept the situation.
Robert E. Lee

What a cruel thing is war: to separate and destroy families and friends, and mar the purest joys and happiness God has granted us in this world; to fill our hearts with hatred instead of love for our neighbors, and to devastate the fair face of this beautiful world.
Robert E. Lee

Bravery is the capacity to perform properly even when scared half to death.
Omar N. Bradley

I am convinced that the best service a retired general can perform is to turn in his tongue along with his suit and to mothball his opinions.
Omar N. Bradley

If we continue to develop our technology without wisdom or prudence, our servant may prove to be our executioner.
Omar N. Bradley

Leadership in the democratic army means firmness, not harshness; understanding, not weakness; justice, not license; humaneness, not intolerance; generosity, not selfishness; pride, not egotism.
Omar N. Bradley

Leadership is intangible, and therefore no weapon ever designed can replace it.
Omar N. Bradley

Ours is a world of nuclear giants and ethical infants. We know more about war that we know about peace, more about killing that we know about living.
Omar N. Bradley

The way to win an atomic war is to make certain it never starts.
Omar N. Bradley

Wars can be prevented just as surely as they can be provoked, and we who fail to prevent them, must share the guilt for the dead.
Omar N. Bradley

We need to learn to set our course by the stars, not by the light of every passing ship.
Omar N. Bradley

Do not needlessly endanger your lives until I give you the signal.
Dwight D. Eisenhower

I deplore the need or the use of troops anywhere to get American citizens to obey the orders of constituted courts.
Dwight D. Eisenhower

If a problem cannot be solved, enlarge it.
Dwight D. Eisenhower

If men can develop weapons that are so terrifying as to make the thought of global war include almost a sentence for suicide, you would think that man's intelligence and his comprehension... would include also his ability to find a peaceful solution.
Dwight D. Eisenhower

It is far more important to be able to hit the target than it is to haggle over who makes a weapon or who pulls a trigger.
Dwight D. Eisenhower

Pessimism never won any battle.
Dwight D. Eisenhower

That was not the biggest battle that ever was, but for me it always typified one thing-the dash, the ingenuity, the readiness at the first opportunity that characterizes the American soldier.
Dwight D. Eisenhower

The most terrible job in warfare is to be a second lieutenant leading a platoon when you are on the battlefield.
Dwight D. Eisenhower

There is no victory at bargain basement prices.
Dwight D. Eisenhower

What counts is not necessarily the size of the dog in the fight - it's the size of the fight in the dog.
Dwight D. Eisenhower

The art of war is simple enough. Find out where your enemy is. Get at him as soon as you can. Strike him as hard as you can, and keep moving.

Joseph Lumpkin

Ulysses S. Grant

To have good soldiers, a nation must always be at war.
Napoleon Bonaparte

The essence of war is violence. Moderation in war is imbecility.
British Sea Lord John Fisher

He who stays on the defensive does not make war, he endures it.
Field Marshal Colmar Baron von der Goltz

If men make war in slavish obedience to rules, they will fail.
Ulysses S. Grant

War is the mother of everything.
Heraclitus

In war, only the simple succeeds.
Field Marshal Paul Von Hindenburg

Si vis pacem, para bellum
(If you want peace, prepare for war!)
Flavius Vegetius Renatus

War is cruelty. There's no use trying to reform it, the crueler it is the
sooner it will be over.
William Tecumseh Sherman

The true test of a leader is whether his followers will adhere to his
cause from their own volition, enduring the most arduous hardships
without being forced to do so, and remaining steadfast in the
moments of greatest peril.
Xenophon

"What a piece of work is man...in action how like an angel!" Well,
boy, if he's an angel, he's sure a killer angel.
Michael Shaara

Knowing is not enough; we must apply. Willing is not enough; we
must do.

Bruce Lee

There is no standard in total combat, and expression must be free. This liberating truth is a reality only in so far as it is experienced and lived' by the individual himself; it is a truth that transcends styles or disciplines.
Bruce Lee

Humble words and increased preparations are signs that the enemy is about to advance. Violent language and driving forward as if to the attack are signs that he will retreat. When the light chariots come out first and take up a position on the wings, it is a sign that the enemy is forming for battle. Peace proposals unaccompanied by a sworn covenant indicate a plot. When there is much running about and the soldiers fall into rank, it means that the critical moment has come. When some are seen advancing and some retreating, it is a lure.
Sun Tzu

Hence, when able to attack, we must seem unable; when using our forces, we must seem inactive; when we are near, we must make the enemy believe we are far away; when far away, we must make him believe we are near... If he is taking his ease, give him no rest. If his forces are united, separate them.
Sun Tzu

The general, unable to control his irritation, will launch his men to the assault like swarming ants, with the result that one-third of his men are slain, while the town still remains untaken. Such are the disastrous effects of a siege.
Sun Tzu

Simulated disorder postulates perfect discipline, simulated fear postulates courage; simulated weakness postulates strength. Hiding order beneath the cloak of disorder is simply a question of subdivision; concealing courage under a show of timidity presupposes a fund of latent energy; masking strength with weakness is to be effected by tactical dispositions.
Sun Tzu

If we wish to fight, the enemy can be forced to an engagement even though he be sheltered behind a high rampart and a deep ditch. All we need do is attack some other place that he will be obliged to relieve.
Sun Tzu

Hold out baits to entice the enemy. Feign disorder, and crush him.
Sun Tzu

You intentionally leave an opening in your defenses or leave some limb vulnerable to attack in order to draw your opponent into making a predictable attack which you can in turn counter. Like baiting a fish hook.
Bruce Lee

You can prevent your opponent from defeating you through defense, but you cannot defeat him without taking the offensive.
Sun Tzu

To be near the goal while the enemy is still far from it, to wait at ease while the enemy is toiling and struggling, to be well fed while the enemy is famished - this is the art of husbanding one's strength. To refrain from intercepting an enemy whose banners are in perfect order, to refrain from attacking an army drawn up in calm and confident array - this is the art of studying circumstances. It is a military axiom not to advance uphill against the enemy, nor to oppose him when he comes downhill. Do not pursue an enemy who simulates flight; do not attack soldiers whose temper is keen. When the soldiers stand leaning on their spears, they are weak from want of food. Do not swallow bait offered by the enemy. Do not interfere with an army that is returning home.
Sun Tzu

In large-scale strategy it is important to cause loss of balance. Attack without warning where the enemy is not expecting it, and while his spirit is undecided, follow up your advantage and, having the lead, defeat him. Or, in single combat, start by making a show of being slow, then suddenly attack strongly. Without allowing him space for breath to recover from the fluctuation of spirit, you must grasp the opportunity to win...

Sun Tzu

Show him there is a road to safety, and so create in his mind the idea that there is an alternative to death. Then, Strike!
Tu Mu quoted in the Art of War by Sun Tzu.

Do not allow the enemy to escape too soon. They will regroup and attack again. Do not seek to destroy or you will arouse their fear of death and the strength and speed that it entails.
Sun Tzu

You can be sure of succeeding in your attacks if you only attack places, which are left open and not defended. You can ensure the safety of your defense if you only hold positions that cannot be attacked.
Sun Tzu

Attack him where he is unprepared, appear where you are not expected.
Sun Tzu

To fight and conquer in one hundred battles is not the highest skill. To subdue the enemy with no fight at all, that's the highest skill.
Sun Tsu

It is easier to block a chambered punch or kick before the attack is launched. Before it accelerates or develops power, one can halt it with a touch. Stop the punch; check the kick, as you move in. Kick the leg as it starts to kick. Strike the arm as it begins to punch. Jam and invade.
Joseph Lumpkin

The non-action of the wise man is not inaction...The sage is quiet because he is not moved... The heart of the wise man is tranquil... emptiness, stillness, tranquility, tastelessness, silence, and non-action are the root of all things.
Chuang Tzu

The good fighters of old first put themselves beyond the possibility of defeat, and then waited for an opportunity of defeating the enemy.

Joseph Lumpkin

Sun Tzu

Though a warrior may be called a dog or a beast, what is basic to his nature is to win.
Asakura Norikage

Success in warfare is gained by carefully accommodating ourselves to the enemy's purpose.
Sun Tzu

There are many ways to describe a warrior. The descriptions may vary widely, so it may be easier to start with what a warrior is not. The warrior is never a brute. He is never reactionary. He is not arrogant or pompous. The warrior should be a balanced and educated person. Most of all, the warrior should have the strength of a strong man, the heart of a compassionate woman, and the open mind of a little child.
Joseph Lumpkin

A good soldier is not violent. A good fighter does not get angry. A good winner is never vengeful. A good employer is humble. This is the virtue of not striving. It is known as the ability to deal with people.
Lao Tzu

The path of the Warrior is lifelong, and mastery is often simply staying on the path.
Richard Strozzi Heckler

The Way of a Warrior is based on humanity, love, and sincerity; the heart of martial valor is true bravery, wisdom, love, and friendship. Emphasis on the physical aspects of warriorship is futile, for the power of the body is always limited.
Ueshiba Morihei

I'm not a good shot, but I shoot often.
Theodore Roosevelt

Accept the challenges, so that you may feel the exhilaration of victory.

General George S. Patton

When I came back to Dublin I was court martialed in my absence and sentenced to death in my absence, so I said they could shoot me in my absence.
Brendan Behan

You can get much farther with a kind word and a gun than you can with a kind word alone.
Al Capone

Victorious warriors win first and then go to war, while defeated warriors go to war first and then seek to win.
Sun Tzu

If you know the enemy and know yourself, you need not fear the result of a hundred battles. If you know yourself but not the enemy, for every victory gained you will also suffer a defeat. If you know neither the enemy nor yourself, you will succumb in every battle.
Sun Tzu

You smell that? Do you smell that? Napalm, son. Nothing else in the world smells like that. I love the smell of napalm in the morning. You know, one time we had a hill bombed, for twelve hours. When it was all over I walked up. We didn't find one of 'em, not one stinkin' dink body. The smell, you know that gasoline smell, the whole hill. Smelled like... victory. Someday this war's gonna end...
Steven Spielberg, Apocalypse Now

If you are going to deal in death, you should be willing to see the truth of it, not some glorious lie. If I have a battle with another sword player, it is between the two of us, our business, our truth. But if you run a planet and you get pissed off at somebody the next orbit over, you each might send a million soldiers to recycling plants. A smart rocket can come from a thousand klicks away to kill you; it doesn't care and it won't be in the least upset that it has blasted you to atomic debris. That's the real horror of modern war, that it is impersonal. Being cut with a sword hurts, and if you are close enough to do it, you can't miss the other's pain.
Steve Perry

One of man's finest qualities is described by the simple word, guts -
the ability to take it. If you have the discipline to stand fast when
your body wants to run, if you can control your temper and remain
cheerful in the face of monotony or disappointment, you have guts,
in the soldiering sense. This ability to take it must be trained-the
training is hard, mental as well as physical. But once ingrained, you
can face and flail the enemy as a soldier, and enjoy the challenges of
life as a civilian.
Colonel John S. Roosma

You remember the Duke of Wellington was talking of the Battle of
Waterloo when he said that it was not that the British soldiers were
braver than the French soldiers. It was just that they were brave five
minutes longer. And in our struggles sometimes that's all it takes-to
be brave five minutes longer, to try just a little harder, to not give up
on ourselves when everything seems to beg for our defeat.
Paul H. Dunn

Let the soldier be abroad if he will, he can do nothing in this age.
There is another personage,- a personage less imposing in the eyes of
some, perhaps insignificant. The schoolmaster is abroad, and I trust
to him, armed with his primer, against the soldier in full military
array.
Henry Peter

No soldier starts a war-they only give their lives to it. Wars are
started by you and me, by bankers and politicians, newspaper
editors, clergymen who are ex-pacifists, and Congressmen with
vertebrae of putty. The youngsters yelling in the streets, poor lads,
are the ones who pay the price.
Francis Duffy

From the time I was twelve years old until I retired last year at the
age of fifty-seven, the Army was my life. I loved commanding
soldiers and being around people who had made a serious
commitment to serve their country.
Gen. Norman Schwarzkopf

The glory of a workman, still more of a master workman, that he does his work well, ought to be his most precious possession; like the honor of a soldier, dearer to him than life.
Thomas Carlyle

Soldiers are sworn to action; they must win; some flaming, fatal climax with their lives. Soldiers are dreamers; when the guns begin, they think of fire it homes, clean beds, and wives.
Siegfried Sassoon

You can always tell an old soldier by the inside of his holsters and cartridge boxes. The young ones carry pistols and cartridges: the old ones, grub.
George Bernard Shaw

ABRUPT, adj. Sudden, without ceremony, like the arrival of a cannon-shot and the departure of the soldier whose interests are most affected by it.
Ambrose Bierce

The Lord gets his best soldiers out of the highlands of affliction.
Charles Haddon Spurgeon

Soldiers usually win the battles and generals get the credit for them.
Napoleon Bonaparte

War loses a great deal of its romance after a soldier has seen his first battle.
John Singleton Mosby

We are parlor soldiers. The rugged battle of fate, where strength is born, we shun.
Ralph Waldo Emerson

RECRUIT, n. A person distinguishable from a civilian by his uniform and from a soldier by his gait.
Ambrose Bierce

Every lover is a soldier and has his camp in Cupid.
Ovid

As an old soldier, I admit the cowardice: it's as universal as seasickness, and matters just as little.
George Bernard Shaw

In the final choice, a soldier's pack is not so heavy a burden as a prisoner's chains.
Dwight D. Eisenhower

Stand your ground. Don't fire unless fired upon, but if they mean to have a war let it begin here!
Capt. John Parker

The more you sweat in training, the less you will bleed in battle.
Motto of Navy Seals

Some Warriors look fierce, but are mild. Some seem timid, but are vicious. Look beyond appearances; position yourself for the advantage.
Deng Ming-Dao

I dislike death, however, there are some things I dislike more than death. Therefore, there are times when I will not avoid danger.
Mencius

Unless you do your best, the day will come when, tired and hungry, you will halt just short of the goal you were ordered to reach, and by halting you will make useless the efforts and deaths of thousands.
Gen. George S. Patton

Those who are first on the battlefield and await the opponents are at ease; those who are last on the battlefield and head into a fight become exhausted. Therefore, good warriors cause others to go to them and do not go to others.
Sun Tzu,

Even though you hold a sword over my heart I will not give up.
Bushido Quote

And if a warrior does not manifest courage on the outside and hold enough compassion within his heart to burst his chest, he cannot become a retainer. Therefore, the monk pursues courage with the warrior as his model, and the warrior pursues the compassion of the monk.
From the Hagakure

Let me tell you what happened briefly. There were 114,000 separate aerial sorties in 42 days ñ one every 30 seconds. Eighty-eight thousand tons of bombs were dropped. Only seven per cent were guided. Ninety-three percent were free-falling bombs that hit where chance, necessity and no free will took them. There were 38 aircraft lost by the US in the slaughter. That number is less than the accidental losses in war games where no live ammunition is even used. No enemy aircraft rose to meet them. When the ground war came there was no ground war. Name one battle. It wasn't a battle, it was a slaughter. General Kelly said when the troops finally moved forward that there weren't many of them left alive to fight. We killed at least 125,000 soldiers and to date 130,000 civilians. We killed as many as we dared.
Ramsey Clark

If a warrior makes loyalty and filial piety one load, and courage and compassion another, and carries these twenty-four hours a day until his shoulders wear out, he will be a samurai.
From the Hagakure

For those that will fight for it....Freedom.has a flavor the protected shall never know.
L/Cpl Edwin L. Craft

From camp to camp through the foul womb of night
The hum of either army still sounds.
William Shakespeare

When the blast of war blows in our ears
Then imitate the action of the tiger,
Stiffen the sinews, summon up the blood,
Disguise fair nature with ill-favour'd rage.
William Shakespeare

Joseph Lumpkin

O God of battles! steel my soldiers' hearts.
William Shakespeare

Sound all the lofty instruments of war,
And by that music let us all embrace;
For, heaven to earth, some of us never shall
A second time do such a courtesy.
William Shakespeare

They come like sacrifices in their trim,
And to the fire-eyed maid of smoky war
All hot and bleeding will we offer them.
William Shakespearc

All quiet along the Potomac to-night,
No sound save the rush of the river,
While soft falls the dew on
The face of the dead,-
The picket's off duty forever.
Ethel Lynn Beers

If they turn on their radars we're going to blow up their goddamn
SAMs [surface-to- air missiles]. They know we own their country.
We own their airspace. We dictate the way they live and talk. And
that's what's great about America right now. It's a good thing,
especially when there's a lot of oil out there we need.
General William Looney

In case you haven't noticed, we dehumanize our own soldiers, not
because of their religion or race, but because of their low social class.
Send'em anywhere. Make'em do anything. Piece of cake.
Kurt Vonnegut

The more you sweat in peace, the less you bleed in war.
Hyman G. Rickover

Weapons are an important factor in war, but not the decisive one; it
is man and not materials that counts.
Mao Tze-tung

In war there is no second prize for the runner-up.
General Omar Bradley

The patriot volunteer, fighting for country and his rights, makes the most reliable soldier on earth.
Thomas J. (Stonewall) Jackson

....an imperfect plan implemented immediately and violently will always succeed better than a perfect plan.
General George S. Patton

Some people live their entire lifetime and wonder if they ever made a difference to the world. Marines don't have that problem.
Ronald Reagan

Superior firepower is an invaluable tool when entering negotiations.
General George S. Patton

The best form of welfare for the troops is first-class training.
Field Marshal Rommel

Peace and war

BLOWIN' IN THE WIND
Bob Dylan 1963

How many roads must a man walk down
Before you call him a man?

Yes, 'n' how many seas must a white dove sail
Before she sleeps in the sand?

Yes, 'n' how many times must the cannon balls fly
Before they're forever banned?

The answer, my friend, is blowin' in the wind,
The answer is blowin' in the wind.

I believe in the doctrine of non-violence as a weapon of the weak. I believe in the doctrine of non-violence as a weapon of the strongest. I believe that a man is the strongest soldier for daring to die unarmed.
Mahatma Gandhi

The purpose of all war is peace.
Saint Augustine

Master the divine techniques of the Art of Peace and no enemy will dare to challenge you.
Ueshiba

Peace cannot be achieved through violence; it can only be attained through understanding.
Albert Einstein

PEACE, In international affairs, a period of cheating between two periods of fighting.
Ambrose Bierce

Nurture strength of spirit to shield you in sudden misfortune. But do not distress yourself with imaginings. Many fears are born of fatigue and loneliness. Beyond wholesome discipline, be gentle with yourself.
Desiderata

There are no fixed limits. Time does not stand still. Nothing endures, nothing is final.
Chuang Tzu

With malice toward none; with charity for all; with firmness in the right, as God gives us to see the right, let us strive on to finish the

work we are in; to bind up the nation's wounds; to care for him who shall have borne the battle, and for his widow and his orphan-to do all which may achieve and cherish a just and lasting peace among ourselves and with all nations.
Abraham Lincoln

Let each one understand the meaning of sincerity and guard against display.
Chuang Tzu Circa

Obtaining victory may be easier than preserving the results.
Joseph Lumpkin

He who joyfully marches to music in rank and file has already earned my contempt. He has been given a large brain by mistake, since for him the spinal cord would fully suffice. This disgrace to civilization should be done away with at once. Heroism at command, senseless brutality, deplorable love-of-country stance, how violently I hate all this, how despicable and ignoble war is; I would rather be torn to shreds than be a part of so base an action! It is my conviction that killing under the cloak of war is nothing but an act of murder.
Albert Einstein

In Nature, things move violently to their place and calmly in their place.
Sir Francis Bacon

When the world is peaceful, a gentleman keeps his sword by his side.
Wu Tzu

Courage is the price that life exacts for granting peace.
Amelia Earhart

What is life? It is the flash of a firefly in the night. It is the breath of the buffalo in the wintertime. It is the little shadow, which runs across the grass and loses itself in the sunset.
Crowfoot, a Blackfoot warrior

The tumult and shouting dies; the captains and kings depart; still stand the sacrifice, a humble and contrite heart. Lord God of Hosts, be with us yet, lest we forget - lest we forget.
Rudyard Kipling

The end of our Way of the sword is to be fearless when confronting our inner enemies and our outer enemies.
Tesshu Yamaoka

ISA 40:31 But they that wait upon the LORD shall renew their strength; they shall mount up with wings as eagles; they shall run, and not be weary; and they shall walk, and not faint.

I count him braver who overcomes his desires than him who conquers his enemies: for the hardest victory is the victory over self.
Aristotle

Out of a martial art, out of combat I would feel something peaceful. Something without hostility.
Bruce Lee

We seek peace, knowing that peace is the climate of freedom.
Dwight D. Eisenhower

A brave and passionate man will kill or be killed. A brave and calm man will preserve life. Of these two types, which is good and which does harm?
Lao Tzu

'Twere best at once to sink to peace, Like birds the charming serpent draws, To drop head-foremost in the jaws of vacant darkness and to cease.
Alfred, Lord Tennyson

A mind at peace, a mind centered and not focused on harming others, is stronger than any physical force in the universe.
Wayne Dyer

A toast to the weapons of war, may they rust in peace.
Robert Orben

All business sagacity reduces itself in the last analysis to a judicious use of sabotage.
Thorstein Veblen

Americans will listen, but they do not care to read. War and Peace must wait for the leisure of retirement, which never really comes: meanwhile it helps to furnish the living room.
Anthony Burgess

An election is coming. Universal peace is declared, and the foxes have a sincere interest in prolonging the lives of the poultry.
George Eliot

As peace is the end of war, so to be idle is the ultimate purpose of the busy.
Samuel Johnson

Atoms for peace. Man is still the greatest miracle and the greatest problem on this earth.
David Sarnoff

Be at war with your vices; at peace with your neighbors, and let every new year find you a better man.
Benjamin Franklin

Democracies are indeed slow to make war, but once embarked upon a martial venture are equally slow to make peace and reluctant to make a tolerable, rather than a vindictive, peace.
Reinhold Niebuhr

Disarm, disarm. The sword of murder is not the balance of justice. Blood does not wipe out dishonor, nor violence indicate possession.
Julia Ward Howe

Establishing lasting peace is the work of education; all politics can do is keep us out of war.
Maria Montessori

Joseph Lumpkin

God is day and night, winter and summer, war and peace, surfeit and hunger.
Heraclitus of Ephesus

I don't want peace that passeth understanding, I want understanding which bringeth peace.
Helen Keller

I intend to leave after my death a large fund for the promotion of the peace idea, but I am skeptical as to its results.
Alfred Nobel

I will not by the noise of bloody wars and the dethroning of kings advance you to glory: but by the gentle ways of peace and love.
Thomas Traherne

If you cannot find peace within yourself, you will never find it anywhere else.
Marvin Gaye

Imagine all the people living life in peace. You may say I'm a dreamer, but I'm not the only one. I hope someday you'll join us, and the world will be as one.
John Lennon

In the arts of peace man is a bungler.
George Bernard Shaw

It is easy enough to be friendly to one's friends. But to befriend the one who regards himself as your enemy is the quintessence of true religion. The other is mere business.
Mahatma Gandhi

It is wise statesmanship which suggests that in time of peace we must prepare for war, and it is no less a wise benevolence that makes preparation in the hour of peace for assuaging the ills that are sure to accompany war.
Clara Barton

Let us ever remember that our interest is in concord, not in conflict; and that our real eminence rests in the victories of peace, not those of war.
William McKinley

Life is pleasant. Death is peaceful. It's the transition that's troublesome.
Isaac Asimov

Mankind must remember that peace is not God's gift to his creatures; peace is our gift to each other.
Elie Wiesel

No country has suffered so much from the ruins of war while being at peace as the American.
Edward Dahlberg

The people of the world genuinely want peace. Some day the leaders of the world are going to have to give in and give it to them.
Dwight D. Eisenhower

Nothing contributes more to a person's peace of mind than having no opinions at all.
G. C. Lichtenberg

Obsessed by a fairy tale, we spend our lives searching for a magic door and a lost kingdom of peace.
Eugene O\'Neill

He does not want war but if war is to be fought it is to be won. True victory is in finding a path that does not lead to conflict. Conflict leads to a lesser victory, which is based on survival and not resolution. Ultimate victory is the saving of all life and honor. This is winning without fighting.
Joseph Lumpkin

The arts of peace and war are like two wheels of a cart, which, lacking one, will have difficulty in standing.
Kuroda Nagamasa

Higher worth is like water. Water is good at benefiting ten thousand beings without vying for position... In dwelling, be close to the land. In heart and mind value depth. In interacting with others, value kindness. In words, value reliability. In rectifying, value order. In social affairs, value ability. In action, value timing. In general, simply don't fight and hence have no blame.
Lao Tzu

Life is a hard battle anyway. If we laugh and sing a little as we fight the good fight of freedom, it makes it all go easier. I will not allow my life's light to be determined by the darkness around me.
Sojourner Truth

An unjust peace is better than a just war.
Marcus Tullius Cicero

Whoever undertakes to set himself up as a judge of Truth and Knowledge is shipwrecked by the laughter of the Gods.
Albert Einstein

Victory goes to the player who makes the next-to-last mistake.
Savielly Grigorievitch Tartakower

In order to make an apple pie from scratch, you must first create the universe.
Carl Sagan

A designer knows he has achieved perfection not when there is nothing left to add, but when there is nothing left to take away.
Antoine de Saint-Exupery

A man's feet should be planted in his country, but his eyes should survey the world.
George Santayana

A society grows great when old men plant trees whose shade they know they shall never sit in.
Greek Proverb

An eye for an eye makes the whole world blind.

Mahatma Gandhi

When I hear somebody sigh, Life is hard, I am always tempted to ask, Compared to what?
Sydney J. Harris

In times like these, it is helpful to remember that there have always been times like these.
Paul Harvey

The loud little handful will shout for war. The pulpit will warily and cautiously protest at first. The great mass of the nation will rub its sleepy eyes, and will try to make out why there should be a war, and they will say earnestly and indignantly: It is unjust and dishonorable and there is no need for war. Then the few will shout even louder. Before long you will see a curious thing: anti-war speakers will be stoned from the platform, and free speech will be strangled by hordes of furious men who still agree with the speakers but dare not admit it...Next, statesmen will invent cheap lies, putting blame upon the nation that is attacked, and every man will be glad of those conscience-soothing falsities, and will diligently study them, and refuse to examine any refutations of them; and thus he will by and by convince himself that the war is just, and will thank God for the better sleep he enjoys after this process of grotesque self-deception.
Mark Twain

I do not want the best to be any more the deadly enemy of the good. We climb through degrees of comparison.
Archbishop Edward White Benson,

The only way to win World War III is to prevent it.
Dwight D. Eisenhower

The final battle against intolerance is to be fought--not in the chambers of any legislature--but in the hearts of men.
Dwight D. Eisenhower

A people that values its privileges above its principles soon loses both.
Dwight D. Eisenhower

There is--in world affairs--a steady course to be followed between an assertion of strength that is truculent and a confession of helplessness that is cowardly.
Dwight D. Eisenhower

Every gun that is made, every warship launched, every rocket fired signifies, in the final sense, a theft from those who hunger and are not fed, those who are cold and are not clothed.
Dwight D. Eisenhower

To map out a course of action and follow it to an end requires some of the same courage that a soldier needs. Peace has its victories, but it takes brave men and women to win them.
Ralph Waldo Emerson

Soldier rest! thy warfare o'er, Sleep the sleep that knows not breaking, Dream of battled fields no more, Days of danger, nights of waking.
Sir Walter Scott

It is a brave act of valor to condemn death, but where life is more terrible than death it is then the truest valor to dare to live.
Sir Thomas Brown

Wars are not acts of God. They are caused by man, by man-made institutions, by the way in which man has organized his society. What man has made, man can change.
Frederick Moore Vinson

Great spirits have always found violent opposition from mediocre minds. The latter cannot understand it when a man does not thoughtlessly submit to hereditary prejudices but honestly and courageously uses his intelligence.
Albert Einstein

To be prepared for war is one of the most effectual means of preserving peace.
George Washington

Since wars begin in the minds of men, it is in the minds of men that the defenses of peace must be constructed.
UNESCO Constitution

When we say, War is over if you want it, we mean that if everyone demanded peace instead of another TV set, we'd have peace.
John Lennon

Peace is more important than all justice; and peace was not made for the sake of justice, but justice for the sake of peace.
Martin Luther

Bravery and cowardice are not things that can be conjectured in times of peace. They are in different categories.
From the Hagakure

Liberty is never unalienable; it must be redeemed regularly with the blood of patriots or it always vanishes. Of all the so-called natural human rights that have ever been invented, liberty is the least to be cheap and is never free of cost.
Robert A. Heinlein

When I got back from the war in 1945, I refused to make war pictures.
James Stewart

Four score and seven years ago our fathers brought forth on this continent a new nation, conceived in liberty, and dedicated to the proposition that all men are created equal.
Abraham Lincoln

War does not determine who is right - only who is left.
Anonymous

When war enters a country, it produces lies like sand.
Anonymous

War would end if the dead could return.
Stanley Baldwin

Weapons are tools of violence; all decent men detest them.
Lao Tzu

What difference does it make to the dead, the orphans and the homeless, whether the mad destruction is wrought under the name of totalitarianism or in the holy name of liberty and democracy?
Mahatma Gandhi

I am not only a pacifist but a militant pacifist. I am willing to fight for peace. Nothing will end war unless the people themselves refuse to go to war.
Albert Einstein

They wrote in the old days that it is sweet and fitting to die for one's country. But in modern war, there is nothing sweet nor fitting in your dying. You will die like a dog for no good reason.
Ernest Hemingway

Strike against war, for without you no battles can be fought! Strike against manufacturing shrapnel and gas bombs and all other tools of murder! Strike against preparedness that means death and misery to millions of human beings! Be not dumb, obedient slaves in an army of destruction! Be heroes in an army of construction!
Helen Keller

Man is the only animal that deals in that atrocity of atrocities, War. He is the only one that gathers his brethren about him and goes forth in cold blood and calm pulse to exterminate his kind. He is the only animal that for sordid wages will march out and help to slaughter strangers of his own species who have done him no harm and with whom he has no quarrel. And in the intervals between campaigns he washes the blood off his hands and works for the universal brotherhood of man with his mouth.
Mark Twain

The citizen who sees his society as democratic clothes being worn out and does not cry it out, is not a patriot, but a traitor.
Mark Twain

We do not admire a man of timid peace.

Theodore Roosevelt

The soldier, above all other people, prays for peace, for he must suffer and bear the deepest wounds and scars of war.
Douglas MacArthur

What the horrors of war are, no one can imagine. They are not wounds and blood and fever, spotted and low, or dysentery, chronic and acute, cold and heat and famine. They are intoxication, drunken brutality, demoralization and disorder on the part of the inferior ... jealousies, meanness, indifference, selfish brutality on the part of the superior.
Florence Nightingale

To give victory to the right, not bloody bullets, but peaceful ballots only, are necessary.
Abraham Lincoln

It is only those who have neither fired a shot nor heard the shrieks and groans of the wounded who cry aloud for blood, more vengeance, more desolation. War is hell.
William Tecumseh Sherman

Humanize war? You might as talk about humanizing hell!
British Admiral Jacky Fisher

You can have peace, or you can have freedom. Don't ever count on having both at once.
Robert A. Heinlein

Let us recollect that peace or war will not always be left to our option; that however moderate or un-ambitious we may be, we cannot count upon the moderation, or hope to extinguish the ambition of others.
Alexander Hamilton

We make war that we may live in peace.
Aristotle

Where there is no peril in the fight, there is no glory in the triumph.

Joseph Lumpkin

Pierre Corneille

In peace sons bury fathers, but war violates the order of nature, and fathers bury sons.
Heroditus

I don't know whether war is an interlude during peace, or peace is an interlude during war.
Georges Clemenceau

The most disadvantageous peace is better than the most just war
Desiderius Erasmus

It is war that shapes peace, and armament that shapes war.
Thomas Fuller

You can no more win a war than you can win an earthquake.
Jeannette Rankin

War may sometimes be a necessary evil. But no matter how necessary, it is always an evil, never a good. We will not learn how to live together in peace by killing each other's children.
Jimmy Carter

War is not its own end, except in some catastrophic slide into absolute damnation. It's peace that's wanted. Some better peace than the one you started with.
Lois McMaster Bujold

The only winner in the War of 1812 was Tchaikovsky.
Solomon Short

Wars teach us not to love our enemies, but to hate our allies.
W. L. George

You can't separate peace from freedom because no one can be at peace unless he has his freedom.
Malcolm X

Never believe that a few caring people can't change the world. For indeed, That's all who ever have.
Margaret Mead

Do not fear your enemies, The worst they can do is kill you. Do not fear your friends, At worst, they may betray you. Fear those who do not care, They neither kill nor betray, but betrayal and murder exist because of their silent consent.
Bruno Jasienski

You can't say that civilization don't advance, however, for in every war they kill you in a new way.
Will Rogers

No more wars, no more bloodshed. Peace unto you. Shalom, salaam, forever.
Menachem Begin

Peace is more important than all justice; and peace was not made for the sake of justice, but justice for the sake of peace.
Martin Luther

Peace hath her victories, no less renowned than War.
John Milton

Since wars begin in the minds of men, it is in the minds of men that the defenses of peace must be constructed.
UNESCO Constitution

As peace is the end of war, it is the end, likewise, of preparations for war; and he may be justly hunted down, as the enemy of mankind, that can choose to snatch, by violence and bloodshed, what gentler means can equally obtain.
Samuel Johnson

Peace and justice are two sides of the same coin.
Dwight D. Eisenhower

With Malice toward none, with charity for all, with firmness in the right, as God gives us to see the right, let us strive on to finish the work we are in, to bind up the nation's wounds.
Abraham Lincoln

"You may not destroy someone else's world unless you are prepared to offer a better one." *Franz Kafka*

"Life is a tragedy for those who feel and a comedy for those who think." *Jean de la Bruyere*

"One should have insight into this world of dreams that passes in the twinkling of an eye."--Hojo Shigetoki (1198-1261)

"Many men feel that they should act according to the time or the moment they are facing, and thus are in confusion when something goes beyond this and some difficulty arises."

"One should not be envious of someone who has prospered by unjust deeds. Nor should he disdain someone who has fallen while adhering to the path of righteousness."--Imagawa Sadayo (1325-1420)

"A man with deep far-sightedness will survey both the beginning and the end of a situation and continually consider its every facet as important."--Takeda Shingen (1521-1573)

"It is bad when one thing becomes two. One should not look for anything else in the Way of the Samurai. It is the same for anything that is called a Way. Therefore, it is inconsistent to hear something of the Way of Confucius or the Way of the Buddha, and say that this is the Way of the Samurai. If one understands things in this manner, he should be able to hear about all says and be more and more in accord with his own."

"A person who is said to be proficient at the arts is like a fool. Because of his foolishness in concerning himself with just one thing, he thinks of nothing else and thus becomes proficient. He is a worthless person."--Tsunetomo Yamamoto (1659-1719)

"To reach a house, you must first enter the gate. The gate is the pathway leading to the house. After passing through the gate, you enter the house and meet its master. Learning is the gate to reaching the Way. After passing through the gate, you reach the Way. Learning is the gate, not the house. Don't mistake the gate for the house. The house is located farther inside, after the gate is passed. Because learning is the gate for the house, don't think books you read are the Way. Books are a gate for reaching the Way." *from the book The Sword and The Mind.*

"The debate is controlled by the extremes, each side shouting out answers and accusations over the heads of the people in between who are kept from formulating questions by the din of the argument all around them. Each paints the other with a broader brush. Each has an arsenal of names and adjectives to deploy against the other side. No one listens. Everyone screams." *Thomas Lynch*

"When presented with two options always pick a third." *Derrick Jensen*

"We find that the age of electronics has converted the human race in large measure from doers to watchers. It seems incontrovertible that great numbers of Americans, especially young Americans, would rather watch than do." *Jeff Cooper*

"...the something you have left out is a hole in your information. Chaos theorists have been quick to point out that both in principle and practice, there will always be missing information, a limitation to our knowledge, a hole in the data...Now we hold chaos itself up to reality and find our overlooked third dimension revealed there, mystery. What we don't know may be sufficiently powerful to overturn in a moment our entire existence and certified knowledge...think of missing information as the trickster of chaos theory...Perhaps at a deep level our reluctance to embrace the missing information has something to do with our anxiety about death - the ultimate missing information, the ultimate unknown that makes all knowledge shrink to nothing. Chaos tells us that the missing information is a window to the whole. In the pit of uncertainty looms our access to creative possibilities. We jump too quickly from the openness of the question to the need for its resolution. But what if what we are seeking doesn't lie in any answer but at the center of the question in the very depths of the missing information?" *John Briggs from the book Seven Life Lessons of Chaos.*

"The road to truth is long and lined the entire way with annoying bastards." *Alexander Jablokov*

"Always listen to yourself. It is better to be wrong than simply to follow convention. If you are wrong, no matter, you have learned something and you will grow stronger. If you are right, you have taken another step towards a fulfilling life." *Hagakure*

"It takes one to know one." *The all-time great comeback utilized by many.*

"A little chaos is always necessary." *Anonymous*

"Nerve succeeds." *Anonymous*

"...Matters of great concern should be treated lightly...matters of small concern should be treated seriously." *Hagakure*

"It is never too late to be what you might have been." *George Eliot*

"The only honorable response to violence is counter-violence. To surrender to extortion is a greater sin than extortion in that it breeds and feeds the very act it seeks to avoid." *Jeff Cooper*

"I didn't surrender neither. But they took my horse and made him surrender. They have him pulling a wagon up in Kansas, I bet." *Chief Dan George from the movie The Outlaw Josey Wales*

"Life of human kind is indeed very short and trifling. You ought to spend your time doing whatever you like to do...If this is misinterpreted, it will harm youth. So I have not spoken of this to the young samurais." *Hagakure*

"The Great Man is he who does not lose his childlike heart." *Mencius*

"Whatever occupies the mind at the time of death determines the destination of the dying...when you make your mind one-pointed through regular practice of meditation, you will find the supreme glory of the Lord." *Bhagavad Gita*

" kind word can warm three winter months." *Japanese proverb*

"The winds of grace are always blowing but it is you that must raise your sails." *R. Tagore*

"There is in all visible things...a hidden wholeness." *Thomas Merton*

"There is something to be learned from a rainstorm. When meeting with a sudden shower, you try not to get wet and run quickly along the road. By doing such things as passing under the eaves of houses one still gets wet. When you are resolved from the beginning, you will not be perplexed, though you will get the same soaking. This understanding extends to all things." *Hagakure*

"Words that come from the heart enter the heart." *Moses Ibn Ezra*

"You need to be alone. Roshi told me. It is the terminal abode. You can't go any deeper in your spiritual practice if you run from loneliness...Resistances are painful, and they deepen as we deepen." *Roshi Katagari and Natalie Goldberg*

"He who will not apply new remedies must expect old evils." *Francis Bacon*

Our bodies are given life from the midst of nothingness. Existing where there is nothing is the meaning of the phrase, "Form is emptiness." That all things are provided for by nothingness is the meaning of the phrase, "Emptiness is form." One should not think that these are two seperate things.
Tsunetomo Yamamoto

"Another name for God is surprise." *Brother David Steindl-Rast*

Joseph Lumpkin

Joseph Lumpkin

CPSIA information can be obtained
at www.ICGtesting.com
Printed in the USA
LVHW080303310321
683048LV00022B/387